THE SOLAR AGE RESOURCE BOOK

D1254606

THE SOLAR AGE RESOURCE BOOK

*by the editors of **Solar Age** Magazine*

EDITOR: Martin McPhillips
EXECUTIVE EDITOR: Bruce Anderson

Library of Congress Cataloging in Publication Data

Main entry under title:

The Solar age resource book.

1. Solar energy. 2. Solar heating.
I. McPhillips, Martin. II. Anderson, Bruce,
1947- III. Solar age.
TJ810.S487 621.47 78-74580
ISBN 0-89696-050-1 pbk.

Copyright © 1979 by SolarVision, Inc.
All Rights Reserved
Library of Congress Catalog Card Number: 78-74580
ISBN: 0-89696-050-1
Published simultaneously in Canada by
Beaverbooks, Pickering, Ontario
Manufactured in the United States of America 2HC380

Designed by Techart Associates

Acknowledgements

The Solar Age Resource Book

Dozens of people were involved in the preparation of this book. Many of them are listed below. In addition, special thanks to our authors for their excellent and provocative articles; to the product manufacturers who saw this *Resource Book* as an opportunity for reaching thousands of people with information about their important products; and to the *Wind Energy Digest* and the *New Hampshire Times* for sharing their information on wind and wood products.

The Solar Age Resource Book

Editorial: by *SOLAR AGE Magazine*

EDITOR: Martin McPhillips

ASSISTANT EDITOR: Donald Sango

EXECUTIVE EDITOR: Bruce Anderson

TECHNICAL CONSULTANTS: Total Environmental Action, Inc.
Winslow Fuller
Douglas Taff

EDITORIAL ASSISTANTS: Janis Kobran
Sean O'Brien
Melanie Wallace

CLERICAL ASSISTANCE: Alice Ferner

Production: by *SOLAR AGE Magazine*

PRODUCTION MANAGER: Deborah Napior

PRODUCTION EDITOR: Deborah Napior

DESIGN, ILLUSTRATIONS: Techart Associates

This Resource Book was made possible by support from *Solar Age Magazine*. *Solar Age* is a monthly publication that provides the facts—dependable, useful, essential information on integrating solar applications with energy-efficient buildings. Each month, *Solar Age* provides a wide range of key articles. Topics such as active, passive, and hybrid solar systems, heat storage and distribution, retrofitting, energy conservation, greenhouses, solar electricity, wind and wood energy, and more. And, regular features cover innovative solar house designs, and latest developments, products, installation advice, business trends, and opinion. It's a magazine that puts hard, factual information into a format that can be immediately used by many types of readers, whether they are experienced in solar or new to the field.

Sure, these days you can read about solar almost everywhere. But when energy-conscious people want to use solar successfully, they turn to *Solar Age*.

Solar Age Magazine
Dept. RB, Box 4934,
Manchester, NH 03108

Table of Contents

Listings

Introduction

Solar energy is a simple and straightforward physical phenomenon, yet the variety of methods that can be employed to convert it into useful energy are staggering in their number, and often in their complexity.

If the many issues of solar technology seem agonizingly complex, an endless mire of conflicting opinion and fact, then simplicity, intelligence and thoroughness must reclaim you from that mire. Sunworshippers, environmental fanatics, politicians, and alternative ideologists aside—the sun will probably make it as an energy source because plain old people get off on the idea. Whatever their reasons—annoyance at oil companies, an urge for self-sufficiency, being backyard tinkerers or technology buffs—the general public likes the idea of energy from the sun. People can do it on their own, their own way; they can compete with their neighbors or have fun with their kids. Nations can decrease the balance of payments deficit by replacing foreign oil with solar heat, and pocket the savings.

One thing Americans hate is to be under the thumb of their monthly bills. When oil heat was so cheap that it was a peripheral and accepted expense, it was alright with almost everyone, but now the typical American is ready to get out from under that bill because it is killing him and because there is no need for it.

Solar energy is the edge of a new economic knife. It is already possible to replace much of one's heating oil with simple solar heating systems. Soon photovoltaic power—the *direct* conversion of sunlight to electricity—will also be widely available. Government spending and private investments in the area of photovoltaics are bringing prices down to a point where they will soon be too inexpensive to ignore. The issues remain complex; the sun remains stubbornly simple.

The use of solar energy runs counter to the ways we generally think of receiving and using energy. Solar radiation is not delivered in trucks like oil or gas. The major reason for *The Solar Age Resource Book* is to show that the ways and means of using the sun for heating one's home are already established. The hardware is on the shelf; reliable and low-cost do-it-yourself methods are available; architects and engineers are learning and using the latest solar ideas—even the government appears to be gradually, if somewhat reluctantly, growing in its awareness of what must be done. There are good and bad ideas. And, the relative value of a solar solution depends largely on what produces the most energy for the least investment for the longest time with as little trouble as possible.

In the case of hardware, we have been able to list around 100 flat-plate collectors already on the market. Although a good enough way to collect solar energy, the flat-plate collector already faces stiff competition from the "passive" school, where a building acts as its own collector and storage unit, with little or no mechanical assistance. Passive solar is as simple as collecting the sun's rays through large south facing expanses of glass or in a low-cost, attached solar greenhouse that will also produce food. In fact, the *solar* greenhouse, because of its durability, low cost and efficiency, must now be considered the "state of the art" for solar heating. For some reason this is a concept that people find hard to comprehend, yet as a collector it leads in most of the important categories. Its aesthetic appeal is another important consideration—greenhouses or "solar rooms" can be beautiful while they heat your home and give you growing space. Douglas Taff's article on the solar greenhouse explains more about it and why it is a good way to go solar. Jeremy Coleman shares his building experiences with attached greenhouses in the Northeast.

Along with the greenhouse articles are two others that describe additional durable, low-cost systems. David Wright, a leading solar/environmental architect, provides the basic drawings and concepts for a "direct gain" solar building module. Bruce Anderson, author of *The Solar Home Book*, and executive editor of *Solar Age* magazine, discusses converting an existing or new building to include simple "thermosiphoning" solar heaters. The basics of these systems are adaptable to many different applications.

Benjamin Rogers will start you toward an understanding of how the weather conditions at your building site can be manipulated to solve some of your design problems, and Michael Riordan addresses storage problems.

Another article by Doug Taff offers a crash course on how to avoid the pitfalls associated with buying solar hardware.

Helene Kassler gives you an update on the ever advancing state of photovoltaic technology and economics.

Allan Frank, editor of the *Solar Energy Intelligence Report*, examines the potential impact of government funding on the rate of emergence of solar technology, and Jeff Cook writes a summary of the fast-emerging issues in solar education and training.

Covering pressing, yet more technical ground, Charles Michal shows how a solar system can be sized with a Scientific Pocket Calculator; Joe Kohler describes the available computer programs designed to simulate and size solar systems; and John Yellott writes of the superior qualities of glass, its precise relationship to solar radiation, and how the two work best together.

Rick Schwolsky outlines the basics for installing a solar domestic hot water system—with good advice for solving tough problems that plague installers. Joe Carter writes an introduction to our listing of wind energy conversion systems for small scale application, and William Hauk assays the reality of wood-burning for the potential stove buyer, leading us to our Wood Stove Buyer's Guide.

Along with the wood stove buyer's guide, we have wind and solar products buyers guides, followed by an alphabetical listing of solar manufacturers. For easy cross-referencing we have included the categories of products they produce.

And closing the book is a listing of solar resource and service companies—architects, engineers, information agencies, and researchers—listed geographically, to aid you in your search for companies who provide solar services.

The basic techniques for using the sun have been available for years. The great surge of interest in solar during the last few years has actually constituted the revival of very old ideas. This latest revival, certainly the most serious effort of all time, has been a matter of refining and gaining knowledge of and confidence in the old ideas.

What is really most striking about the current advances is that they make us seem rather foolish for not having gotten down to it earlier. It was that famous phantom "the energy crisis" that led oil companies to drastically increase their prices that led to the realization that the sun is really a prime, super-abundant, global energy source.

With our mixture of both technical and simple introductory—but practical and to the point—articles we hope to take the interested layperson and the solar professional to a point of interface. We believe that because of the decentralized nature of solar, this interaction is a necessity—the greater the general knowledge—both technical and philosophical—the less confusion and the quicker things get done.

During these opening moments of the proliferation of a great new group of solar technologies you are privileged to take part in the initial process of selection. Your choices should be based on the facts, of which the most pressing are performance, cost and durability.

Martin McPhillips

Martin McPhillips
Editor

THE SOLAR AGE RESOURCE BOOK

"Don't fight the site." In summer, the trees grow cooperative leaves to shade collector; in winter they drop away to let the sun through. Designed by Fuller Moore, College Corners, Ohio.

Microclimate: "Don't Fight The Site"

by Benjamin T. Rogers

Buck Rogers is a consulting Mechanical Engineer and the person who first coined the term "passive" in reference to natural solar energy systems.

Man does not conquer nature. Intelligent man diverts the overwhelming thrust of natural forces to serve his needs.

Solar design is very site specific. Subtle local environmental conditions can make or break a project. These local conditions have been lumped into a package of factors that we call microclimate. The sensitive designer diverts the favorable factors to his benefit; in like fashion he subverts the unfavorable factors that would compromise his objectives.

When a building sits comfortably on its site, there is a good chance that its occupants will be content. This is basic.

If the designer does his homework, the energy flows that occur in nature will be in harmony with the design. The habitat will work with these flows rather than ignoring or antagonizing them. "Don't fight the site," make it work for you.

In a simplistic yet primary sense, one can think of a habitat as an expensive solar collector with integral storage. If the design is thoughtfully executed with sensitivity and respect for all the forces and flows that exist at the site, the result will be a unique solar collector—a collector that you can live in.

Doing a thorough job with homework is the big problem. Architects and designers are visual creatures. The architectural program may follow function as far as floor areas and circulation patterns are concerned; but the form as it exists in the mind of the designer too often follows fashion. The architectural precept "form follows function" has been subverted by the limited approach of the practitioner.

The flow that any designer in the solar idiom thinks of first is the solar flux, the sunshine or solar energy that falls on the proposed site. Millions of dollars have been spent on collector design and analysis, hundreds of papers have been published, many (like Hottel & Woertz) have become the basis of a whole technology; books have been published, some very good (Duffie & Beckman), and some not so good. It might appear that the use of solar energy for heating buildings is an open book. The designer need only to apply well established principles to develop a satisfactory system. This is not true.

It is a common opinion that solar energy is a diffuse or thinly distributed source, and that this property poses big problems in solar design. Used to dealing with the condensed energy of fossil fuels, we try to apply fossil fuel technology to solar applications and cannot get a fit. System size and cost get out of hand. We pay a gross premium for clumsy design. Perhaps the problem lies in our basic logic. Perhaps we know all the answers, but in the meantime the questions have changed. We had better question some accepted precepts.

Solar energy is a concentrated, not a diffuse, source. If we take the total yearly production of coal in the United States, reduce it to Btu and divide this figure by the area of the United States in square feet, we find that coal represents less than 170 Btu per square foot per year. The same typical square foot will see about 300,000 Btu of solar energy. We have been confused by an elaborate technology and a complex transportation system that has developed within a fossil fuel society.

It is commonly assumed by solar designers that the first step in developing a solar heated habitat is to make sure that the building is a tight, low heat loss, thermal envelope. This is a quality that can easily be justified by a heat balance. Building losses must be provided for, the provider being the solar collector area, and each square foot of collector has an associated cost. This is straight forward calculation, but the philosopher might well look at this argument, and speaking from within his protective cloak of technical ignorance, ask: why is so much emphasis placed on conserving a nondepletable resource? The question would, of course, remain unanswered. Something subtle is missing, for within our present technological ritual, there is no answer.

Life-cycle costing and other systems of economic analysis have been called on to justify the use of conventional stereotyped solar systems. The response of the financial community has been less than enthusiastic; witness the aborted solar features of the new Citicorp building in New York. True, in the term they will pay out, but in the term we will all be dead. Who wants to wait? Innovative technology needs to be introduced, but the majority of the profession has doggedly plugged along applying conventional practice to what is in fact a whole new ball game. Old habits are hard to break.

Each specific site has its own spectrum of advantages and disadvantages. The trick lies in putting together solutions based on the qualities of the site itself. Site adaptation has been well addressed by a number of workers. Architect Richard Crowther[1] and others have published good books devoted largely to this aspect of environmental design. Rather than catalog a large number of factors, I will offer a few examples from my experience that will demonstrate the sort of reasoning that I think makes sense. Some are examples of my own mistakes.

I once added a fairly large living room to the modest log cabin in which I live in northern New Mexico. To deal with the harsh summer sun that is intensified by the altitude (6,000 feet), I designed a pitched roof with natural ventilating channels. Heated air tends to rise and the flow that is produced by this effect is referred to as natural convection; likewise, flow produced by a fan is called forced convection. In this case, solar radiation, first struck a light weather roof, and the natural convective path below removed most of the heat and discharged it through a high vent before it had a chance to encounter the building insulation. The scheme worked fine, but the whole system has now been abandoned. The ventilating channels have been filled with vermiculite insulation. How did this come about? Well, a deciduous tree that was small when the project started has now grown up to shade the roof in the summer. Special measures are no longer needed to keep the space comfortable. Such a simple solution as the intelligent use of plantings is an important part of any site adaptation scheme. I did learn something from the exercise—a lush growth of weeds shades my greenhouse during the summer. I leave them alone until October when they are dry and easily pulled and piled with a rake.

There is a magnificent hacienda near Taos, New Mexico that I have studied for a number of years. This adobe building is sited on a ridge overlooking the Ranchos de Taos Valley. Esthetically everything is right, functionally everything is wrong. There is an orchard in the yard, irrigated by the community acequia that runs along the ridge above the place. Centuries of irrigation have resulted in a perched water table, so the site is always lush. But adobe is hygroscopic, that is, it readily absorbs and retains moisture; further, its insulating properties or thermal conductivity are a strong function of moisture content. A damp adobe wall will lose heat many times faster than a dry wall. The house is always cold, even in the summer when evaporative cooling is also a factor, and I would classify it as uninhabitable in the winter. A classic case of the wrong material on a promising site, or the right material on the wrong site. Site adaptation involves not only the surface features and the microclimate; one must also look below the surface.

In rough terrain, particularly where there are nearby mountains or high hills, there is a characteristic pattern of winds during the summer months. During the day the sun warms the valley floor and walls, which in turn warm the air, making it lighter as it heats up; the result is a flow of air up the valley. This is frequently referred to as a thermosiphon effect and is another example of natural convective flow. At night after things have cooled down, the flow reverses, and the cooler air from the higher elevations flows down the valley. Simple enough in concept, but in the real world the mechanism is most complex. Mass transport of moisture is a factor, and most valley-mountain systems, if examined in detail, will be found to be a semi-closed microclimate existing as an independent environmental entity. Evapotranspiration (a big word that can in a sense be thought of as the "breathing" of plants) puts moisture into the warm air that is flowing up the valley during the day. When this air reaches the cool higher altitudes of the upper valley, much of the moisture that it is carrying condenses and a portion of this water enters the stream system that characteristically flows down the valley. Thus, a stream gauging station located in the lower reaches of the valley could easily measure the same bit of water over and over again during a given growing season. This is one reason that phreatophyte control, conceived as a water conservation measure, frequently produces surprising results, usually adverse. (Phreatophyte control is a term used by water "conservation" people so that you won't know what they are up to; read as: "stream bank defoliation!")

Adaptive architecture can take advantage of these natural flows, but only with great care. During the day, the up valley air flow can be used to induce natural ventilation through a building. If the air flows through selective plantings before entering the building, the evapotranspirative cooling of the plant will cool the entering air. A neat system.

Taking advantage of the evening down valley flow can be pretty tricky. This flow usually begins as a thin sheet of the heavier cool air flowing along the ground, while the gross up-valley flow continues above it. So, in a given location we have down flow close to the ground and up valley flow above it. The cool air sheet thickens with time, until at some point later in the evening all air flow is down the valley. Nature, as usual, tries to advise one about this temperature inversion. Look to fruit trees and other early blooming vegetation. In my orchard there is a clearly defined line below which the chill mountain air has killed the low blossoms, and above which the air has not dropped below freezing.

Trees and other vegetation tell one a great deal about a site. Even for a site without weather records one can estimate the design requirements by matching the growth to a location for which good records do exist. Willows warn of a high water table or a local moist condition that might

1. *Sun/Earth by the Crowther Solar Group, etc.*

Make the site work for you, as designer David Wright did with his Karen Terry House II, by using the earth as shelter for the north side and exposing the south to the rays of the sun.

not be suspected. Vegetation will often indicate the chemical or mineral nature of the land; an experienced botanist or forester can be of great help when these factors need to be evaluated.

I once sited a sewage lagoon at the toe of a sunlit valley wall. The thermosiphon induced by the hillside produces air flow across the lagoon surface and up the hill. Aerobic decomposition is encouraged by a steady supply of fresh air across the water surface, and any odor resulting from anaerobic digestion is swept away and diluted without becoming a nuisance. The design has worked well for a number of years.

The best way to evaluate the microclimate of a site is to observe it in detail for a full year. Most designers cannot afford to spend the time to do this, but on the other hand, if the habitat is to be occupied for generations perhaps such effort is entirely justified. Should those who come after us be subject to mistakes we have made in our haste? We often speak of progress, but much of what we do simply is not progress.

I have been following the development of a greenhouse that was sited with great care. This structure is built into a south facing hillside. The site study continued for a full round of seasons. A lady we will call Marlyn decided on a rather unique approach to environmental accommodation. Basically, the idea was that if the given environment was too harsh, one could create an environment, in this case by building a big barn of a greenhouse, inside which a habitat that was compatible with the created environment could be constructed. One would step out ones' door and have two seasons available. The greenhouse season, with growing plants and a chaise lounge, or the bright sun, snow, and sparkling cold of a typical New Mexico winter at 7,000 feet. The whole project was to be primarily solar heated.

Marlyn spent the best part of a year testing promising sites on a large tract of land. This involved a lot of winter days with a good book on sunny south slopes, and a lot of below freezing nights in a sleeping bag. She did her work so well that the inner dwelling is simply delineated, rather than isolated, from the primary greenhouse. Isolation was not required. Backup heat for the living area is provided by a wood stove, and heat stratification (warmer air will hug the ceiling area) keeps the sleeping loft comfortable after the fire goes out. The structure is about 65 × 20 feet, and the budget involved $6,000 in cash and a lot of help from friends and persons interested in the project. A large part of the success of the habitat has resulted from the loving care that produced a quality site study before the first batter board was set.

Another flow of energy that tends to be neglected by designers is so called night sky radiation. As Budyko and others have pointed out, the Earth, considered as a free body in space, is in near thermal equilibrium, i.e., Earth loses about as much heat as it gains. If this were not so we who inhabit Earth would be in peril, the temperature would slowly increase or decrease until living things could no longer survive. As with any body undergoing radiant energy exchange, this is a continuous two way process. During the day the incoming solar energy is largely visible light, with a substantial component falling in the infrared and a lesser component in the ultra violet. Some of this radiation is reflected, some of it is adsorbed, and some of the adsorbed energy appears as heat, warming the adsorbing surface. The heated surfaces re-radiate in the infrared, and much of this energy escapes through the atmosphere to outer space. At night the predominant energy flow is from Earth, as the surfaces that have been heated during the day radiate to the cool night sky, thus the term night sky radiation.

Night sky radiation was pretty well understood in ancient times. It was used to collect water from the atmosphere for domestic use, and to irrigate crops. The most spectacular application was probably to produce ice in areas of the Middle East where temperatures seldom dropped below freezing. The technique involved using massive structures that were shaded to allow minimal heat gain during the day, and to keep the water pools that were to be frozen insulated to prevent heat gains from the surrounding earth. At night the pool surfaces would see little heat gain from the surroundings, but would lose heat by radiation to the sky. Temperatures would drop and ultimately ice would form. These structures are known as "ice walls" and "ice pits." There is abundant archaeological evidence of their use.

The physics of the process is quite straightforward. If a body of water is exposed to a clear sky on a quiet night, and if it is well insulated from its surroundings, it will radiate to the sky and be cooled. When the water reaches the dew point (that temperature at which moisture will start to condense out of the humid air) the pool will begin to collect moisture. If the air is dry, or stratification has depleted the moisture content in the air at the surface of the pool so that the dew point of the air in that region is below freezing, conditions are favorable for the formation of ice. Twenty Btu per hour per square foot is probably a typical energy flow rate.

Site and architecture can sometimes be combined to create a microclimate by design. The main building of the classic Taos Pueblo is an example. The site is flat, no nice south facing hillside to take advantage of, yet the architecture creates one. The building faces south with each story stepped back one room module. In winter the low angle rays of the sun heat the vertical faces. Natural convective flow develops up the wall and across the first roof top work area. This cross flow is induced by the up-flow on the next wall, and so on to the top of the building. On sunny winter days the Taos people spend a lot of time on these comfortable work areas. The ancient Taos designers indulged in lavish use of their non-depletable solar resource without recourse to a lot of solar hardware. One wonders how the contemporary aerospace oriented solar designer would deal with the problem!

Sometimes the most innocent set of circumstances will produce a microclimate that is worthless. An example that comes to mind is a solar collector sited on the bank of a stream. Many streams are partially fed by springs in the stream bed that introduce a flow of water at ground water temperature. The combination of relatively warm water and very cold air produces a fog that effectively blocks the sun from the collector surface. This most often occurs on very cold days, the days when the collector needs all the help it can get.

The trick in making effective use of microclimate, as I have said, lies in putting it all together. Let's examine some ways of pulling together some of the factors already discussed.

A south slope is a good site for a habitat. We can carefully use the natural convection up the slope to wash our collection system covers with warmer air and thus reduce the temperature differential that the system must deal with. If the system is a greenhouse, located below the dwelling, natural convection can again be put to work to circulate warm air throughout the building. A well designed greenhouse is, in a sense, a contrived microclimate that will not freeze. Why not put our collectors for heating domestic water inside the greenhouse? By doing so we avoid all the fuss with anti-freeze, heat exchangers, and code problems. In this example we have used natural convection in three ways. We have improved the performance of our collectors, provided for natural air circulation throughout the dwelling, and put the water heater in the ideal location to establish a convective loop between heater and storage tank. Further, by creating a microclimate in the greenhouse, we have developed an environment that allows us to use a simple water heater rather than a complicated non-freezing unit. One should also note that pumps and blowers are not mentioned or required. Simplicity can be the height of elegance.

A well designed dew pond can be a source of cold water for summer cooling. I have already mentioned dew ponds: basically a pond well insulated to prevent heat gains from the earth that is cooled by night sky radiation. Dew ponds are site specific, but if conditions are favorable they present a means of zero energy primary cooling. The quantity of energy required to make use of the chilled water will depend on the ingenuity of the designer.

Vegetation can have a profound effect on the microclimate of a site. Make use of what exists, and consider contrived plantings to be as much a part of the environmental design of a dwelling as the fabric of the building. A whole spectrum of potential applications exist: shading, cooling, humidity control, and seasonal wind flow control, to mention only a few.

Microclimate, natural or contrived, may well be the solution to the present contradiction that exists in the contemporary practice of solar design. We are dealing with a non-depletable resource; we should be free to use it lavishly. But present practice is oriented toward capital intensive solar hardware. The high cost of these systems dictates adherence to the energy conservation ethic. We must conserve a perpetual self-renewing energy resource. How foolish! A highly mechanistic solar cooling system can air condition a building at great expense. A planting of alfalfa can cool a whole valley and provide a valuable salvage resource at the end of the cooling season. The cost of this collector is reckoned in dollars per acre rather than dollars per square foot.

I have no quarrel with the retrofit oriented solar hardware designer trying to solve todays problems. But, in the long run, solar hardware is the next generation of dinosaurs! ☼

Dr. Douglas Balcomb's house, Santa Fe, New Mexico

The Solar Greenhouse

by Douglas Taff

Doug Taff is Executive Vice-President of Parallax, Inc., Hinesburg, Vt, and a technical consultant to Solar Age Catalog.

All greenhouses are, of course, solar. They have windows that admit sunlight. The big difference with the *solar* greenhouse is that the heat gained from sunlight, rather than escaping soon after it arrives, is captured and held for nighttime use.

The one essential step involved is storage. You have got to be able to store some of the solar heat; this is the first imperative of the *solar* greenhouse. Before discussing storage methods, however, it would be wise to look at the reality of greenhouses that are not *solarized.*

They use fuel with abandon. Some five to fifteen billion Btu can be consumed per year just to maintain one acre of commercial tomato production. A hobby greenhouse will kill you in the same way. The fact that they produce home grown vegetables pales in light of fuel bills that average hundreds of dollars per year. In 1976 over one thousand acres of home greenhouses were sold in the United States, and there is probably four to six times more acreage in existing home greenhouses than the commercial variety. That's 30 to 45 thousand acres of food production capacity flawed by unholy fuel consumption. These days, with the cost of energy rather steep and solarization of a greenhouse, especially a smaller one, being a relatively easy step, to permit such consumption is like spitting into the wind.

Solar greenhouses are energy self-sufficient, and whether commercial, hobby, or home-sized they are really quite different creatures than conventional greenhouses. By adding heat storage systems and insulation, and by using the latest window glazing materials, a solar greenhouse might not only become self-sufficient, it could also produce heat for any house it is attached to. One of the best, most attractive ways of using this idea is to design the greenhouse as an integral part of a new house. Dr. Douglas Balcomb's house in Santa Fe, New Mexico and Parallax III in northern Vermont show that this approach is adaptable to different climates. (See photographs.) The Balcomb house makes use of a thick adobe wall separating the greenhouse from the rest of the building to absorb and store heat. As the greenhouse becomes hotter and the warmer air rises naturally up to the roof, fans and ducts carry it away to tons of rock held in a bin beneath the floor. Heat will be stored there for nighttime use by the entire building.

Parallax III has a similar storage system and is designed to gain from 52 to 65 percent of its heat from the

Table 1. Fuel reduction contributed by an attached solar greenhouse.***

City	65 Degree Base	Solar Heat Produced by Greenhouse* (kW-hrs)	Heat Loss of Greenhouse (kW-hrs)	Ratio of Heat Gain to Heat Loss	Fuel*** Reduction (%)	Savings **** ($)
New York, NY	4871	5652	2007	2.8	32.1	324.85
Boston, MA	5634	5592	1856	3.0	28.5	227.90
Burlington, VT	7865	4476	2931	1.5	8.4	77.25
Philadelphia, PA	5251	5452	1548	3.5	32.0	206.91
Baltimore, MD	4654	4818	1414	3.4	43.5	156.58
Chicago, IL	6155	4993	2159	2.3	19.8	130.36
Springfield, IL	4561	5754	1821	3.2	37.0	173.05
Milwaukee, WI	7205	5965	2735	2.2	19.2	125.97
Denver, CO	6283	7897	1996	4.0	40.3	224.24
Dayton, OH	5597	4803	2042	2.4	21.1	99.40
Cincinnati, OH	4870	5003	1356	3.7	32.1	124.00
Duluth, MN	10000	6809	3968	1.7	12.2	

* Energy available after transmission and reflection losses subtracted
** Based on 55° nighttime setback and materials as described in Table 1
*** Dwelling is assumed to use 2.33 kW–hrs. per degree day (base 65). This quantity of heat is typical of an average U.S. home.
**** Value of energy is based on available electrical costs during January, 1976

Dr. Balcomb's house and Parallax III demonstrate attractive ways of designing the greenhouse as an integral part of a new house.

greenhouse. The use of large storage bins in both houses prevents waste from occuring when, at the peak collection hours, the greenhouse begins to overheat. By ducting the excess to storage bins the heat gain will increase, because a hot greenhouse will lose heat to the outdoors more rapidly.

A solar greenhouse can easily be added on to an existing house or building. The Holdridge house, also in Vermont, uses four 55 gallon drums filled with a total of 208 gallons of water as storage mass—called "thermal mass"—to hold heat collected by its greenhouse addition. The water drums are located in the greenhouse itself to be heated directly by sunlight. There are also two blowers: one at ceiling height blows warm air from the greenhouse into the main house, and the other takes cooler air from the floor level of the house into the greenhouse to be warmed. A differential thermostat continuously senses the greenhouse temperature and checks it against the house temperature. If the temperature in the greenhouse is 5° warmer than the house, the blowers turn on and bring heat into the house. Overheating in the greenhouse is prevented by supplying surplus heat to the house throughout the heating season. Table 1 shows the approximate contribution that an attached solar greenhouse can make to a single family house under the weather conditions of various American cities. (This table does not, however, represent the maximum heat producing potential. Thermal blankets will increase the amount of heat by cutting night losses.)

From any point of view solar greenhouses make a lot of sense. They can provide the basic necessities of life—food and heat—and also provide a warm, livable space. Everyone who has an image of what a solar greenhouse is sees it a little bit differently. To some it's a food growing facility; to others it's a low cost solar collector. It can be almost anything its designer wants it to be.

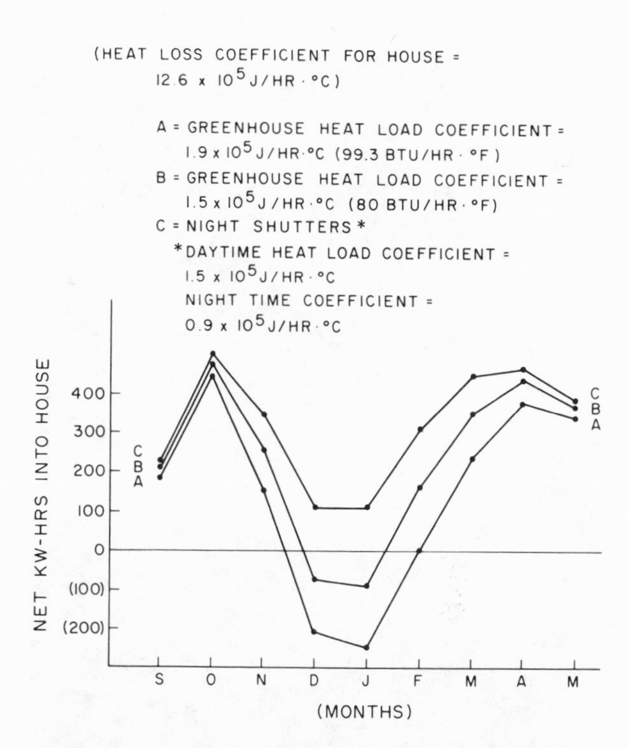

Fig.1. Simulated monthly performance of a prototype greenhouse in Burlington VT, used to heat a house.

For a completely fresh perspective, consider what a solar greenhouse could mean to people in developing countries, a Peruvian farmer in the foothills of the Andes for instance. It's a means of extending his growing season. It's a way of providing fresh vegetables and a source of energy to warm his home. It's a way of saving 60 to 80 percent of his yearly income that would normally be spent on food and fuel. To a sociologist or economist the

Table 2. Yearly vegetable production in a 10 ft. by 12 ft. greenhouse.

Vegetable	Yield (lbs)	Value ($)
beans	20	13.80
beets/turnips	14	4.20
chard/spinach	16	6.24
cucumbers	68	19.72
lettuce	81	39.69
onions	24	6.48
peppers (all kinds)	12	8.28
radishes	9	3.60
squash (summer)	40	7.60
tomatoes (small var. potted & hanging)	184	145.36
TOTALS	468	254.97

The thick adobe wall separating the greenhouse from the rest of the building absorbs and stores heat.

greenhouse might represent a means of stimulating self-sufficiency and self-help programs.

Consider these same benefits applied to the American consumer. His standard of living might be higher than the Peruvian farmer, but the economics of food and fuel still dictate, to a great extent, the quality of his life. No sensible person argues against fresh vegetables. At least admit that home grown tomatoes are inevitably better and fresher tasting.

Few people realize how much food a small inexpensive greenhouse can produce and how beneficial the savings can be if heat is removed from the operating expenses. Table 2 illustrates reasonable annual yeilds that could be expected from a small 10 by 12 foot greenhouse using a typical mix of vegetables prized by Americans.

Mid-winter growth was stressed in this chart so the numbers are not inflated by summer produce. The true value of the vegetables is well over the $250 indicated because of decreased spoilage, increased quality and better nutrition.

The heat producing capacity of a solar greenhouse is somewhat more complex. In most cases designers have been working on intuition rather than hard data. Therefore, the results of the study reported here are only preliminary.

Let's examine the yearly solar collection efficiencies and capacities of window-wall systems versus active collector systems that have been intensely monitored by A. O. Converse and his team at Dartmouth College.

Burlington, Vermont, the location we will use for our comparison, is located at 45° N latitude. Its winters are characteristically cold (8,000 degree days) and cloudy (24 percent of possible sunshine in December). Active systems are not generally cost-effective in this climate because of the heat losses resulting from the large temperature difference (Delta-T) between the absorbing surface of the collector and the outside air. The maintenance and high initial costs of active collectors dictate a high fixed cost for any solar heat delivered to the house. Windows (i.e. as in greenhouses), on the other hand, operate at a much lower temperature difference during daylight hours, but nighttime heat losses must be controlled if they are to be effective collectors. Window-walls are only moderately expensive, inherently long lived, and have low maintenance costs. Active systems installed in areas north of 45° N latitude consistently demonstrate design and life-cycle cost problems directly related to cold and cloudy climates. It is therefore extremely important to create a data base that analyzes the thermal efficiencies of several window-wall/solar greenhouse systems and compares them to active collector designs both in terms of their total yearly solar heat delivery capacity and their life-cycle. Table 3 illustrates a study for Burlington. Listed are six window systems, as well as six active collector systems. Each is designed with heat storage. The active systems have been monitored in actual operating and the data presented has been adjusted for Burlington's climate.

The data indicates quite clearly that for window-walls, efficiencies, delivery capacities, and lower costs are accomplished by adding additional layers of glazing, night drapes, and inclining the glazing to optimize sun angle. A solar greenhouse-type design is obviously the way to proceed. The active system data is presented and lists the type of system, glazing detail, tilt angle, absorber plate design, and coating.

This table indicates two important things: 1. For Burlington, Vermont a double glazed window system, with adequate storage and a nighttime drape will be as effective a heat collector as most active collector systems, but at a lower cost. 2. The same system as above, but inclined to take advantage of the sun's geometry was shown to perform better or equal to any of the active systems tested—even those containing solar assisted heat pumps. Such inclined, double glazed, and draped walls are commonly found in solar greenhouses. They are twice as cost-effective as solar assisted heat pump systems and almost three times better than conventional active collector systems.

The present costs of electric and oil heat in Burlington are $14.64 per million Btu and $5.10 per million Btu respectively. The active systems are barely competitive with electric heat even if maintenance (not factored into the chart) is ignored. Solar greenhouse systems, especially if night draped, are competitive even with present day oil prices. If one also accepts the idea that such systems will last much longer than 20 years (there's hardly anything that can go wrong with them) and that during their life maintenance will be minimal, then their life cycle cost appears even more inviting.

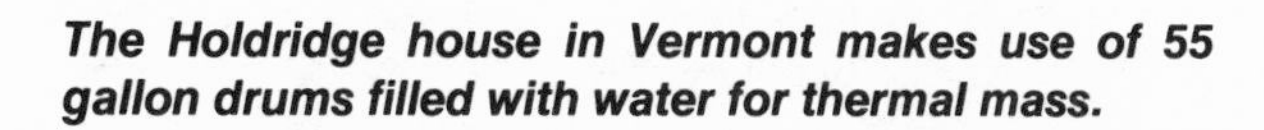

The Holdridge house in Vermont makes use of 55 gallon drums filled with water for thermal mass.

Table 3. A comparison between effective collectors.*

COLLECTOR SYSTEM	EFFICIENCY (%)	BTU/FT²-YR (Kcal/m² yr)	$/10⁶BTU-YR
WINDOW–WALL			
SINGLE GLAZED (u = 1.13, t = .88)	0	negative (———)	———
DOUBLE GLAZED (u = .55, t = .77)	8.7	12,722 (34,502)	22.60
TRIPLE GLAZED (u = .38, t = .68)	17.4	25,189 (68,312)	14.64
DOUBLE GLAZED PLUS NIGHT DRAPE (u = .55/.17, t = .77	27.3	40,163 (108,922)	9.96
DOUBLE GLAZED AT 60° (u = .55, t = .77)	24.8	50,620 (137,281)	5.86
DOUBLE GLAZED AT 60° PLUS NIGHT DRAPE (u = .55/.17, t = .77)	39.6	80,928 (219,477)	4.94
ACTIVE COLLECTOR			
AIR SYSTEM – DOUBLE GLAZED (vertical, finned, selective)	29.2	42,908 (116,367)	18.06
AIR SYSTEM – DOUBLE GLAZED (vertical, corrugated, paint)	28.4	41,764 (113,264)	17.79
AIR SYSTEM – DOUBLE GLAZED (60°, flat plate, paint)	21.5	43,797 (118,777)	12.50
LIQUID SYSTEM – SINGLE GLAZED (45°, flat plate, paint)	28.8	57,615 (156,252)	13.02
LIQUID SYSTEM – SINGLE GLAZED PLUS HEAT PUMP 45°, flat plate, paint)	44.2	88,510 (240,039)	10.59
LIQUID SYSTEM – DOUBLE GLAZED PLUS HEAT PUMP (45°, flat plate, paint)	37.7	75,507 (204,775)	9.84

*DATA SET FOR BURLINGTON, VERMONT

Taking this preliminary data one step further and including other cities within the United States, it can be easily shown that small (8 by 12 foot aperture) solar greenhouses can be effective solar heating systems. Look again at Table 1, where, incidentially, the effect of thermal drapes are not factored in. The addition of drapes can have a significant effect in the coldest, cloudiest cities. In Burlington the drapes (R-4) essentially double the net heat output (Table 3). This can also be shown graphically (Fig. 1) by simulating the net heat output by a month as a function of three different greenhouse construction methods. The greenhouse is assumed to have 96 square feet of glazing and 108 square feet ot floor area. Decreasing the heat loss by adding insulation and cutting air infiltration and by ultimately adding thermal drapes has an immediate, beneficial effect on the yearly net heat output to the house.

Food and fuel are basic human requirements. The promise of a solar greenhouse lies in its use of simple, universally available technologies to balance the debits and realities of this earth. There's probably one in your future.

BIBLIOGRAPHY

1. Gillett, D.A., "Solar Powered Greenhouses," Proceedings of the First Annual Conference of the New England Solar Energy Society, Amherst, MA., 537–550, (1976).
2. Lawand, T. A., "The Design and Testing of an Enviromentally Designed Greenhouse for Colder Regions." *Solar Energy,* 17, 5, 307–312, (1975).
3. Marlboro Solar Greenhouse Conference Proceedings, Marlboro College, Marlboro, VT., (1977).
4. Michal, C., "Glazed Area, Insulation and Thermal Mass in Passive Solar Design," Proceedings of the First Annual Conference of the New England Solar Energy Association, Amherst, MA., 295–301, (1976).
5. Nash, R. T., et al., "Thermally Self-Sufficient Greenhouses," Proceedings of the 1976 Annual Meeting of the American Section of the International Solar Energy Society and the Solar Energy Society of Canada, Winnipeg, (1976).
6. Taff, et al., "Design Considerations, Theoretical Predictions, and Performance of an Attached Solar Greenhouse Used to Heat a Dwelling," Proceedings of the 1977 Annual Meeting of the American Section of the International Solar Energy Society, section 33–1, Orlando, FL., (1977).

A Direct Gain Solar Cabin

by David Wright

David Wright is an Environmental Architect based in Sea Ranch, Calif.

Fig. 1. Do-it-yer-self solar module.

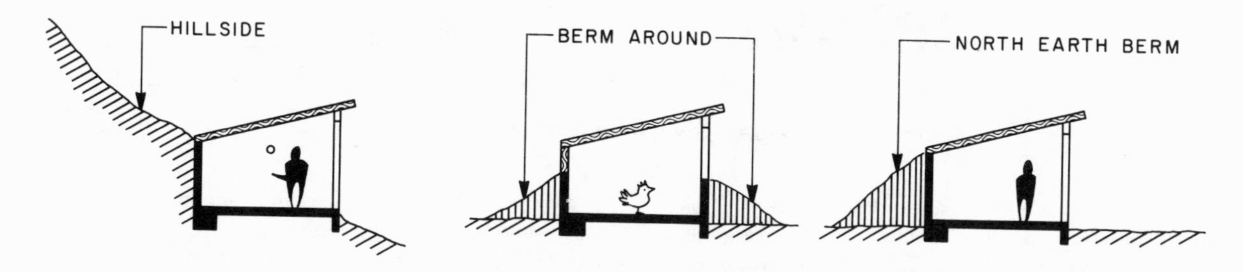

Fig. 2. Solar module variations.

Solar energy jargon can be terrifying to the beginner. It is easy to get knocked off the track by the various concepts and terms before you actually do something.

If you have more regard for "hands-on" than book learning, build a solar workshop or weekend cabin. You will have something for your learning effort and, rather than moving basic concepts around in your head, you can move them around at the building site.

The design presented here is a prototype that can be built with many variations. I'm going to give you the basic guidelines and drawings. You can choose the specific materials and add your dose of imagination.

This basic passive design uses the "direct gain" concept that, like a solar greenhouse, turns the building itself into a solar collector and storage unit. The essential requirements are few, but they should explain to you how sunlight must be courted to keep a building warm.

Solar Collection is accomplished by an oversized window, which in fact amounts to a "window-wall." This faces south and admits winter sunlight into the cabin.

Once you have the sunshine inside you need *Heat Storage* to keep the energy there for nighttime use. This is done with materials—called thermal mass—that can absorb and hold heat (water and masonry are used here).

With the heat collected and absorbed, keeping it within the cabin requires a little *Thermal Integrity*. This means that you don't cheat on insulation and weatherproofing. You must build a heat envelope.

The final requirement is the keeper of the others; it is *Control Flexibility*, the creative use of shading and moveable insulation to prevent overheating as well as nighttime heat loss.

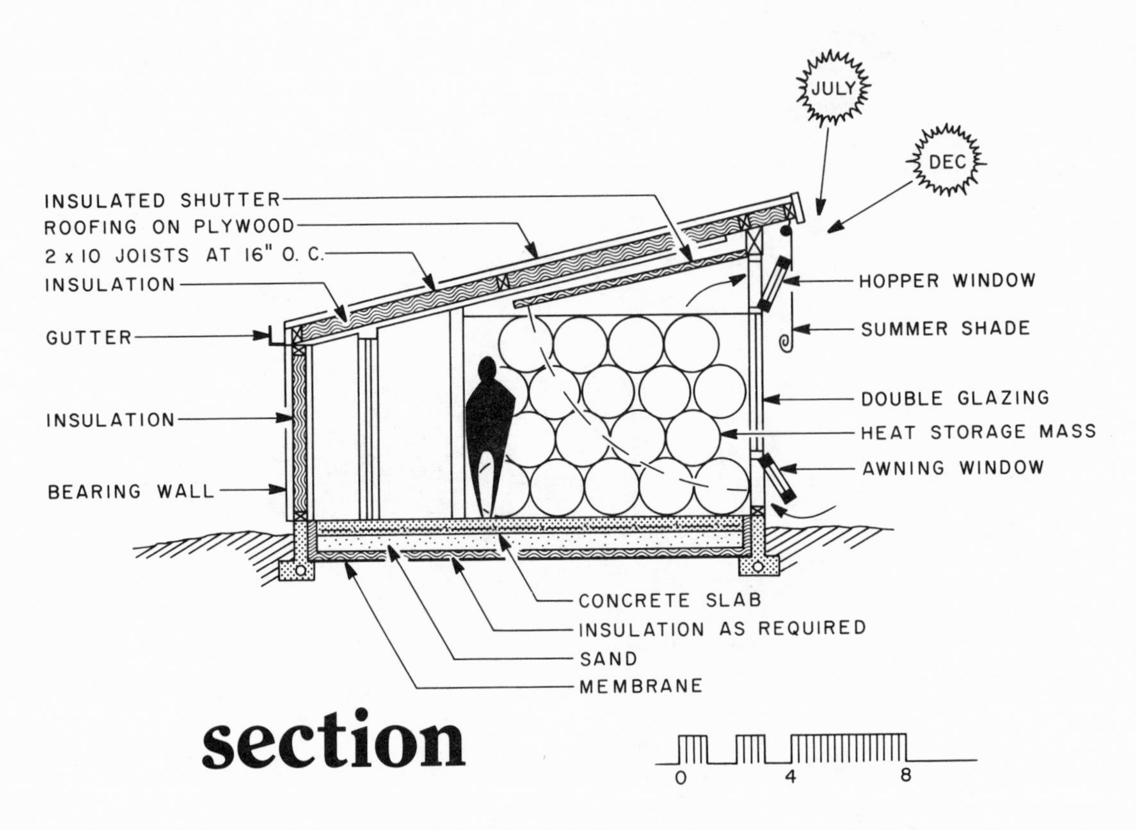

Fig. 3. Solar module section.

The "greenhouse effect" is one of the main principles behind collecting solar heat. Sunlight arrives in the form of shortwave radiation. These waves pass easily through the window-wall and strike a surface. As the surface material heats up, it re-radiates some of its heat in the form of longwaves. Most of this longwave energy will not readily pass back through the glazing. Much of the original energy is trapped in the space in the form of heat. It is stored in the thermal mass and air, escaping back to the outside at a much slower rate than it came in. The idea is to keep it where you want it, as long as you need it.

Passive designs can be adapted to local building techniques, architectural styles, and materials. Salvaged or scrounged materials can often be used. You are not obligated to use equipment or materials that are marketed by manufacturers for the building industries. On a grass roots level, you can build a structure for a minimal capital outlay. This too is energy conservation. I recommend, however, that you use good durable glazing materials (glass, greenhouse fiberglass, and the like) and quality insulation (fiberglass, treated cellulose, or plastic foams). These are generally worth the investment because they are effective, safe, and long lived.

A wood frame system is illustrated here. It is simple and effective, and can also be adapted to make use of post and beam, poured concrete, masonry, adobe, log, rammed earth, and so forth. The fiberglass insulation (R-19 walls and R-27 ceiling) can be replaced by equivalent values of any other insulation material (wood, treated cellulose, pumice, polystyrene, polyurethane, rock wool, etc.). It is good to build well and seal all cracks. Do the job once. Build for permanence.

Consideration of the building code is important. Usually, local building officials are concerned with the health, safety, and well being of the general public. It is always a good idea to comply with their regulations. They might not be "tuned-in" to natural solar design, but they can offer important information and will probably be interested in how your design works. Local conditions vary considerably, and building department officials can help you determine specific data such as soil bearing values, earthquake problems, depth of frost-line, and other characteristics that should be taken into account.

Remember that with passive solar design it is difficult to completely "blow it". If you only get 60 percent solar heating, that's far better than a conventional building does. I tend toward design "overkill" when possible. With enough flexible insulation devices, sun shades, and thermal mass, it is easy enough to control gain and loss to attain about 80 percent heating in most climates.

As a "rule of thumb" for climate zones where the outside 24 hour low average temperature is above 30° F most of the year, a good ratio of window-wall collection area to floor area is about 1 to 3. This will normally be enough to maintain an inside temperature above 60° F, provided proper operating technique is used. When the sun goes away for several days or when the outside temperature is expected to drop below 10° F, close the insulating shutters and use an auxiliary heating source (woodstove, gas fired heater) to stay comfortable.

The thermal storage mass can be sized for the number of days of "normal" cloud cover. Use a total of 150 pounds of stone or concrete, or 30 pounds of water per square foot of window-wall area. When possible, storage mass should be located where the sun will strike it directly. If the sun does not strike it directly, use approximately four times the calculated mass to be safe. The heat energy is distributed to mass outside the direct gain surfaces by heat stratification, thermal convection currents, and longwave bounce or re-radiation. This indirect heat distribution is difficult to calculate, yet it occurs automatically. All mass within an insulated structure tends to seek thermal equilibrium. Whether one heated mass can "see" another and radiantly transfer its energy or not, the heated mass will try to give, and the cooler mass will attract heat, sharing the energy until they are all the same temperature. Sometimes this occurs quickly, sometimes not. The factors that affect this heat transfer process depend on the nature of the mass, shape, color, relative temperatures, and exposure of the interior materials in relation to one another. It is difficult to calculate this transfer, but common sense will often tell you what to expect.

In this design the thermal storage mass is the concrete floor slab and the water-wall room partitions. The floor slab is in an ideal position for absorbing direct sunlight all day long. The heat energy will be driven down into the concrete and absorbed. A warm floor is one of the great pleasures of a direct gain solar house.

The water-walls are built perpendicular to the solar window and will receive heat on each side, first in the morning, then in the afternoon. Masonry will also work well in this position. The higher portions of the wall that are not directly hit by sunlight will absorb scattered radiation and the heat rising to the ceiling. This partition, when it does lose its stored heat, will lose it back into the room and not through an exterior wall.

Posts and a bearing beam carry the weight of the roof and anchor the storage mass. If you anticipate adding

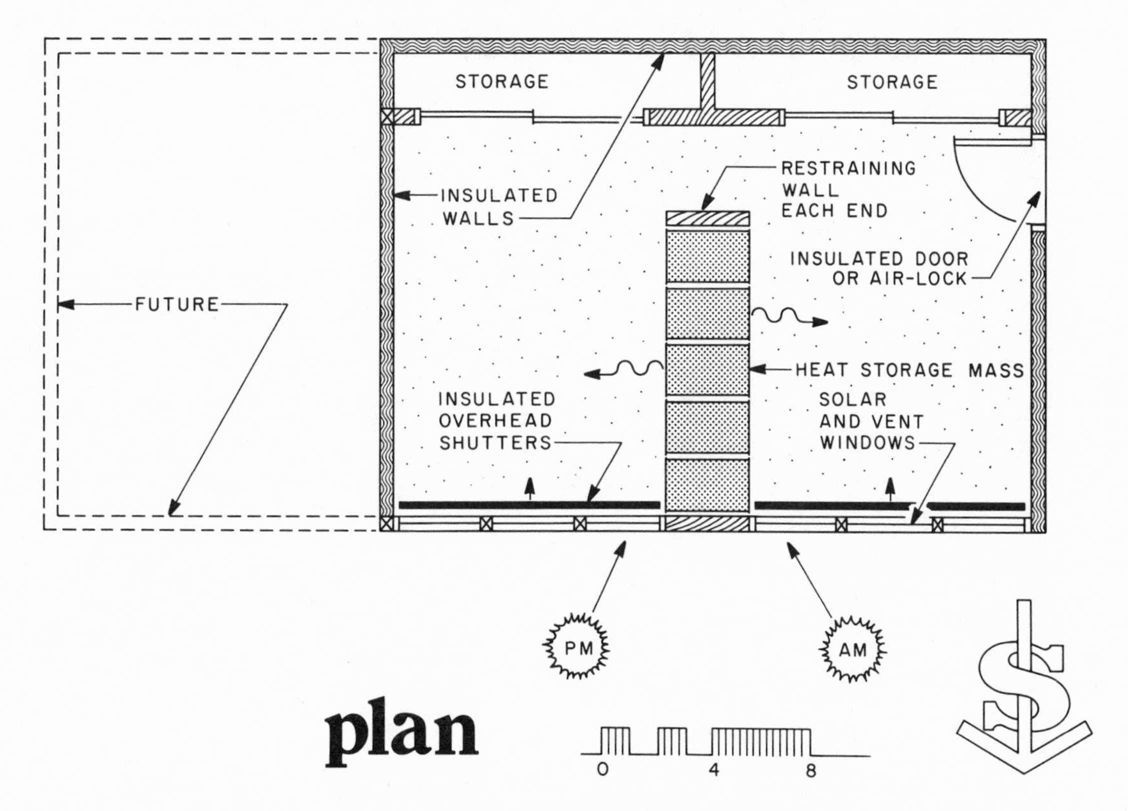

Fig. 4. Solar module plan.

another room in the future, build the adjacent wall as a post and beam frame. Then, when you are ready for the addition, remove the exterior siding and insulation panels, use them for the new outside wall, and install thermal mass in their place. This system may be expanded as many times as you like.

For cold climates with over 6,000 degree days of heating and where the frost line is deep, insulate down to the top of the footing and under the floor slab. It is also good practice to increase the thermal mass to store as much heat as possible when the sun is available. This can be done by increasing the floor slab thickness and by adding additional water containers. In extremely cold climates, above 8,000 degree days of heating, the cabin can be sunk into the ground and earth bermed on the north or weather sides. This increases the insulation, reduces heat loss profile, and greatly reduces infiltration loss.

When experimenting, allow room to increase the thermal mass. I like to rough calculate, or "eye-ball" the design, then try it for a winter season. Then, if you need to insulate better or add storage, this can be done without messing up the usable floor area. The water storage vessels can be fiberglass tubes or boxes, oil drums, paint cans, glass jugs, or just about anything else than won't leak. If the appearance of these vessels is not to your liking, cover them with a dense conductive plaster. The plaster must bond to the vessel to assure conductive transfer. An attached metal covering will radiate and conduct heat to the containers, and some vertical air spaces will allow convective transfer within the wall.

The trick to direct gain systems is to allow for flexibility. The weather is seldom "normal." Certain years will have a long winter, or a warm fall, or a cold spring. If you size the overhang for total summer shading, you might be sorry. I like to allow enough roof overhang or beam extension to accommodate outside vertical shades of bamboo, snow fencing, canvas, or the like. Then you can shade when needed as much as you like. With properly placed low windows and high vents you can have adequate air circulation anytime. It may be necessary to vent excess heat at times during the winter. The roof overhang should keep rain and snow off the glass and away from the vents.

The overhead hinged shutters shown here are perhaps the simplest way of attaining a high "R" value, while at the same time permitting out-of-the-way placement when open. Many other solutions (beadwall, bifold shutters, sliding doors, insulating curtains, etc.) may be suitable depending on their functions, aesthetics, and economics. Any such device should be easy to operate and built to last.

If you have a need for a greenhouse, guest room, cabin, study, or shop, and you would like to experience natural solar space conditioning, this module is an easy and economical way of "doing it." Be the first on your block to have one! ☼

Thermosiphoning Solar Air Heaters

by Bruce Anderson

Bruce Anderson is the Executive Editor of Solar Age Magazine, author of the Solar Home Book, and President of Total Environmental Action, Inc., Harrisville, N.H.

Thermosiphoning is the natural movement of fluids—air or water—due to differences in their temperatures. When, for example, hot air shoots up a chimney from a fireplace, it does so because of its relative lightness (bouyancy) compared to the heavier cooler air in the room and outside.

The sun's energy beating down on solar collectors can be used to create this chimney effect. The sun warms the fluid in the collector, causing it to expand, become lighter, and rise. If the collector is placed below the point where the heat is needed, the heat rises naturally to that point without the use of fans or pumps.

Hundreds of thousands of solar water heaters, many dating back to the late 1800's, were of this design (Fig. 1) The collector sits below the storage tank. As the solar heated water rises by thermosiphon action to the storage tank, cold water from storage flows down to the bottom of the collector. This cycle will continue all the while the sun shines.

Thermosiphoning will also occur at night, but in the opposite direction, establishing a cooling cycle. If, for example, the storage tank is below the collector (instead of above it as in the previous example), during the night the cool water in the collector will flow down to the storage

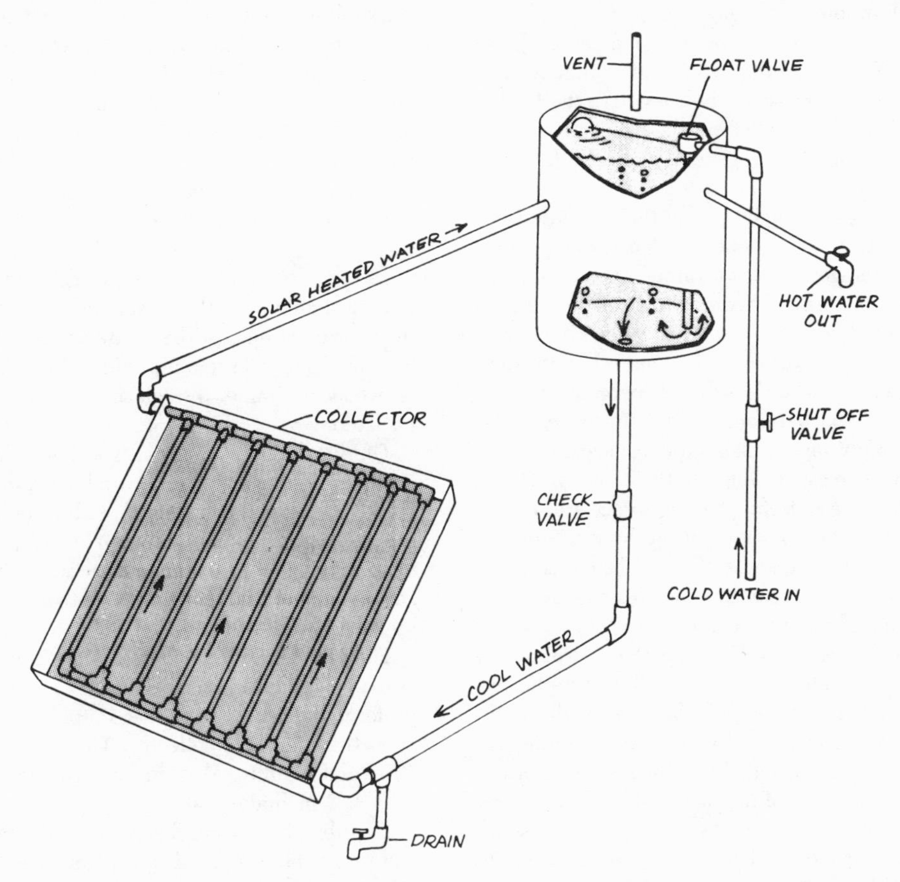

Fig. 1. A Thermosiphoning water heater with separate collector and storage tank.

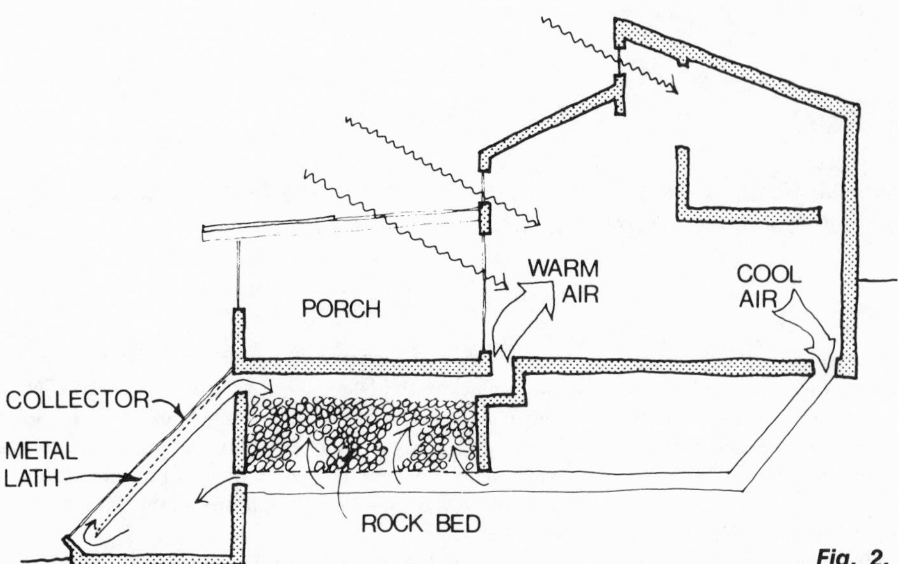

Fig. 2. Thermosiphoning rock bed

tank. Warm storage water will flow up to the collector to be cooled. However, the conventional design (Fig. 1) prevents this nighttime cooling action, that in most solar designs, is undesirable.

Fig. 2 shows the most common adaptation of the thermosiphon principle to the heating of buildings. The collector is again at an elevation below where the heat is needed. Solar heated air rises from the collector into the house through ordinary floor registers. Cooler air from the house returns to the collector.

If the house is warm enough the registers are closed and the solar heated air travels from the collector to a plenum above the rock storage bin, down through the rocks, and back to the collector. At night, warm air rises through the rocks and into the house, by-passing the cold collector.

Heat storage is not necessarily required with solar collectors, particularly when less than 20 percent or so of a building's heat is expected from the sun. In this case the solar features simply serve to keep the furnace off during the day, and in some cases, well into the evening. This is usually true when the collectors have a surface area of less than 15 percent of the floor area of the house. Sunlight entering through large expanses of south-facing glass can also keep the furnace off the entire day; this may overheat the house, even during cold weather.

Converting your home to solar, through construction of this air heating collector, is not a terribly complicated affair. Millions of homes have walls that face south. Many of these walls can be converted with relative simplicity. Heat storage, as I mentioned, is not necessary in many cases because virtually all of the heat will be used by the house when the sun is shining. As with most air-heating collectors, this wall mounted design can be built easily into the walls of both new and old buildings.

Materials

The materials needed to build this collector (Fig. 3) are glass or plastic glazing, the absorber plate, rigid insulation, interior wall finish, and the required trim including air grills and backdraft dampers. Sealants, glazings, and other materials must be able to withstand summer heat, because the dampers are kept closed and the collectors get quite hot.

Tempered double-pane insulating glass is used here as the glazing. By using stock sizes manufactured for sliding glass doors, economy of installation and availability for purposes of replacement are assured. The supports for the glazing should have the least amount of connection to the thermosiphoning absorber plate as possible, to keep heat loss by conduction to a minimum. This design makes use of wood to frame the glazing, providing a bit more insulation.

The insulating core consists of a standard type of roof insulation supplied by several manufacturers. These boards are usually 3 by 4 feet, with a nominal thickness of 1½ inches and a nominal R-value of 6.7. The mineral board side of this urethane/mineral composite faces the absorber plate; the flammable urethane foam is protected by the gypsum board.

The absorber plate is corrugated metal siding, a readily available building material complete with fasteners and preformed closure strips that simplify the construction of an airtight, durable collector. The ribs give this absorber structural stability. Alcoa makes a ribbed 8 inch siding that uses .032 inch aluminum.

The absorber is purchased as mill-finished aluminum and must be prepared for painting with an etching-cleaner. The recommended finish is a thin coat of flat

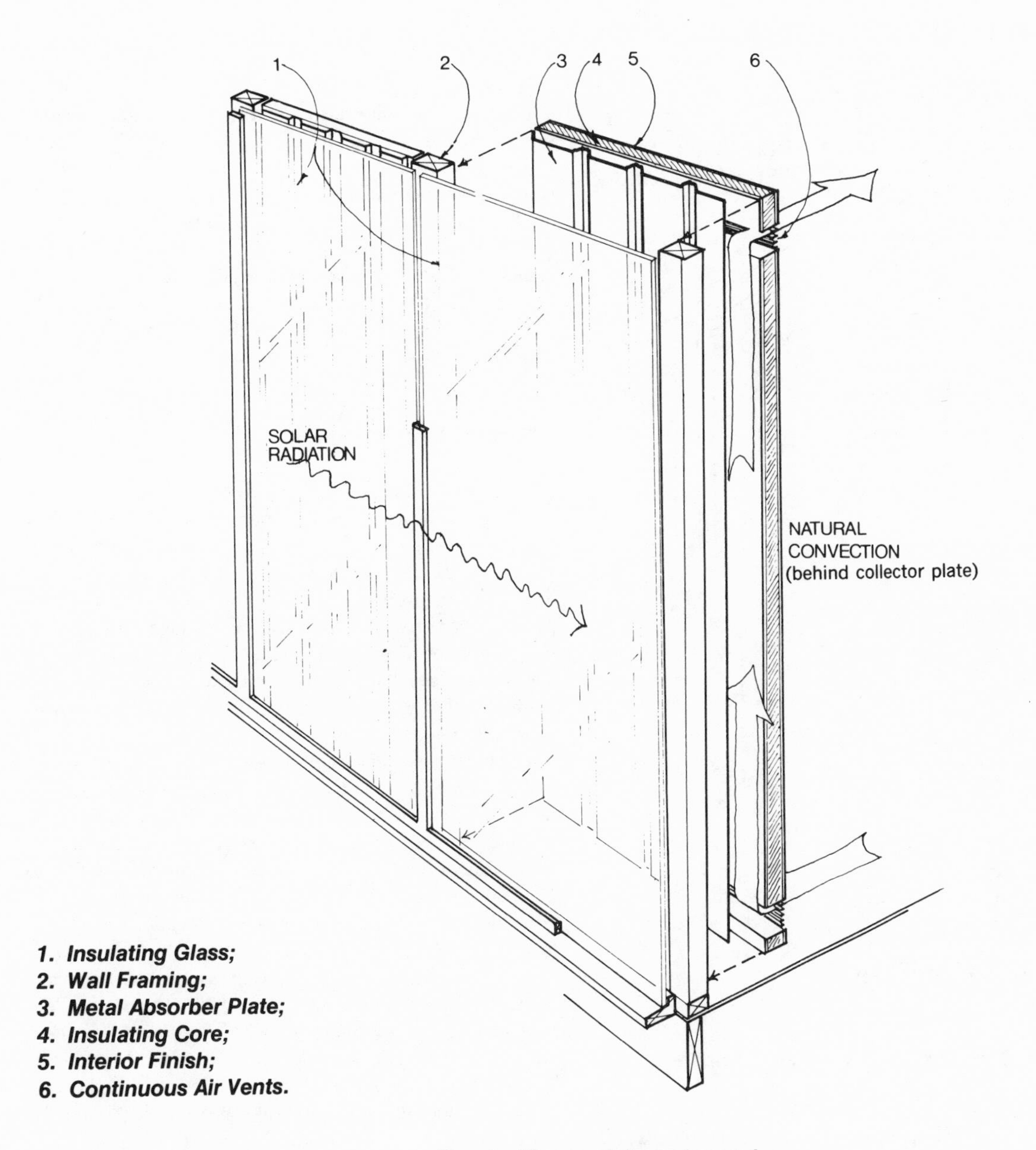

Fig. 3. Thermosiphon air panel

black enamel over a suitable primer. "Nextel" brand 'Black Velvet' by the 3M Company is a suggested choice.

Heat is vented from the panel to the house through a full-width register that has a manual air-flow control. Backdrafts, which occur at night or during very poor weather, are prevented by the use of lightweight (1 mil) plastic film one-way dampers. These must be specially made. The plastic film is attached to a punched or die-cut 24 gauge galvanized sheet with double size adhesive tape (Fig. 4).

Design

The dimensions of the collector are based on getting it to fit into a somewhat typical wood framed wall. Doubled 2 by 4 inch studs, 3 feet on center replace conventional framing. The collector structure should not be relied upon for overall wind bracing of the building frame. Good building practice requires the use of plywood panels or diagonal bracing.

This three foot module permits the use of 34 by 90 inch insulating glass panels. The 90 inch size creates a floor to

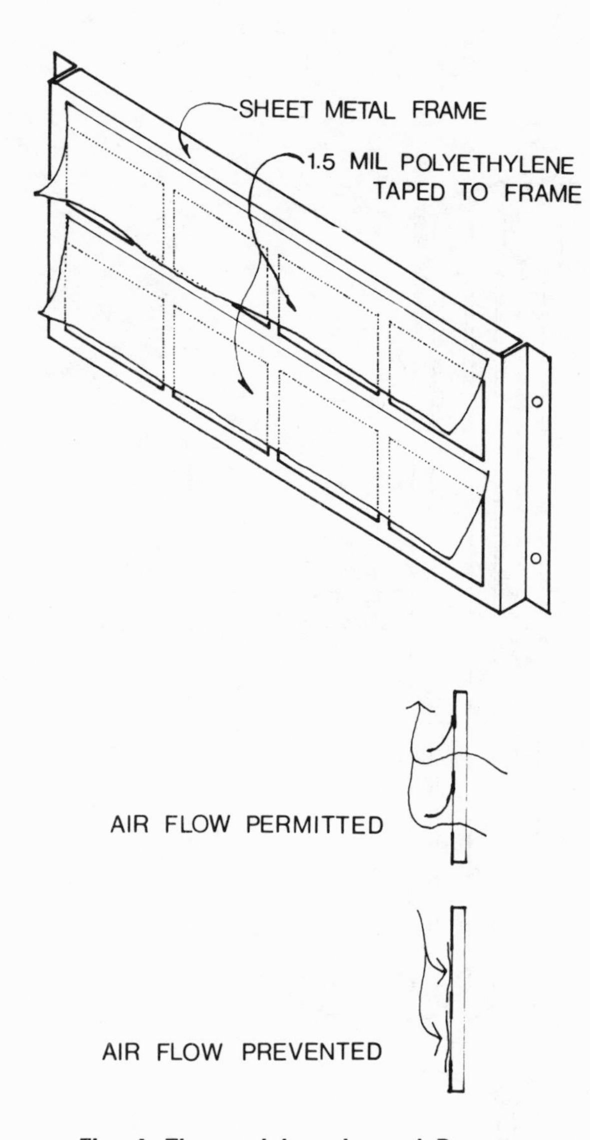

Fig. 4. Thermosiphon air panel: Dampers

ceiling panel that has an overall depth, from the black absorber to the interior finish, of 3 inches. This leaves a one-half-inch clearance to the exterior stud surface to which the glazing is attached.

Construction and Installation

These light-weight (3 to 4 pounds per square foot) panels are best made in the shop, with the glazing and interior trim (vent grills and dampers) added on site. If the seal between the sides of the panel and the abutting construction is adequately caulked, a temporary weather-skin results, and installation of the glazing need not be rushed for fear of damage by the elements.

Costs

First costs for thermosiphoning air heaters are dependent primarily on variations in labor costs. The major materials used are widely available from a number of competitive manufacturers.

The cost of materials is in the $3.75 to $4.25 per square foot range. Estimates of installed cost are about $11 per square foot. Because the panel replaces the conventional wall in light-frame construction, which usually costs $3 per square foot, the net cost of the solar aspects of the panel is $8. *Operating costs are non-existent.*

Thermal Performance

At times of little or no sun, air flow in the thermosiphoning air heater is low to non-existent (reverse flow being prevented by the one-way dampers). But under sunny conditions air flow is rapidly established. This design has a probable maxium flow rate of 2 cubic feet per minute per square foot, with an average temperature rise from inlet to outlet of 95° F. The resulting 90 Btu per square foot per hour is an average collection efficiency similar to that of low tempatures flat-plate collectors used in standard active systems.

The chief factor in overall performance is the ratio of the heating load to collector area. Effective performance deteriorates rapidly as system size is increased because without storage the space will overheat, creating waste. Estimates of useful energy output range from 30,000 to 120,000 Btu per square foot per heating season. This is equivalent to the energy from ½ to 1½ gallons of fuel oil (or 10 to 40 kilowatt hours of electricity). The high numbers in this range are typically associated with very poorly insulated houses in cold climates (Madison, Wisconsin); the low numbers with buildings with low heat loads in warm climates (Atlanta, Georgia; Ft. Worth, Texas); 50,000 to 80,000 Btu per square foot each heating season in moderate climates—like Boston—for total solar contributions between 40 and 60 percent of the total load, respectively—are reasonable estimates for preliminary design.

This wide fluctuation in estimated performance is due mainly to the lack of stabilizing thermal mass that stores excess heat. To increase system performance, especially in well insulated (low-load) buildings, increase the building's heat storage, e.g. double the thickness of gypsum board. Overheating can occur whenever collector areas are large enough to provide over 30 percent of the seasonal heating load. This can be controlled somewhat by the strategic use of thermal mass—which can be anything from waterbeds, to masonry room dividers, to Etruscan urns filled with pebbles. This collector remains, however, one of the most simple, effective, and economic ways of using solar now.

Experiences With Attached Greenhouses

by Jeremy Coleman

Jeremy Coleman is a designer and builder from Marlboro, Vt., who specializes in solar and energy conserving design and construction, including solar greenhouses.

A number of attached solar greenhouses were erected in southern Vermont during the 1978 building season. Although the geometries and glazing approaches vary, the greenhouses demonstrate certain similarities. They tend to be of a uniform size (typically about 8 by 16 feet), since 16 feet is the available length along the south building wall and 8 feet is the distance that can usually be obtained horizontally from the building.

Figure 1 is a schematic section of three greenhouses attached along the eave of a house. Each example attempts to solve the problem of building out from the south line with a minimum of vertical travel.

The second example, which is optimum from the standpoint of solar gain, required stepping down 24 inches (three steps) into the greenhouse. If the grade had been slightly higher, this would not have been possible.

A disadvantage of stepping down into the greenhouse is that a trap for cold air is created at night—cold falls down off the glazing onto the plant beds. The cold air, however, can be drained into a partially heated basement. Alternatively, a small fan can be used to break up the temperature stratification in the greenhouse.

Another purely geometrical consideration is height above grade. Extra snow cover occurs below the glazing due to snow falling off the face of the greenhouse.

The greenhouses in the Vermont area functioned well over the winter. However, growth from December through February was relatively unproductive due to low light levels. Almost all the greenhouses ran "cold"—temperatures were allowed to fall at night and during cloudy days, maximizing solar heat gain but minimizing productivity.

Heat vs. Light

This conflict—plant growth vs. solar gain—is a paradox of solar greenhouse design. Plant foliage and direct storage mass "compete" somewhat for light, and partial glazing of the end walls reduces net solar heat gain. Nonetheless, an 8 × 16 foot double-glazed solar greenhouse, with 50 percent end wall glazing and 120 net square feet of south glazing, will produce a net heat gain of at least 4.5×10^6 Btu in a 7,000-degree-day climate, assuming that all heat overflow can be stored or used in the attached dwelling, and that the average temperature of the greenhouse is 65°F over the

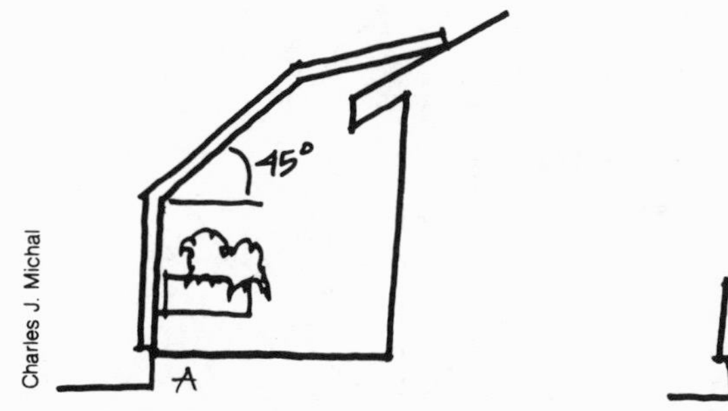

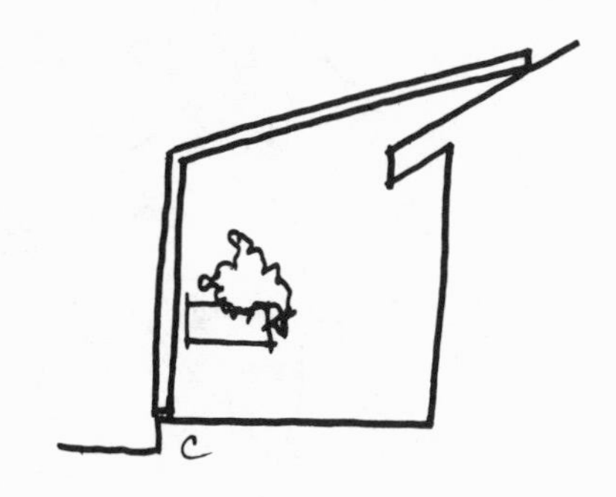

Fig. 1. The floor of greenhouse A is defined by the existing floor level of the house. The beds are raised, and little vertical extension is left for the roof turn and glazing span. In greenhouse B the 60° roof slope and lower roof overhang push the floor down three steps below the house floor level. This is ideal from a heat-producing perspective but creates cold growing beds. In greenhouse C the high grade and low eave produce vertical, roof-level glazing. The angle is too low for good heat gain, but the sun room provides excellent conditions for plant growth.

course of the heating season. If the greenhouse is run "cold," if overheating can be avoided, and if a reflective thermal curtain is used, heat production of over 12×10^6 Btu is possible. Some 75 percent or more of the greenhouse heat load occurs at night—so the advantages of a thermal curtain are obvious.

Greenhouse builders are coming down off the learning curve, and some construction and operating lessons have been gained over the past year.

Condensation

Historically, commercial greenhouse construction has evolved to handle the large amounts of condensation that form on the glazing and framing members, especially at night and on cold days, when the glass is cold. Dripping water forms spots on the leaves of some commercial crops, such as African violets, and reduces the market value of the crop. Greater concern is the rot in the structure promoted by water and warm atmosphere. For this reason cypress, redwood, and cedar have been widely used in greenhouse construction. Eaves have often been made of iron. Today, aluminum tubing and extrusions are used with polyethlene houses. Fasteners, too, must be carefully chosen: brass, hot-dipped, stainless-steel, and aluminum screws have all been used. (Rot-resistant woods, especially redwood, are corrosive to most metals.) With this history in mind, many builders have questioned the wisdom of using fasteners and framing materials common to the house construction trade—but no one knew how serious condensation problems could be.

Surveys appear to conclude that condensation is not really a severe problem in attached greenhouses. The amount of condensation depends on the percent of relative humidity and the temperature of the exterior surface on which condensation forms. Two factors create an entirely different situation in a vented, attached greenhouse compared with a free-standing greenhouse. One, air from the greenhouse is constantly flowing into the dwelling, reducing relative humidity in the greenhouse. This is particularly true during sunny periods, when bed evaporation and plant transpiration are at a maximum. Two, attached greenhouses are usually double- and occasionally triple-glazed. Thus interior surfaces are warmer than they are in the old commercial glass structures, and the dewpoint temperature is not reached so easily. Among the greenhouses in the Vermont area, only one has significant condensation buildup, and it is single-glazed.

Since condensation buildup in double-glazed houses is small, framing members are not in any structural danger. A possible exception is the sill member below two separate layers of glazing.

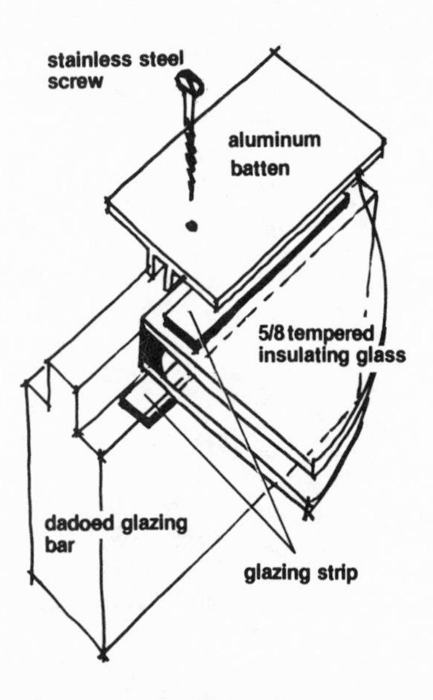

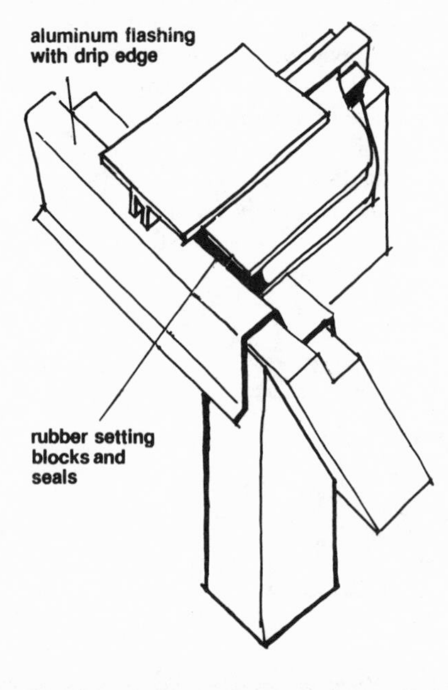

Fig. 2. A glazing detail for 5/8 inch tempered insulating glass, such as a 34 x 76 inch patio door. (Non-tempered insulating glass should not be used in any non-vertical application.) The lower layer of glazing tape helps level any irregularities in the glazing bar. A 3/16 inch space should be left between the sides of the glass and the glazing bar, 1/2 inch at the top of the glass. The flashing must be set on the eave before the glazing bars are set. The glazing batten should be tightened down only enough to compress the upper layer of glazing tape.

Zinc chromate screws that hold down these aluminum battens corroded in one month. Although "gooping" the heads with silicone can prevent corrosion, it is better to use stainless steel, aluminum, or brass. (The screw on the left is brass.)

If condensation is not really a problem in the attached greenhouse, it is still prudent to think about what happens to the moisture that is vented into the attached dwelling. Especially in modern vapor-tight construction, excessive humidity in the house can lead to moisture buildup in wall and ceiling insulation, which could ultimately be worse than condensation in the greenhouse. Care must be given to venting of insulated framing cavities. Some Farmers Home Administration (FHA) offices recommend that ceiling vapor barriers be omitted, so vapor pressure can be relieved into an open attic where it can be easily dissipated. Excessive vapor buildup has not been observed in any of the greenhouse and residential structures in the Vermont area, but since these problems can occur in houses without an attached greenhouse, it seems obvious that an attached greenhouse could exacerbate a marginal situation.

Condensation does occur between the inner and outer glazings, except in the case of sealed insulating glass. It does not appear possible to eliminate this condensation, but it is possible to minimize the problem. The condensation builds up at night or whenever the glazing is cold and is vaporized when the sun warms up the space between the glazing. The remedy is to seal the inner glazing as tightly as possible (to keep as much vapor as possible out of the glazing cavity) but to allow the outer glazing to breathe.

Two ⅛ × 2 inch weep holes for a 24 inch glazing cavity will allow the vapor to be vented but will not contribute significantly to convective heat loss. A glance at the weep holes on the bottom of an aluminum combination storm window tells the story. The sill at the bottom of the glazing should slope down and out towards the weep holes. Here is the place where condensation can accelerate rot. If the sill is not made of rot-resistant wood, it should definitely be treated with pentaphenol or clear Cuprinol preservative.

Storage

It is often said that an attached solar greenhouse should have a heat storage capacity of at least 20 Btu/°F per square foot of glazing. For an 8 × 16 foot attached greenhouse with 120 square feet net south-facing glazing, this means 2,400 Btu/°F of storage—about five 55 gallon oil drums filled with water. This would require 10 linear feet of wall space on the north side of the greenhouse (*i.e.*, they fit) and would form a 36 inch high shelf for plant beds. Before putting in the drums, however, it is worth questioning whether storage is really necessary.

The case for forgoing storage is based on the idea that space in the greenhouse would be better used for storage of garden materials. Futhermore, the thermal mass must be directly irradiated by the sun to be most effective.

The primary function of storage is to prevent overheating and the loss of usable thermal output. If heat transfer from the greenhouse to the dwelling is good enough to prevent greenhouse overheating, and if the heat load of the dwelling (including solar gain) can absorb the influx of solar heat, then overheating is not a problem—storage is not needed.

If the residence is insulated with, for example, R-11 walls and an R-20 or less ceiling, and if solar gain is not great (20-40 sf for an average-sized house, for instance), the house can likely use the greenhouse heat most of the time. Moving enough air into the house, however, may not be easy. Even allowing for heat storage capacity of the moist plant beds and for evaporation and transpiration of water in the greenhouse, it will still be necessary to move on the order of 600 cfm out of the greenhouse (about 5 cfm/sf south glazing).

Even with good venting to the house, a small fan is required to move this quantity of air. In one greenhouse with no storage and with a sliding glass door and a double-hung window venting air to the house, temperatures peaked at 104°F in March (clear skies, 50°F ambient). However, a small window-type exhaust fan running at low speed (about 1,000 cfm) reduced the peak to 85°F (same conditions), which was only 10°F higher than the temperature of the house—an entirely adequate performance. Such a fan can be easily operated by an inexpensive cooling thermostat.

Another function of storage is to provide on-site heat delivery at night. However, nighttime heat loads are much smaller than daytime cooling loads, so it is much easier to keep the greenhouse warm at night by convection from the house. In a tightly-built double-glazed dwelling, a centrally-placed door left partly open at night will keep the greenhouse warm enough for many cold-weather vegetables. An 8 × 16 foot greenhouse in Brattleboro, Vt., fell below 50 °F only once all winter with the venting door open all the time. Since most people like to run their greenhouses "cold" to reduce heat loss and maximize net solar heat gain, the door need be left only partially open.

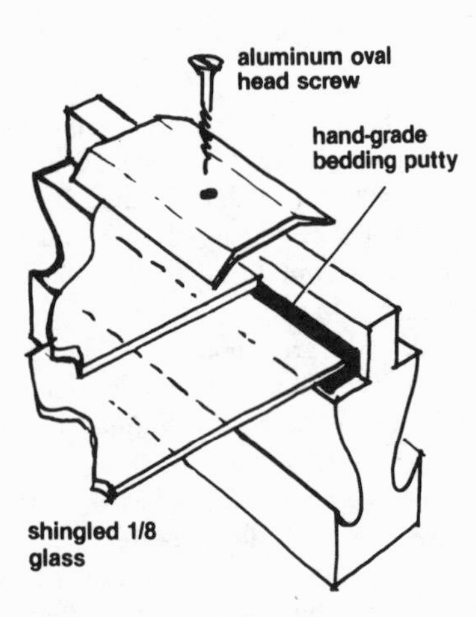

Fig. 3. A detail of shingled 1/8 inch glass in a traditional greenhouse glazing system. The glass is set in hand-grade bedding putty and is capped with an aluminum bar cap and aluminum oval-head screws. The bar caps also serve to prevent the glass pieces from sliding downwards.

Thermal mass is not useless—often it is a necessity. On any well-insulated house with solar orientation, thermal mass is required to soak up heat produced in an attached greenhouse. Even if overheating is not a problem, greenhouse mass can reduce net daily heat loads in the fall and spring when daytime ambient temperatures are high (and the house needs no heat) and nighttime temperatures are low. For a greenhouse that maximizes both solar heat gain and plant production, remote storage in a rock bin or containerized water storage system is an attractive alternative. Cornerstone Foundation in Brunswick, Maine, has done some work with remote storage using water-filled plastic containers obtained as scrap from the U.S. Navy.

Glazing

Glazing possibilities for the attached solar greenhouse include small glass panes, large panels of insulating glass, acrylic (and perhaps polycarbonate) sheet, fiberglass reinforced polyester (FRP) sheet, polyethylene film (UV treated), and special double glazings, such as Acrylite SDP, Tuffak Twinwall (polycarbonate), and FRP panels (Kalwall). No matter what glazing system is used, it must be installed properly to prevent leaks or, worse, damage from unforeseen forces.

Glass for the attached greenhouse comes in two general forms: large panels (⅝ inch or thicker; patio door sizes—34 × 76 inches—are popular and cheap) and smaller ⅛ inch thick panes recycled from old greenhouses.

The insulating glass employs glazing tape and silglaze. The flashing detail is a drip edge to allow the water to fall clear of the eave of the greenhouse.

The battens for the silglaze application are an extrusion available from Kalwall for its FRP panel system. The T-shape gives the batten rigidity, where 1½ × ⅛ inch aluminum strips will tend to fold under the inward pressure of the screws and the outward pressure of the glazing tape. Wooden battens have the necessary rigidity but weather severely—especially at a 60° or 45° tilt angle—and hence are likely to require frequent maintenance.

It is important to use aluminum or stainless steel screws for the aluminum battens. Zinc chromate screws will corrode almost immediately.

A greenhouse is often situated under an eave below a large roof expanse that sheds water from rain and melting snow onto the greenhouse glazing. This puts an additional burden on the glazing integrity, especially in winter, when dripping water freezes and thaws against the weather seal. A gutter along the top of the greenhouse would help remove most of the extra water. A dark-colored solar gutter might stay free of ice and snow at least during sunny winter periods. Obviously, icicles or sliding snow (metal roofs) can be a problem.

The Storage Problem

by Michael Riordan

Michael Riordan teaches solar design at Cabrillo College and is a co-author of ***The Solar Home Book.*** *He is a partner in Cheshire Books.*

The sun isn't always shining when we need it most, so a capacity for storing heat is almost always needed in a solar heating system. Fortunately, more solar heat than a building can use is generally available when the sun is shining. By storing this excess, a solar heating system can provide energy as needed—not according to the whims of the weather.

An understanding of heat storage begins with an understanding of the nature of heat. Quite simply, heat is the motion of the atoms and molecules in a substance—they're twirling, vibrating, and banging against one another. Any common material stores heat as its temperature rises. Its atoms and molecules vibrate more rapidly, and it feels warmer to the touch. This sensible heat is subsequently released when the molecular vibrations decrease and the temperature falls. The walls and floors of a building can be used to store heat in this way—if the room temperature is allowed to fluctuate. A tank of water or a bed of rocks commonly provides sensible heat storage in active solar heating systems. Unfortunately, it takes high temperatures or large volumes of material to store enough heat to warm a building for a few cold days.

Most materials that change phase from solid to liquid absorb a large amount of heat when they melt and release it as they solidify. When a pound of ice melts at 32° F, for example, it absorbs 144 Btu; the water molecules vibrate so rapidly that they break loose from the crystal lattice and roam freely about in the liquid phase. When the water later freezes and the molecules return to the lattice, this latent heat of fusion is released to the surroundings. But ice is a poor prospect for solar heat storage because its melting temperature is too low.

Substances that melt near room temperature are much better candidates for solar heat storage media. Glauber's Salt or sodium sulphate decahydrate, a fairly common eutectic, or phase-changing, salt has been used for this purpose.

Incorporation of heat storage into a building without incurring undue cost or drastically limiting the space available for occupants is one of the more intriguing

Table 1. Annual solar heating fraction for masonry and water walls.*

City	Annual Percent Solar Heating		
	Solid Wall	Trombe Wall	Water Wall
Santa Maria, Cal.	98.0	97.9	99.0
Dodge City, Kan.	69.1	71.8	77.6
Bismark, N.D.	41.3	46.4	49.8
Boston, Mass.	49.8	56.8	60.0
Albuquerque, N.M.	84.4	84.1	90.8
Fresno, Cal.	82.4	83.3	85.5
Madison, Wisc.	35.2	41.6	43.1
Nashville, Tenn.	60.7	65.2	68.2
Medford, Ore.	53.3	56.1	59.0

*All results are for a thermal storage mass of 45 Btu per °F per square foot of glass—approximately 18 inches of concrete or 8.6 inches of water. The Trombe wall differs from the solid masonry wall in that thermocirculation is permitted during the day. Source: J.D. Balcomb, J.C. Hedstrom, and Robert D. McFarland.

problems of solar design. There are many possible choices for a heat storage medium, but these practical considerations have restricted the designer's options to a few common materials like water, rocks, brick, concrete, and Glauber's Salt. Within these limitations, the design approaches are legion. What follows is a survey of the principle heat storage methods currently in favor.

Passive Systems

The heat storage in a passive solar system is generally incorporated into the building components. Structural elements like massive masonry walls and floors are common; containers of water in the south wall or on the roof have been successful. Large volumes are necessary because storage temperatures are low—usually less than 100° F. The storage materials are often placed in direct sunlight and double as solar absorbers. Heat transfer to the rooms relies on convection, natural conduction, and thermal radiation, with an occasional assist from a fan or blower.

Let's face it, you've got to build floors and walls anyway, so it's worthwhile to spend a little extra time and money to design them to retain solar heat. If room temperatures are allowed to fluctuate, say from 55° F to 65° F over the course of a day, these building elements can absorb excess heat on sunny days for evening or nighttime use. Depending on the choice of building material, the expanse of south-facing glass, and the temperature swing allowed, fairly large quantities of heat can be stored. For best results, the heat capacity of your building material—the number of Btu absorbed by a cubic foot during a 1° F temperature rise—should be large. Concrete and stone, for instance, store almost twice the heat per unit of volume as do pine boards. By far the best heat storage materials are water and steel—but the former is hardly a structural material, and the latter makes sense only in buildings with severe structural loads. Considerations of cost, availability, and ease of construction all point to the use of masonry materials like adobe, brick, and concrete.

A 6-inch concrete floor slab under a 1,200 square foot house can store 192,000 Btu over a 10° F temperature rise—almost enough to carry a well-insulated house through a night of freezing temperatures. If the aesthetics bother you, brick or adobe make good substitutes. In any case, the floor should be exposed to as much direct sunlight as possible and should be well insulated around its perimeter. In regions with high water tables, you may need to insulate under the floor slabs. Exterior masonry walls can also provide heat storage if they are insulated (at least 2 inches of polyurethane or polystyrene) on the outside.

Masonry south walls can be used for heat storage and as collectors if a double layer of glass covers the outer surface, with an air space between glass and wall. The most common approach is the thermal wall made famous by Felix Trombe and Jacques Michel in Odeillo, France. A series of ducts along the top of the wall permits solar-heated air to rise between glass and wall into the room behind the wall. Another series of ducts along the bottom of the wall allows cooler replacement air to flow into the space. The rest of the solar heat migrates through the masonry with a time lag characteristic of the wall thickness and conductivity. Wall thicknesses as high as 18 inches and as low as 8 inches have been recommended.

Table 2. Specific heats and heat capacities of common building materials.

Material	Specific Heat (Btu/lb/°F)	Density (lb/ft³)	Heat Capacity (Btu/ft³/°F) No Voids	Heat Capacity (Btu/ft³/°F) 30% Voids
Water	1.00	62	62	43
Steel	0.12	490	59	41
Rock or Stone	0.21	170	36	25
Concrete	0.23	140	32	22
Brick	0.20	140	28	20
Oak	0.57	47	27	19
Sand	0.19	95	18	13
Pine or Fir	0.67	27	18	13
Clay	0.22	63	14	10

The final word probably awaits the results of an extensive series of simulation analyses being done at Los Alamos Scientific Labs.

A major drawback of a glazed masonry south wall is its low conductivity. Temperatures in excess of 140° F can build up on the solar surface, causing large heat losses and lower collection efficiency. While they are not structural elements, containers of water used as substitutes for the masonry wall do not have this problem. Convection currents in the water carry the solar heat rapidly away from the collection surface, keeping it much cooler. In addition water, with a heat capacity twice that of most masonry materials, can store more heat in the same volume. Recent computer simulations at Los Alamos have shown increases of 1 to 6 percent in the annual solar heating fraction if a water wall is used instead of a masonry wall in the same climate.

The extra cost of containers is by far the dominant expense of water walls, but imaginative solutions abound. They range from Steve Baer's famous drumwall to vertical storage tubes fabricated from fiberglass-reinforced polyester (e.g., Kalwall), to galvanized steel culverts welded shut and turned on end. About 4 gallons of water are recommended for each square foot of glass, varying with climate and solar heat requirements. If a water storage wall is properly sized, the water temperature will range from 70° F to 100° F, exceeding 100° F only rarely. Over such a temperature range, 1,000 gallons can store 250,000 Btu—quite an appreciable figure. The containers need not sit directly behind the glass: they can be used as room partitions or placed against back walls and illuminated by skylights. All they need is good exposure to the winter sun.

Containers of water on the horizontal roof of a building can also be used to store solar heat. Thermal radiation carries the heat to the rooms directly below. Very even heating occurs and water temperatures as low as 70° F can be used to heat the building. Pioneered by Harold Hay, this roof pond approach is best suited to arid climates where snow buildup is rare and summer cooling important. Unfortunately for heating, the sun is low on the horizon during winter, and the daily insolation on a horizontal surface is least when it's most needed. Consequently, roof ponds must cover large areas—often the entire roof of a house. Moveable insulation must cover all this area during winter nights. One of the major problems encountered in early roof pond systems was the large heat losses around imperfect seals in this moveable insulation.

During the summer, the roof pond can be cooled rapidly by thermal radiation to the cool night sky characteristic of arid climates. By day, the insulation is rolled back in place and any excess room heat is absorbed in the pond for release later that night.

One of the most successful solar houses in this country, Hay's Skytherm house in Atascadero, California, has used roof ponds to provide 100 percent solar heating and natural cooling since 1973. During a test year (1973 to 1974) the roof temperature always remained between 66° F and 74° F without auxiliary heating or cooling. Pond temperatures frequently rose to 85° F during a winter day. With an average pond depth of 8 inches, the total volume of 7,000 gallons of 85° F water stores more than 875,000 Btu—enough to keep the house warm through four cloudy January days.

Hybrid Systems

There are many other ways to include additional heat storage in a passive system. Gravel beds or containers of water located inside the interior walls or under the floor are just a few examples. Such auxiliary heat storage volumes enhance the overall storage capacity and extend the system's carrythrough. Almost always, you will have to use external power for fans and blowers that send the solar-heated air from the collection surface to the storage volume. Without direct sunlight, auxiliary storage needs a continuous flow of warm air past the storage medium for effective heat transfer. Although there are exceptions, external power needs increase when you segregate the heat storage volume from the solar collection surface. There is also added expense for ductwork and insulation. But you do gain more flexibility in locating the storage volume(s) and can open up more space for occupancy. Such a system is no longer "passive" in the strictest sense. The term "hybrid system" has been coined to categorize those systems that are passive in their basic approach yet do not shy away from using a small amount of external power to help move the heat around.

Hybrid systems open up a whole new stomping ground for the ingenious and resourceful. In one approach, the ground under a concrete floor slab is insulated to a depth of 2 feet. Ducts through this ground storage carry excess solar heat from the collector or the rooms bring excess solar heat during the day. For a 1,200 square foot house, such an auxiliary storage volume can hold 384,000 Btu over a 10° F temperature rise. That's two days carry-through for a well-insulated house during freezing weather. Rocks or sand have sometimes been substituted for earth. In another approach, drums, cans or plastic bags filled with water are placed in the insulated crawl space just beneath a house. Small containers of water could even be placed between the floor joists or inside the walls in woodframe house construction. In all these approaches, solar heated air must be blown through or past the auxiliary storage; stored heat returns to the rooms via natural convection and thermal radiation or a small fan can be used. Only a small amount of external power is needed.

Active Systems

The storage medium you choose depends a lot on the collector you use. Liquid-heating collectors usually require liquid storage—large tanks of water are the rule. Most air-heating collectors need a storage medium consisting of rocks or small containers of water or eutectic salts. Both rocks and water are cheap and plentiful, and the technology of their use is well understood. With its high heat capacity, water can store almost five times the heat as the same weight of rock. With 30 percent void space between the rocks, a storage bin needs two and one-half times the volume as a tank of water storing the same amount of heat over the same temperature rise. Rock storage bins do not have the corrosion and containment problems of water tanks, and they can be built more cheaply. On the other hand, water costs next to nothing while rock can be more than $10 a ton delivered. Either way, it's not difficult to spend $500 to $1,000 for a heat storage unit capable of storing 500,000 Btu.

Good waterproofing products and large sheets of plastic have simplified the once formidable task of containing a large volume of water. Galvanized steel tanks were once a common choice, but they are prone to leaks. Glass linings and fiberglass tanks helped alleviate the corrosion problem but increased the initial cost. Nowadays, poured concrete tanks with vinyl plastic liners are finding increasing use. Other new approaches use wood or metal frames to support the plastic—eliminating poured concrete.

Rock storage bins are simpler than water tanks; a very adequate bin can be built from bricks or concrete blocks. With appropriate bracing, you could even use plywood and 2 by 4 lumber. Dr. George Löf succeeded, in his own house, with vertical fiberboard construction. Because of its great volume and weight, putting a rock storage bin inside the building is a problem. Locating it in a crawl space or under a concrete floor slab are two excellent solutions.

In general, the solar-heated fluid from the collector should enter the storage tank or bin at its top. Heat stratification keeps the topmost areas the warmest, so that heat can be withdrawn at the highest possible temperatures for distribution to the rooms. Air or water returning to the collector should enter the storage tank or bin at its top. Air or water returning to the collector should exit from the bottom of the bin or tank, where temperatures are the coolest. Sometimes, with liquid systems, it's necessary to use heat exchangers to isolate the fluid in the collector or distribution loops from that in the water storage tank. No such devices are necessary with rock storage.

Harry Thomason has developed and patented an ingenious approach to heat storage that uses the best features of rocks and water. He surrounds a tank of solar-heated water with a bed of fist-sized rocks. Not only do the rocks provide extra heat storage, but they also act as a cheap, effective heat exchanger. The heat from the tank diffuses slowly through the rocks, to be carried to the rooms by a circulating airstream. Because of excellent heat transfer to the air from the large total surface area of the rocks, storage temperatures as low as 75° F can still be used effectively. The use of lower storage temperatures means the collector can operate at lower temperatures—hence more efficiently.

Storage Temperature and Size

The higher the temperature a storage medium can attain, the smaller the storage bin or tanks need to be. For example, 1,000 gallons or 134 cubic feet of water can store 167,000 Btu as its temperature rises from 80° F to 100° F and 334,000 Btu from 80° F to 120° F. It takes 334 cubic feet of rock (assuming 30 percent voids) to store the same amounts of heat over the same rises.

Offhand, you might be tempted to design for the highest storage temperature possible in order to keep the storage size down. But for most space-heating systems high storage temperatures would seriously degrade both collector efficiency and overall system performance. The storage unit must still be hot enough to feed the baseboard radiators, fan coil units, radiant panels or other heat distribution apparatus. In general, the upper limits on the storage temperature are determined by collector performance and the lower limits by the constraints of efficient heat distribution.

In residential systems, the storage should be large enough to supply a house with enough heat for one average January day. For a well-insulated 1,200 square-foot house, that's 200,000 to 600,000 Btu (600 to 1,800 gallons of water or 12 to 36 tons of rock over a 40° F temperature rise) in most U. S. climates. The storage volume should also be large enough to store all the solar heat collected in a winter day, without driving the storage temperature excessively high. Taken together, these two criteria usually require that 1½ to 4 gallons of water or 70 to 200 pounds of rock be used for each square foot of collector surface.

Larger storage volumes can be used to attain several days' storage capacity, and the collector will operate more efficiently at the lower inlet temperatures occasioned by having so much storage. Up to a point, the extra money spent on the storage medium and container (and on larger heat distribution equipment) can be recouped by the savings realized by using fewer collectors. Just where is that point? Some designers like to think that 10 gallons of water or 400 pounds of rock per square foot of collector are not excessive. A recently-built solar house in Waltham, Massachusetts, achieves 100 percent solar heating with a little more than 13 gallons of water storage per square foot. Unfortunately, its designer later concluded that the huge system just wasn't economically justifiable.

Eutectic Salts

Eutectic, or phase-changing, salts are just about the only real alternative to rocks and water for heat storage in an active solar heating system. Very large amounts of heat can be stored as the latent heat of fusion of these salts. A pound of Glauber's Salt, the most widely studied and used, absorbs 104 Btu as it melts at 90° F and about 21 Btu as its temperature rises another 30° F. To store the same 125 Btu over the same rise would require about 4 pounds of water or 20 pounds of rock.

Much smaller storage volumes are possible with eutectic salts. Consequently, they offer the designer unusual versatility in locating the heat storage. Closets, thin partitions, structural voids, and other small spaces become likely heat storage bins. To some extent, we can return to the passive concept of integrating heat storage volumes into the normal building components. And because the storage temperature remains close to the room temperature, especially with Glauber's Salt, little insulation is needed around the storage volume.

With so many things to recommend them, why haven't eutectic salts seen widespread use for years? Until recently, the major problem was the separation of components after partial melting had occurred, rendering the salts useless after several cycles of melting and freezing. Another problem has been supercooling of the solution. Dr. Maria Telkes has recently solved these problems; more than 1,000 cycles have been tallied and further experimentation is upping the count.

The remaining hurdle to full-scale use of eutectic salts is an economic one. Eutectic salts capable of storing a million Btu can be prepared for as little as $200, but the cost of containers can quadruple that price. Rigid plastic tubes seem to offer the best short-term hopes of solving the problem. With luck, the next decade should see real commercialization of eutectic salts for storage.

State of the Art

With sufficiently large storage volumes and very thick insulation, it's possible to store solar heat for long periods of time—weeks and even months. With 17,000 gallons of water in a basement tank surrounded by 2 feet of insulation, the first MIT house, built in 1939, could store heat in the summer for use in winter. This tank could hold more than 10 million Btu at 195° F, the maximum temperature observed. No auxiliary heating was needed over the two full heating seasons the house was in operation. But the MIT research team concluded that 100 percent solar heating was an uneconomical goal at that time.

More recently, Harold Hay's success demonstrates that there are climates and designs that make 100 percent solar heating a reasonable goal. The primary advantages are elimination of auxiliary heating systems and the resultant simplification of the entire solar heating system. The initial costs of a backup furnace could be applied instead to building larger storage units.

Storage system design has improved gradually over the forty years since the first MIT house was built. With rigid board and plastic foam insulation, it's possible to use large volumes of earth around a building for long-term heat storage. Some designers have tried large pools of water either in the basement or in an attached greenhouse. Confined aquifers not too deep underground may soon be used for long-term heat storage by community-scale solar heating projects and total energy systems. The growing use of heat pumps has contributed, making long-term coolness storage (with water as ice) as well as long-term heat storage feasible.

Further developments in packaging eutectic salts in concrete seem to hold great promise for heat storage in both active and passive systems. Suntek Research Associates of Corte Madera, Calif., is developing Thermocrete—blocks of foamed concrete impregnated with the salts. Stratification and supercooling do not occur. The very fine concrete particles act as seeds for crystal growth when the temperature falls to the freezing point of the salt. Moreover, the concrete acts as a cheap structural container. If Thermocrete blocks can be mass-produced inexpensively and sealed reliably, they will provide entirely new avenues of approach to heat storage design—walls or floors receiving direct sunlight, stacks of hollow-core Thermocrete blocks for ample heat storage in an active system, interior walls and partitions built from hollow-core blocks through which solar-heated air could be blown. The building itself could store millions of Btu with only a small temperature rise. Long-term heat storage would be the rule, not the exception.

References

F. Trombe *et al.*, "Some Performance Characteristics of the CNRS Solar House Collectors," in *Passive Solar Heating and Cooling*, Conference and Workshop Proceedings, Albuquerque, New Mexico, May 18–19, 1976, p. 201.

Bruce Anderson, "Designing and Building a Trombe Wall," in *Solar Age*, August, 1977, p. 25.

J. D. Balcomb, J. C. Hedstrom and R. D. McFarland, "Simulating as a Design Tool," in *Passive Solar Heating and Cooling*, p. 238.

J. Douglas Balcomb, James C. Hedstrom and Robert D. McFarland, "Thermal Storage Walls for Passive Solar Heating Evaluated," in *Solar Age*, August, 1977, p. 20.

H. R. Hay and J. I. Yellott, "A Naturally Air-Conditioned Building," in *Mechanical Engineering*, Vol. 92, No. 1, 1970, p. 19.

Mark Hyman, Jr. "Trying for 100 Percent," in *Solar Age*, January, 1977, p. 24.

Kurt Wasserman, "An Interview with Maria Telkes," in *Solar Age*, June, 1976, p. 14.

Charles F. Meyer and David K. Todd, "Conserving energy with heat storage wells," in *Environmental Science and Technology*, Vol. 7, No. 6, June 1973.

Day Charoudi, Charles Tilford, *et al.*, paper presented to the Winnepeg ISES Conference, 1976.

Photovoltaics: Readying A Technology

by Helene S. Kassler

Helene Kassler was formerly the Assistant Editor of SOLAR AGE magazine in Harrisville, N.H.

The magic of photovoltaic power—in goes sunlight, out comes electricity—makes it the most romantic solar technology.

Solar cells make electricity without burning fuels. Massive transmission lines or cooling towers are unnecessary. Photovoltaics work most efficiently at the site of use—they are the liberation forces in a world of evermore centralized energy production. They should first of all, free us from bondage to large utility companies.

The first photovoltaic cells, produced by Bell Laboratories in 1954, were a laboratory curiosity—superfluous when compared to cheap and plentiful fossil fuels. They didn't find real use until the 1960 space program was established. It was a match made in heaven. Weight, *not* cost, loomed as the major concern.

In June of 1977, however, the respected Congressional Office of Technology Assessment released a year and a half long study that jolted the energy community. If fuel costs continue to rise at the present rate, they concluded that electric and heat-producing solar systems for the home could compete with conventional systems in ten to twenty years—even without technological breakthroughs.

The announcement took many people by surprise. Until then, photovoltaics were generally seen as an energy source for the year 2025—too expensive for today—awaiting a technological miracle. True, in 1959, NASA's first silicon cells cost $2,000 a peak watt. But by early 1976, the government's first large purchase for down-to-earth uses averaged $21 a peak watt. And in late 1977, further purchases averaged less than $11 a peak watt.

Already, photovoltaics have found their way to the fringes of real competition with existing energy sources. In remote areas, solar cells are less expensive than supplying new power lines or fossil fuel generators. Mountain-top communication relays, railroad switches, pipeline and bridge corrosion prevention systems, and offshore buoys are all being economically powered by solar cells. The OTA study concluded that a multi-million dollar business potential exists, even today, in developing countries, for photovoltaic systems. In the capital cities of the Third World, utility rates are often 20¢ to 25¢ a kilowatt hour. In remote areas, if electricity is available at all, rates can reach 45¢.

How have these changes occurred? Two factors worked together. Fuel costs have skyrocketed, a trend that seems to have become a destiny. But the photovoltaic industry (until 1976 a handful of predominately small, underfinanced, independent companies) kept working to improve solar cell efficiencies and production techniques—and succeeded. Today, the quiet sidestep into the photovoltaic industry by the major oil companies attests to the rising popularity and potential of this power source.

Looking to the future—The Department of Energy's (DOE) goal of $2 a peak watt by 1982 and 50¢ a peak watt by 1986 have been called attainable and in some corners, conservative. At $2 a peak watt, photovoltaics translates into electricity at 40¢ a kilowatt hour; at $1, 10¢. And at 50¢ a peak watt, solar cells will approach utility rates of 6¢ and 8¢—competitive with rates in many cities today. Some people predict cheaper prices sooner—especially if the government does some considerable buying.

The future of photovoltaic economics may not rest only on their electrical output, but also on their ability to serve in "total energy systems" where both electricity and heat are produced simultaneously. The idea is logical. Solar cells are expensive to produce. You can, however, combine them with low-cost concentrators that throw more sunlight on each cell, achieving greater output for a smaller cost. Because these systems tend to overheat, water pumped across the back of the panels collects "waste heat" that can be used to supply industrial, space or water heat. This has no effect on the electricity supplied by the cells, and is mainly a bonus. Some companies see this as the residential solar system of the future, a true home-owner-owned utility. Many phototoltaic and concentrator companies, both large and small, are aggressively working on research or prototypes.

But what is the status of photovoltaics today—technologically, economically, and industry-wise? Single crystal silicon solar cells, those thin round wafers are presently the standard bearers—the only cells commercially available in significant quantities anywhere in the world. Two companies sell cadmium sulfide cells (to government test facilities, laboratories), but these have cell efficiencies that hover around at 5 percent while single crystal silicon is over 10-13 percent.

But silicon cells are expensive today. While photovoltaic technology is well understood, production techniques are not advanced. They are responsible for much of the expense. Other types of cheaper photovoltaic substances are under serious investigation—and could come to dominate all or part of the future market. But for today, single crystal silicon solar cells are it.

The most oft-repeated phrase concerning single crystal silicon photovoltaics is that they need no major technological breakthrough to achieve cost-effectiveness and widespread use. What they do await is the coming of mass production, assembly lines, and automation. Presently solar cells are mostly hand-made; labor accounts for 35 percent of the cost. Moreover, the energy required to make them is high. Roughly 7,000 kilowatt hours of energy goes into manufacturing a 1 peak kilowatt cell with an energy payback period of four years.

But some people in the industry claim it needn't be that high. Dr. Joseph Lindmayer, founder and president of Solarex Corp. has an idea for a solar breeder factory similar in theory to the nuclear breeder—it produces more fuel than it consumes. The solar cells manufactured by the factory go right up on the plant's roof. Scientists at Solarex estimate that a concentrating solar breeder could reduce the energy payback down to one to two years. Solarex is presently building two 2-kilowatt prototypes with DOE funding.

Places for cost reductions abound: silicon purification, silicon wafer production, cell production, module manufacturing. The goals are for less waste in production,

Practical uses of photovoltaics today. Florida Power Corp. is using photovoltaics to prevent galvanic corrosion from eating through a mile-long 15kV underground cable in St. Petersburg, Florida.

greater automation, or both. Labor, materials, and energy are seen as the greatest barriers for near-term cost effectiveness and widespread use. And this is the area where the government, industry, and research laboratories around the world are putting much of their effort.

Solar cell manufacturing is at present one of the most tedious and laborious industries in the modern world. Knowing something about the silicon solar cell production routine provides an understanding of the problems faced by today's industry.

The process starts by converting sand into 99.5 percent pure "metallurgical grade" silicon that costs 10¢ to 20¢ a pound. Not pure enough for solar cells, the silicon must reach "semiconductor grade" with fewer impurities than one part per billion, at a price of $144 a pound. The metallurgical silicon is converted to a gaseous or liquid form for purification and production of the electronic grade chemical. This high quality silicon accounts for $3.50 out of each cell's present $11 cost.

If you ever wondered why solar cells are round, the next step explains it. Most photovoltaic silicon is grown as a single large crystal. A rotating silicon crystal seed is dipped into molten silicon that is "doped" with calculated amounts of boron. As it is slowly withdrawn, the seed grows into a large single crystal of silicon—3 or 4 inches in diameter and shaped like a salami. In this stage, much of the molten silicon is wasted, left unused in the crucible.

Next, ultrathin diamond saws or wires cut the salami crystal into wafers. Here too, waste takes its toll. Up to 50 percent of the material ends up as silicon dust on the carving floor. Impurities are then diffused into the top of each wafer to provide a junction that separates the electrical charges. Electrical contacts attached to the top and bottom of the cell provide a circuit through which the electrons can flow—electricity. An anti-reflective coating added to the top layer minimizes the energy lost due to reflected sunlight.

The result is one silicon solar cell. One cell that must be interconnected with others, mounted in a panel, and encapsulated in glass or plastic—more labor, more materials, more energy.

Single crystal silicon is not the only substance gifted enough to produce the photovoltaic effect. Many companies, universities, and research laboratories are aggressively working on alternatives, seeking a less expensive, or more efficient solar cell—or both. More detailed descriptions are available in two *Science Magazine* articles, Volume 197, 29 July, 1977 and Vol. 199, 10 Feb., 1978. Efficiencies of the alternative cells should match or surpass the 10 percent to 13 percent of the standard-bearing, commercially-available silicon. Most of these are thin films that require far simpler production techniques; silicon alternatives are being investigated too.

National Weather Service station in Long Island Sound uses NASA-Lewis array.

Cadmium sulfide/copper sulfide, among the more frequently mentioned prospects, are being developed by Solar Energy Systems (a Shell Oil Co. subsidiary), Westinghouse, Photon Power Corp (French National Oil Co.), University of Delaware, Monosolar Inc., and Rockwell International. Rather than slow and tedious growth of crystals, thin-film is glass coated with cadmium and copper sulfide by a variety of techniques, either spraying, wet chemical, or vacuum deposition. Although not as plentiful as silicon, far less of the compounds are needed for this much simpler production process. The result is a less expensive solar cell. But its efficiency is also less: 5 percent compared to silicon's 11 percent. In turn these cells require greater area to offer the same output as silicon solar cells. And they degrade when combined with concentrators. Moreover, cadmium compounds are highly toxic under certain conditions (when ignited). At present, Solar Energy Systems and Photon Power Co. are closest to manufacturing these cells commercially. When they will begin, however, is not yet known.

Gallium arsenide, another photovoltaic hopeful is far more expensive than silicon, but has a far greater theoretical efficiency—especially when coupled with concentrating collectors. Its efficiencies of 19 percent remain undiminished even at 1700 X the sun. Even unconcentrated gallium has a 15 percent efficiency. These cells hold great promise as part of total energy systems that would provide both electricity and heat. They maintain high efficiencies even at the high temperatures generated in concentrating systems—16 percent efficiencies at 100° C, 12 percent efficiencies at 200° C. Gallium arsenide is something of a rare material, and is also toxic. IBM in Hopewell Junction, N.Y., Varian Co., of Palo Alto, Calif., and Hughes Research in Malibu, Calif., are working on the gallium arsenide option.

And the list continues. There are even other silicon photovoltaic prospects. Among the more well-known is the "edge-defined ribbon growth" developed by Dr. A. I. Mlavsky of Mobil-Tyco Labs. His single crystal silicon, grown in a die, produces a thin silicon ribbon in a rectangular shape that minimizes waste and makes more efficient use of space in arrays. Similar work is being undertaken by IBM, of Hopewell Junction, N.Y.

"Amorphous" substances are another highly touted possibility. While single crystal silicon has crystallographically-ordered atoms, amorphous substances are of disordered structure, yet can still supply the sunlight-freed electrons. Most work in this area is on an "alloy" of hydrogen and silicon. In general, the amorphous materials can be much thinner and absorb more light than crystalline silicon; production would be far easier too. Present efficiencies are at 5.5 percent, but may reach 15 percent. Workers on the hydrogen/silicon compounds include William Paul at Harvard University, Jan Trauc at Brown, and researchers at RCA Corp. Stanley Ovshinsky at Energy Conversion Devices Inc., of Troy, Mich., works with non-silicon amorphous compounds.

Thin film polycrystalline silicon is also under investigation. Rather than growing a single silicon crystal, these cells are made by depositing a thin film of silicon crystals onto a substrate. They are cheaper to produce. But irregularities must be controlled—they absorb the light-freed electrons.

Some work is on modifications of the standard silicon solar cell. A joint project of National Patent Development Corp. and MIT involves a vertical juncture cell. Standard cells have one junction capable of producing approximately ½ volt—regardless of size. The vertical juncture cell has many junctions sandwiched together, much like a butcher-block table. The junctions are then parallel to incoming light—output from each multi-junction cell can be as great as 10 volts.

Thermophotovoltaics, another avenue of work, could dramatically increase the efficiency of silicon cells—perhaps up to 30 to 50 percent. Richard Swanson at Stanford designed a system that recycles the sunlight and shifts its spectrum so that more light is used by the cell.

Then too, there is the work on the multi-color or multi-band-gap solar cell containing two or more cells stacked on top of each other. The top cell is designed to convert high frequency light to electricity; the next cell converts middle frequency light to electricity.

Assessments or comparisons of photovoltaic "hopefuls" with the standard-bearing single crystal silicon is difficult. While many hopefuls are more efficient or less expensive to produce, they are also rarer. Single crystal silicon may be expensive but it comes from sand, one of earth's most plentiful substances. Sand shortages or embargos are rather unlikely. Silicon is also a well-known and researched substance—thanks to years of work by the booming semi-conductor industry. Few of the other substances are as well understood. Yet the prospect of highly efficient yet cheap cells is enticing, as demonstrated by the prodigious numbers of oil and other large corporations now researching off-beat photovoltaic substances.

The federal government is also busy at work on a master photovoltaic program that offers short to long-term goals and projections for costs, efficiencies, industry output, production techniques, research and development and demonstration projects.

Their major effort (as of mid-1978) is a three-pronged, multi-agency research and development program involving NASA-sponsored Jet Propulsion and Sandia Laboratories, and several divisions of DOE.

Jet Propulsion Laboratories (JPL) manages the Low Cost Solar Array project. Dedicated to reducing the cost of flat-plate solar arrays, their research includes projects to provide lower-cost, high-grade silicon; improve ingot-growing, wafer-slicing, silicon ribbon-growing; and develop mass-production processes for cell fabrication and encapsulation. JPL now works on non-silicon systems as well. As Paul Maycock, Acting Assistant Director for Photovoltaics in DOE explained, "Anything that's ready for manufacturing, JPL works on. After that, for systems development, the project goes to Sandia."

Sandia Laboratories is in charge of DOE's second area of concern—the Concentrating System Development Program. There's beauty to the logic of concentrators—replace expensive photovoltaic materials with cheaper tracking concentrators and acquire the same electrical output for less money. They work with several types of concentrators with concentrating ratios greater than 5 to 1, including tracking (fresnel, parabolic trough) and non-tracking (Winston compound parabolic collector, and v-trough), and photovoltaic arrays (silicon, gallium arsenide). Their goal is greatly increased efficiencies.

For now, though, concentrators system prices rank in the same range as flat array photovoltaics. DOE recently contracted with 17 companies for concentrating systems. The budget estimate for these systems averaged about $15 to $16 a peak watt installed, higher than the flat array average of $11 a peak watt.

Once JPL awards photovoltaic funding for a field project, Sandia works to assure the most favorable use of the system. They work on optimal design, balancing efficiencies with cost effectiveness, and storage.

The third area of DOE work, Advanced Materials Research and Development will be run by SERI starting in Fiscal Year 1979. This program covers work on cadmium sulfide, gallium arsenide, thin film polycrystalline silicon, and amorphous silicon. These cell types are all less expensive to produce than single crystal silicon, but their efficiencies are less too. The hope is to raise their efficiencies above 10 percent and then move on to mass production.

So far, large photovoltaic purchases by JPL are the government's most visible activity. Many people see massive purchases as quite important. They reason that large solar cell buys will funnel dollars into the industry and stimulate the market—as the volume increases costs will drop. Companies will finally have the needed capital to automate their plants, or at least move more toward mass production.

And costs have dropped. As previously mentioned, the first silicon solar cells in 1959 cost the space program $2,000 a peak watt. In early 1976, ERDA's first photovoltaic purchase of 46 kilowatts averaged $21 a peak watt (all figures in 1975 dollars). Later that year, prices averaged $15.50 in a 130 kilowatt purchase. Most recently, DOE's 1977 purchase of 190 kilowatts averaged $11 a peak watt.

And finally, the Mississippi County Community College, in Blytheville, Arkansas, slated for 1979 completion received a DOE grant to photovoltaic power the school with a concentrating system. They contracted with Solarex, the largest solar cell manufacturer, for 50,000 silicon solar cells to generate 362 peak kilowatts. The panels averaged $6 a peak watt—$3 for the actual cell modules, and $3 for the optical concentrating system.

Where do most government-purchased cells wind up? After testing at JPL, the arrays are tested in field applications. Forest Service lookout towers, Department of Transportation road-side warning systems, and Department of Defense communication and radar systems, Air Force Station in Mount Laguna, California—all have federally-purchased photovoltaic arrays.

Yet the government could do far more. The National Energy Plan includes a Federal Photovoltaic Utilization Program to fund installation of photovoltaic systems for federal buildings. If the plan is ever passed, such projects would offer triple benefits; added stimulation for the photovoltaic industry, reduced fossil fuel consumption, and accessible demonstration projects that would be visited by more people than at offshore buoys, midwest irrigation ditches, or mountaintop relay stations.

BDM Corp. of McLean, Va., completed a study for the Federal Energy Administration (now part of DOE) in 1977 recommending a greater purchasing effort by the federal government to help reduce photovoltaic costs. They concluded that a cumulative market of 500 megawatts by the 1980's is the only way to reach federally-projected cost goals. They urged government purchases. Four hundred million dollars invested in 152 megawatts of solar cells over 5 years could bring costs down to 75¢ a peak watt in 1983—in advance of DOE's projected date. Moreover, these cells would save the government $1.5 billion to $2 billion in 20 years.

BDM labeled the Department of Defense (DOD) as one of photovoltaics' major potential markets. Although DOD has made some purchases, they could do far more, according to BDM. Solar cells have tremendous potential in field use where their size and weight offer major advantages over generators. BDM forecasted a 10 megawatt market in DOD when prices drop to $10 a peak watt.

The future of photovoltaics is hard to predict. The possibilities are richly diversified. In the near future, silicon cells will probably remain the norm—improved efficiencies and production techniques reducing their costs. Beyond that, anything is possible. Any one of the prospects could win hands-down. It could be an exotic yet efficient substance, an inefficient but cheap compound, easily-produced single or polycrystalline silicon, a concentrator-enhanced cell,—or any combination of the above.

Often it is the little problems that are generally ignored. The materials developed to bring theory to reality were more often than not chosen out of ignorance and an astounding disregard for common sense. The simple fact of the matter is that selecting 'the appropriate' material for any collector, absorber plate, storage tank or solar greenhouse, etc., isn't a simple problem.

Avoiding The Classic Solar Pitfalls

by Douglas Taff

Doug Taff is Executive Vice-President of Parallax, Inc., Hinesburg, Vt.

It is the nature of the sun to heat. Its mission, as some people would view the future and the sun, is to heat with gleaming arrays of solar collectors, solar greenhouses, solar ponds, solar plantations, etc., etc., etc. The development of solar engineering and material science was thought to be straight forward; all the "big" problems could be solved.

Unfortunately—there is a worm in this apple—it was the "little" problems that were generally ignored. Theoretical solar application somehow became divorced from practical engineering. The materials developed to bring theory to reality were more often than not chosen out of ignorance and an astounding disregard for common sense. Within the realm of material selection, scientific arguments and biased sales pitches melt together into a significant mass of confusion. The simple fact of the matter is that selecting "the appropriate" material for any collector, absorber plate, storage tank or solar greenhouse, etc., isn't a simple problem. It can't be solved by finding the appropriate table in the nearest handbook or government publication.

The First Law of Solar Design states, "The whole is always less than the sum of its parts." It is the weakest link that ultimately decides heating capacity, longevity, and ultimate cost-effectiveness. Murphy, the high-priest of blunder, said it: "In any field of scientific endeavor, if something can go wrong, it will. If it can't go wrong it definitely will. And, when you think the worst is over, it isn't."

How, from all this confusion, do you select properly from premanufactured collectors, materials to build collectors with, or from all the accessory of a complete system? Well, it isn't easy, but there is a method that has saved many a budding entrepreneur as well as consumer.

The FIRST RULE: Disregard everything you have ever learned about glazing, metals, insulation, heat transfer fluids, pumps, switches, piping etc. Five percent was false to begin with, 50 percent was hearsay, and the remaining

45 percent wasn't based on your application anyhow. Become a skeptic and don't take anything at face value. Remember, it's your cash, your credit, and your reputation. Don't gamble.

SECOND RULE: Read everything you can find on comparative materials testing. Manufacturers' data only comes from a single point of view and it's not necessarily complete. Comparative testing can quickly uncover the not-so-obvious, sometimes hidden truths. An excellent source of data in this regard is the *HUD Intermediate Minimum Property Standards/Supplement for Solar Heating and Domestic Hot Water Systems.*

THIRD RULE: Try to find materials that are similar or identical to those used in systems that have a year or more of homeowner tested experience. Homeowners are keen observers and are blatantly vocal about things that go wrong. Testing done by independent laboratories, universities, and the government, although exceedingly accurate, can be misleading because maintenance and tinkering are often ignored as important parameters.

FOURTH RULE: Make it elegant, but always keep it simple.

FIFTH AND CARDINAL RULE: Never impose unique and/or costly design prerequisites on the system by the selection of a single component.

The last rule is by far the most important and yet it is the most difficult to see until you've already failed to keep it. A friend of mine lives in a house built over 200 years ago. Five years ago nine collectors were installed for the purpose of heating his hot water. The system is highly efficient and has supplied the majority of his domestic needs. However, the design dictated certain prerequisites that on one hand insured its overall thermal efficiency, but also created abominable maintenance problems. If it had been sold to an average consumer, it would have been another in a long line of disasters that led to the statement in one national newspaper that solar hot water heaters have been shown useless. As it happened, it wasn't sold to a consumer. It was created from scratch as an experimental installation and its contribution of hot water will forever be insignificant compared to the lessons it taught.

In designing the system he had to come to grips with the problems of cost and efficiency. His solution was to use aluminum absorber plates covered by two layers of glazing. The collector had been well tested under university and industry conditions. It was even well acclaimed, everything appeared right. However, he had already transgressed the fourth rule. By choosing a collector with an aluminum absorber, he was forced to select plastic plumbing between it and a copper heat exchanger (in order to avoid galvanic corrosion). Dielectric fluids were not yet available for heat transfer. Connecting an aluminum collector to plastic (CPVC) pipe is a simple job. No problem right? Wrong! Hoses are necessary and at 200° F the smell of melting plastic can bring heart failure to any proud "solar" owner.

Moving to high temperature plastics cured the leaking hose problem but it never answered or eliminated the galvanic corrosion created between the absorber and the heat exchanger. Consequently, silicone oil (a dielectric fluid) was used to solve the corrosion problem. But, by doing this an additional problem was created. Because of the new fluid's low surface tension, it slid out of the threaded joints, out of expensive, high temperature plastic fittings, and again created the leaky hose syndrome.

At the same time, because of the increased viscosity of the fluid and its lower heat capacity as well as the heat transfer coefficient, pump horsepower had to be increased and the heat exchanger enlarged. It doesn't take long to leak $40 worth of silicone, especially in a system not designed for it. The next step was to replace the silicone with a mixture of paraffin oils. Everything proceeded smoothly for several weeks. Then the solenoid valves used for tracking the collectors dissolved. (This is an east-west, roof-mounted collector system that tracks by sensing thermal loads.) The packing came out of the gate valves and the good old faithful expansion tank's bladder dissolved in oil. Two months and several bottles of detergent later, the oil was finally and totally removed from the system. In the interim two more expansion tanks and three additional solenoid valves had also been dissolved or destroyed.

Now the system operates to its original specifications, with a glycol based fluid, high temperature hoses, and with a limited (but efficient) lifespan because of its multi-metal nature. It will never function cost-effectively because the addition of preordained constraints had violated the Fifth Rule.

My friend is actually me. The system is on my house and has provided five years of data necessary to save me from many latter "progressive" or highly technical "solar wonders." Some may think it was built too early for what it was designed to do. I don't. It was and is "elegant" as dictated by the Fourth Rule, but it is in no way simple.

There is nothing wrong with aluminum absorber plates, high temperature plastic fittings, silicone or paraffin oils or glycol mixtures. They can be used effectively in dozens of situations. But they must be used in conjunction with a system designed for and responsive to them. Creating a workable solar heating system is not, like some advertising would lead you to believe, just the assembly of parts into a neat "prepackaged" heater. The whole is not the sum of its parts. Any system having a single weak point is no better than that weakness. Compromises must be made if any measure of "cost-effectiveness" is ever to be achieved. Compromises, however, that defy the Fifth Rule only prove Murphy's Law correct.

In weighing the value of a material the most important factor is its effective lifespan. Maintenance costs can negate any benefits earned if and when premature failure occurs. When buying critical, capital intensive, solar hardware, guarantees should be provided for a period of twenty years or longer. Specific "ware" items can be scheduled into the cost of operating the apparatus. It is impossible to save on energy costs if collectors are replaced or critically modified every ten years. It simply is not long enough! Pump seals, on the other hand, are ware items and could be easily replaced as standard maintenance items. Tanks should be trouble free. Period. Two thousand gallons makes one awful, expensive mess.

Failures do occur. And, designing out potential problems is critical to any operation. In the last five years I've replaced two pumps, nine interior cover glazings, three expansion tanks, four 120 gallon glass lined steel storage tanks, numerous fittings, gauges, sensors, and gallons of heat transfer fluid on a single solar domestic hot water heater. It is true the failures were enhanced by the experimental nature of the installation, but the real question is: should they have occurred at all?

Passive and hybrid systems will relieve some of the future pressure to develop effective active systems at all cost. Hopefully the "little", nagging, and ultimately important questions will be answered quickly. We've come a long way but the market is still one based on "buyer beware." We've got a way to go.

A solar heating system can only be judged on its cost-effectiveness, and cost-effectiveness is a function of an immense number of variables. There are four component areas that decide the eventual performance of the system. Any mistake within one of these will cause the economic failure or premature aging of the system. *Collectors, heat exchangers, heat transfer fluids,* and *insulation* are the critical components. Listed below are observations and anecdotes concerning the common sins of solar design. Hopefully, they will help keep the past from becoming the future. Systems can be designed cost-effectively, but only if you recognize the quicksand before you have fallen prey to it.

ANECDOTES AND MORTAL SINS ASSOCIATED WITH COST-EFFECTIVE SOLAR DESIGN.

Collectors

Solar collector glazing should be clean, durable, have a high transmissivity, and be stable under all temperatures up to 400° F.

Fiberglass reinforced polyester is ideally suited for do-it-yourself application because of its light weight, resilience, cost and durability. It does, however, tend to show a high rate of heat expansion. This, coupled with its rigidity, can result in tricky framing details and wrinkling. Thermal degradation may be a problem in some cases, especially above 200° F.

The *acrylics* are beautiful materials because of their high transmissivity and durability. They are, however, subject to sagging if heated to above 200° F and should be used only as the outer glazing in double glazed collectors. The advantages of *glass* are obvious. However, weight, falling objects, and thermal shock are significant problems. *Polycarbonate* has all the advantages of glass except for lower transmissivity and higher cost. This is an excellent material where breakage is a problem. Thermal stress should be avoided, and only UV stabilized material can be used. *Teflon*™ is the ideal inner glazing material because of its high thermal stability and high transmissivity. Its lack of rigidity is a disadvantage.

New and more effective glazing materials are critically needed for both collectors and passive systems.

Absorber plates must show a high heat transfer rate, be coated with a long-lived thermally stable absorptive paint or surface, and be compatible with common plumbing systems. Corrosion problems must be met and dealt with. Selective surfaces are important if high temperatures are to be reached in the collector, but the ultimate concern lies in how an efficient and cost-effective absorber plate can be integrated into an entire system. This is the crux of the problem. You must not impose unique or costly design prerequisites on the system by selection of the absorber plate.

Manufacturers should be making collectors that are easily connected to associated manifolds and plumbing, etc. The use of short, hidden or odd sized nipples without threads, barbs or compression hardware is unreasonable. If possible, collectors should be designed with internal manifolds that plug from one collector to the next. If you do have to deal with a tube connection to the plumbing which cannot be soldered directly, use only silicone hoses. Vinyl, nylon, and polypropylene are all subject to progressive failure due to thermal degradation and heat transfer fluid attack.

If collectors are to be mounted on a roof, the manufacturer should be required to supply hardware and plans of how they are to be mounted and weatherproofed. Roof mounting can be tedious and costly. It's your house and you will have to pay the consequences if a roof system is not designed properly. Leak damage is not usually covered in collector warranties.

Heat Exchangers

Heat exchangers are not designed into systems for the sole purpose of freeze protection. In areas with hard water they provide a barrier to the eventual "liming" and destruction of the collector. Because they are generally made of rigid or spiral wound copper tubing, they are less subject to damage than collectors if "deliming" with an acid wash becomes necessary. Collectors are also subject to localized hot spots and generally operate at a higher plate temperature. Higher temperatures accelerate liming; consequently, if liming is to occur, it can only be effectively controlled within an exchanger. If the water used for domestic purposes is softened by typical sodium exchange resins or if the water is acid by nature, then a problem can arise in the collector via direct corrosion. Any use of water in aluminum or steel without the appropriate inhibitors, of course, causes corrosion. But, if corrosive water is used directly in a copper collector failure will also occur. Therefore, it is best to always isolate the collectors from the storage system by means of a proper heat exchanger and working fluid. Automatic drain-down domestic hot water systems should only be used in areas with ideal climatic and water conditions. Solenoid valves, usually associated with such systems constantly fall prey to Murphy's Law, and as explained above the water problems are not insignificant.

Beware of one major problem with heat exchangers. If they're not properly sized to the collector array, the total system efficiency will be dismal. An excellent rule of thumb is 1 square foot of heat exchange surface for every 2 square feet of collector. This is the absolute minimum in most cases, especially if the heat exchanger is built into the storage tank. In one test of heat exchange capacity we matched a 10 square foot spiral wound coil exchanger located inside a 65 gallon storage tank against a 40 square foot counter flow heat exchanger feeding (via a pump loop) a 120 gallon tank. A single 117.5 square foot collector array supplied both. Because of the plumbing arrangement, the heat transfer fluid first had to pass directly through the internal coil/65 gallon tank before it could enter the counter flow/120 gallon system. Consequently, the internal coil/65 gallon tank had a greater advantage because it always received the hottest fluid directly from the collector array. In one 8 hour period the 65 gallon tank increased in temperature from 65° F to 93° F while the 120 gallon tank went from 65° F to 127° F. In terms of Btu the 10 square foot coil was only able to store 15,197 Btu. The 40 square foot counter flow exchanger stored 62,124 Btu or 4 times more heat. The limiting factor in this system *was the exchanger*, not the collector. And because of this limitation the large expensive collector could never produce the yield for which it was designed. Heat exchangers can determine the cost-effectiveness of an entire system.

Heat Transfer Fluids

Watch out, there's a lot of "snake oil" being sold to "protect" collectors. The problem of choosing an appropriate heat transfer fluid is complicated by the requirement that the fluids be either potable or, if they aren't, that a double walled heat exchanger be used. Double walled systems are expensive. Consequently, many manufacturers are gearing to produce materials based on "food grade chemicals" (fluids that when diluted by water are potable). Propylene glycol is one such material, silicones are another. Ethylene glycol, aromatics, and freons cannot be made non-toxic. Paraffin based oils are unlikely or impossible (depending on formulation) to ever be made potable. Consequently, designing the system for an adequate single wall or double wall heat exchanger predetermines your choice of fluid. From there on, it's betting your system against fluids that can either easily destroy or save it.

In choosing between various fluids watch out for problems associated with toxicity, high viscosity, and low specific heat and heat transfer coefficients. (One manufacturer's advertising included the claim that their fluid had the ability to gain temperature far faster than water, even on cloudy days. What they failed to mention in their advertising was that this property was due almost solely to a low specific heat and had nothing to do with efficiency. What they claimed as a positive benefit was really a negative attribute.) If there's a leak, will the fluid dissolve the roof? Some will. One fluid even took the soles off my shoes. Can threaded joints be sealed effectively? How about pump seals, expansion tank diaphrams, gate valve packing, etc? What accessory costs are going to be incurred? What's the scheduled maintenance and will the manufacturer back you up all the way or will he disappear in true "snake oil" tradition? Are detailed specifications and instructions available and are testing reports offered freely or kept proprietary?

People have lost money, reputations, and installations by choosing hastily. You might lose your shirt, or at very best the soles off your shoes. Don't guess.

Insulation

You must insulate all pipes associated with solar collectors. Some installers haven't in the past. And the brand new storage and hot water heater tanks are already well insulated—right? Not very likely. If the temperature outside your home was −5° F and inside it was 65° F, and if you only had 1 to 2 inches of fiberglass insulation in your walls, I doubt you would consider your home "insulated." A 70° F temperature differential between any storage tank or hot water heater and its surroundings is common. The majority of complaints about poor performance by solar domestic hot water heaters are to some degree related to inadequate insulation of storage, plumbing, or the hot water booster. What's the sense of solar heating water if the heat is then lost from the tank? A typical 80 gallon electric hot water heater will lose 3.5 to 4.0 kilowatt hours of heat per day through its walls. At 5¢ per kilowatt hour that's a minimum of $5.25 per month. Most of the larger (80 to 120 gallon) solar storage tanks have similar rates of heat loss. The significance of this lies in the fact that a solar hot water heater supplying up to 100 percent of the needs of two people in a region where electricity costs 5¢ per kilowatt hour, would only save $8.35 per month on their electric bill. This is after installing a $1,000 plus solar hot water heater. Yet, insulating the tank with an additional layer of fiberglass to a total R factor of 20 or better would save them an additional $4.00 per month right off the top. Any increase in operational efficiency is 50 percent. If the solar heating system supplies less than 100 percent (as in most cases), the ratio of the heat loss to the heat supplied increases. Performance dips and solar heated water is simply being used to heat a closet or basement rather than the purpose for which it was intended. In the same manner, adding more people or insulating it properly decreases the ratio. Insulation, therefore, is one of the most important factors in guaranteeing a healthy return on investment and providing a cost-effective system. It cannot be ignored as insignificant.

Government Funding: Critical Choices

by Allan Frank

Allan Frank is the Editor of Solar Energy Intelligence Report, Silver Spring, Md.

- If the government were to fund solar energy the way it funds nuclear energy, the economics would be dramatically altered toward solar.
- Solar has received more government aid in the first few years of serious funding than did nuclear.
- Solar gets only a minute amount of the total energy budget.
- Solar funding cannot be quickly increased several fold without serious problems of mismanagement and waste.

Which of the above statements are true? They all are. They are also all highly misleading. The federal government is increasing its funding at an incredible pace, and this has opened up a number of other wells from which solar is drawing. The question of how much money can be pumped into a program, and how fast, is being asked by scores of Congressmen and countless numbers of their constituents.

The history of solar spending—which appears to be rising as fast as any other research and development budget in peacetime history—is that liberal spenders have been pushing the levels up as fast as dollar counting conservatives will let them. In fiscal year (FY) 1970, total spending for solar, mostly through the National Science Foundation, was a mere $100,000. Two years later, in FY '72 it grew to $2 million. In the midst of FY '78 the government was committed to spending more than $410 million, plus hundreds of millions of dollars more for conservation programs that are broadly and deliberately worded to include work for solar. For FY '79 Carter has asked for $4.5 hundred million in Budget Authority, but in the hands of the congressional "Solar Alliance" that is likely to be increased, perhaps to the level of $5 or $6 hundred million. Figure 2 details the year by year growth of solar funding, with a more specific breakdown for the latest years.

In addition there are $98 million in the National Energy Plan earmarked for more photovoltaic device purchases, not to mention several billion dollars for tax credits, low-interest loans, loan guarantees, state grants, and more.

So this all means that solar is getting plenty of money, right? That's hard to say. The total energy budget for FY '78 includes $10.6 billion for the Department of Energy (from which solar draws almost all of its money), plus a few hundred million dollars for the Nuclear Regulatory Commission, and several hundred million for the Interior Department. The total energy research and development budget is about $6.7 billion (the rest of the Department of Energy money goes to oil, natural gas, electricity regulation and the Strategic Petroleum Reserve, for the most part). Solar as shown in Figure 1, is really only a small part of the total.

Can more money be spent on solar? Probably. In the past, the federal government has been blamed with being unable to spend all the money that Congress gives it for many areas, including solar. But in FY '77 virtually all the money for energy was spent. Getting rid of the money is relatively easy; not getting rid of an appropriation can get Congress angry enough to cut back spending the following year. Doing it wisely is something else. Getting far more

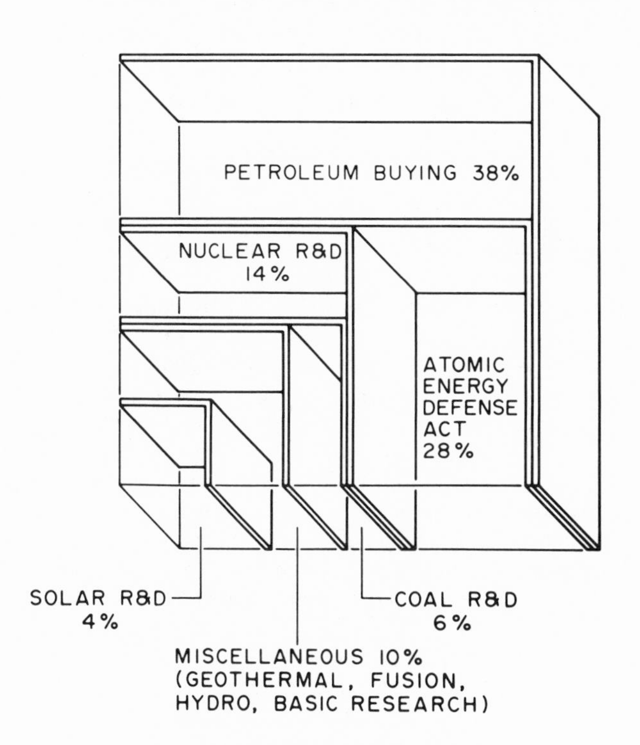

Fig. 1. The Total DOE Budget

money than expected risks wasteful duplication. As the Department of Housing and Urban Development (HUD) has been finding out, getting more money than planned also means staggering work loads in reviewing proposals, raising the possibility of inadequate review. That may not be the case, but given the congenital defects of most bureaucracy it is certainly possible.

Richard E. Balzhiser, Director of Fossil Fuel and Advanced Systems for the Electric Power Research Institute (EPRI) makes an important point: "Funding levels alone are not a valid indicator of priorities." Speaking of the Department of Energy Solar Working Group (SWG), which has prepared a report on how the government should best spend its solar money, he said that, "for the most part, solar is still in the R & D phase, the least costly phase to pursue. Premature funding of the more expensive pilot and demonstration phases could risk early failures, both technically and economically, which could have adverse impact on solar's ultimate role." Nuclear and coal budgets are much bigger than the solar budget because, says Balzhiser, "when one begins to talk about megawatt pilot projects, one quickly reaches the tens and hundreds of millions of dollars for a single project. The solar budget has increased by a factor of about 2,000 over the last seven years. I find it hard to believe that continued growth at that rate will lead to prudent use of the funds or any significant increase in the rate at which solar energy contributes to our energy needs."

More important than asking *how much* can be spent, is to ask whether the current amount is enough to stimulate both the market for solar products as well as the industry to make them. While a research and development project provides money for companies to come up with better products, a demonstration program provides the ability to test them once developed. At the same time, federal R & D money brings out private money.

The market potential for solar energy systems is nothing short of astounding. From the point of view of supply, Denis Hayes of the Worldwatch Institute and Solar Lobby, sees solar as capable of providing 40 percent of world energy needs by the year 2000 and an incredible 75 percent by 2025. In his book *Rays of Hope: Transition to a Post-Petroleum World*, Hayes asserts that, "Such a transition would not be easy or cheap, but its benefits would far outweigh the costs and difficulties." He notes that a recent study of industrial heating demands in the United States, ". . . concludes that about 7.5 percent of the heat is used at temperatures below 100° C and 28 percent below 288° C . . . However, if solar preheating is used, 27 percent of all energy for U. S. industrial heat can be delivered under 100° C and 52 percent under 288° C." Hayes makes a further claim that, "the energy that could reasonably be harvested from organic sources each year probably exceeds the energy content of all the fossil fuels currently consumed annually."

From the business perspective, solar is seen developing an almost $1 billion market by 1985, according to International Resource Development, Inc. (IRD) of New Canaan, Conn. IRD also claims that the structure of the industry would look "very different" in seven years as a "shakeout" of manufacturers occurs. The IRD report contends that initial growth will be heavily dependent upon state and federal government subsidies and tax incentives.

Officials within the Department of Energy (DOE) believe that for every solar home demonstrated by the government, another ten will be built by industry. If DOE demonstrates 5,000 residential units as it has already planned (with more to come), industry is expected to take care of another 50,000. This, in fact appears to be happening already. Two groups in Binghampton, N.Y., Domusol and BFG Corporation, are planning solar housing communities that could number in the hundreds or thousands by themselves. Another group in Colorado, Perl Mack, is working on hundreds and maybe thousands of housing units of its own. A group made up of National Homes, Revere Corporation, General Electric, and Dow Chemical is setting up 450 model homes around the country that could lead to thousands of new solar homes. Dozens of other builders are working on plans varying widely in scope, so it seems possible that the one to ten ratio could hold up. Whether greater spending would have brought this about any faster is uncertain. DOE understands this and is already planning to phase out its solar heating program, and will phase out its cooling program as soon as that technology reaches the stage of readiness now enjoyed by heating.

Recently, Arthur D. Little, Inc. (ADL), a Cambridge, Mass. consulting firm, prepared a quantitative analysis of the heating and cooling market and how it is likely to be affected by various incentives, such as the National Energy Plan. Anywhere from 749 thousand to 7.2 million residential heating and cooling units will be installed by 1985—President Carter has set 2.5 million as the national goal—and from 1.3 million to 12.5 million will be installed by 1990, says the report. This equals annual sales of 87 to 882 thousand units in 1985 and from 144 thousand to 1.2 million units in 1990.

Other calculations by ADL show total collector sales of from 62 to 697 million square feet in 1985 and from 103 to 1,177 million square feet in 1990. Total value of the sales is placed at $2.2 to 15.8 billion in 1985 and from $3.7 to $26.4 billion in 1990.

ADL says that the cost to the government of the incentives that help produce these figures would be from $451 to $5,587 million in 1985 and $509 to $6,887 million in 1990. This level of collector use, at least the high estimates, would reduce by a large equivalent oil con-

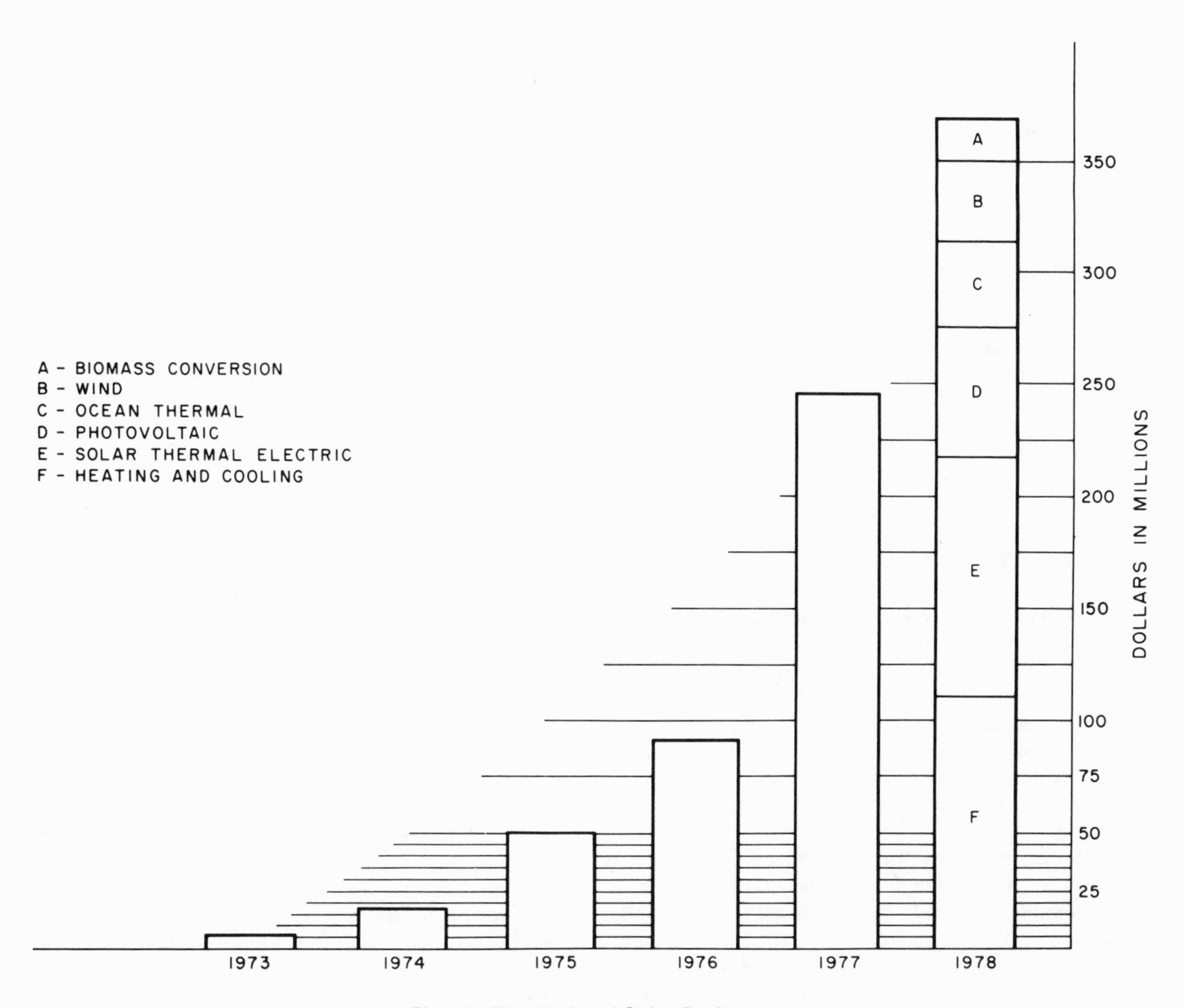

Fig. 2. The National Solar Budget

sumption in the U. S. If the collectors save oil imports, the benefits to the government would be considerable, quite possibly as great as the costs.

These costs of incentives do not apply to heating and cooling alone; in fact the companion report to ADL, prepared by Midwest Research Institute, Kansas City, Mo., says that heating and cooling will account for about one-quarter of the total energy savings of all solar options. ADL and MRI use a conservative estimate of the energy savings that the collectors could provide; the Department of Energy uses models that are 60 percent more optimistic, including the option that the collectors may be replacing electric heating, which is used in about half of the new homes in the U. S. This would triple the oil equivalent savings because electric heating is highly inefficient.

However, if one uses DOE's formula (normalized to one square foot of collector saving 3.1×10^{-4} barrels of oil equivalent per day) and remembers that the expenditures for the incentives may produce savings four times as great as that for heating and cooling alone (as MRI and ADL contend), the savings are considerable. Admittedly, these are tenuous caveats.

Calculations show that in 1985, the collectors in use could save 19,170 barrels a day or almost 7 million barrels a year at the low end of the projection. At the current $12.70 a barrel price for foreign oil, not counting transportation from overseas nor the great potential for a further price increase over the next seven to twelve years, that translates into savings of $243,464 per day or $88.9 million per year in 1985, also at the low end of the projection. At the high end of the 1985 projections the savings are

215,512 barrels a day or 78.7 million barrels per year, and $2.7 million per day or $999 million per year.

The combined calculations for 1990 show potential energy savings of from 31,848 to 363,928 barrels a day or from 11.6 to 132.8 million barrels a year. That translates to from $147.5 to 1,687 million per year. Remembering that if heating and cooling accounts for only one-quarter of the total savings accrued from the incentives, total savings may well surpass the costs. Also, these savings and costs are calculated in terms of single years, such as 1985 or 1990. But after the incentive is spent once, the per-year oil equivalent savings keep piling up year after year.

While most emphasis has been placed on heating and cooling, there is little doubt that massive purchases of photovoltaic cells will bring their prices tumbling. When the government made its first major purchases of solar cells in 1976, 46 kilowatts (electric) cost $21 per peak watt. Later that year, 130 kWe ran for $15 per peak watt. Last year, 190 kWe came with a price tag of just under $11 per peak watt. Cheapest cells were going for $9.49 per peak watt.

Photovoltaic systems that use concentrators have been built for much less. Sandia Laboratories in Albuquerque, N.M., built a Fresnel lens system rated at $3.50 per peak watt, and Solarex Corporation of Rockville, Md., has signed a contract with DOE for a silicon concentrator system of 240 kWe that is worth about $3.30 a peak watt, including profit. It will be connected with a heat recovery solar heating and cooling system for Mississippi County Community College.

More dramatic are the expectations for advanced technologies such as thin-film processing, ribbon growth, and amorphous silicon. DOE wants to get the price down to $1–2 per pW in the early 1980s and to 50¢ per pW by 1986, again in 1975 dollars. And its plan calls for a price of only 10–30¢ per pW and an annual production rate of 50 gigawatts (electric), the equivalent of 50 fullsized nuclear power plants.

The government's plan calls for increasingly large purchases to stimulate production capacity, which in turn allows industry to experience a learning curve price drop that brings in more business. All of this will also pay for experimental work in new techniques that, it is hoped, will leapfrog vast levels of operating cost and get the marketable price down even faster. Some entrepreneurs believe they can come up with cells as cheaply as 10¢ per pW or even less, and in great quantities.

It seems government funding might play a key role in advancing solar by bringing prices down much faster than the modest proposed purchases would tend to indicate. There is evidence that truly prodigious spending plans could do the job quickly and directly. A study conducted last year by BDM Corp., McLean, Va., contends that if the Federal government were to install 152 megawatts of photovoltaic devices on its remote facilities, direct net benefits would be perhaps twice the initial investment. Considering the total energy demand by the Federal Government, 152 MWe is but a tiny portion, equal to about one-seventh of a nuclear power plant.

Investment of less than $500 million—half for the arrays and half for development work, installation, maintenance, battery storage and replacement, and other costs—in 1975 dollars for the 152 MWe is seen returning $2 billion gross over the 20 year lifetime of the systems plus their period of installation, staggered over five years. Using a 10 percent discount rate over this period, the benefit would be $500 million in net discounted present value.

This assumes that the Department of Defense replaces one-fifth of its gasoline generators and that escalation of fuel, operation and maintenance costs would be equal to the general rate of inflation. More realistically, the report said, an escalation rate of 2 percent plus inflation would yield net benefits of $1 billion.

According to the study, this Federal Photovoltaic Utilization Program would cut solar cell costs (in 1975 dollars) to $3.50 per pW in 1979 when 7 MWe would be installed; and would drop prices to $0.75 per pW with 35 MWe installed in 1983. Since these are annual averages, actual array prices might slip to as low as $0.47 to $0.67 per pW.

Some analysts are claiming that the Department of Defense should determine that solar cells were a "military necessity" as it did for transistors. They say that this would put the government in a position to spend whatever was necessary to learn what it needed about photovoltaics, purchase whatever it could and do so at a rate far faster than would suit the private sector's needs. If solar cells reached the state of readiness the military wanted, they would be ready for wide use on the private market, at a low price.

Thus, it seems that there are several options for the government to take in solar cell development. Funding research in new techniques will help industry find ways to mass produce the devices and bring costs down while eliminating problems. Purchases build up an indistrial capacity that brings prices down to attract private sales. Massive purchases, according to the most optimistic, will bring prices plummeting by themselves. Commitment by the military might provide the framework for a program of rapid purchases and implementation.

If the Defense Department is willing to buy solar cells only where it is economical and not as a matter of military necessity, their purchases will, of course, be less.

In other recent market surveys large impacts were also projected for different types of solar energy. Windmills, for example, are seen as capable of providing more than one trillion kilowatt-hours of electricity in 1995 by both Lockheed California Co. (which has since dropped out of the windmill business) and General Electric Co. According to Lockheed, this achievement assumes 78,900 large scale units would be in place with a capital investment of $158.4 billion. But this would save 2.14 billion barrels of oil per year (5.8 million bbl per day) worth $25.65 billion at $12 per bbl.

Similarly, GE has projected that the U. S. has 1.07 trillion kWh of potential for windmills. Speaking at a Washington, D.C., conference, John A. Garate said, "A total of 313,500 (wind energy conversion systems) units would be required (assuming 1,500 kWe units) with a combined generating capacity of 470 gigawatts . . . In terms of USA projected energy demand in the year 2000, this output corresponds to 14 percent of the total demand (of 7.9 trillion kWh)."

So it seems that the government can play a singularly important role in determining the price of solar equipment and the viability of a nascent industry. It also seems that at the current, relatively modest level of commitment, prices are beginning to crumble in the areas one would expect to be ready for the market first. Massive expenditures, almost universally seen capable of causing prices to plummet in at least a handful of areas, will require major changes in the way the government has been thinking because the risks are proportionately larger and the promised benefits appear less certain to decision-makers, if only because the investment would be so high.

Whether the government is willing to take so radical a step as to buy the industry the capacity to have a major "quad impact" as the analysts say, is far from certain. While the industry is beginning to manage its costs, it is likely to need those giant purchases eventually to compete with conventional energy sources that have had so long a head start on solar. Only the government can spend so much so fast, but most experts believe there will be much waste in doing so. Still, solar should have the opportunity to compete before the United States becomes locked into using another set of resources. And it may. It took 20 years before nuclear power plants were producing more energy than it took to mine uranium, mill it, enrich it, fabricate it into fuel and build the plants to burn it. For solar to advance to the same relative point just may require immediate infusions of staggering levels of federal funds.

solar age
solar age

Solar Education: The Roots of a Profession

by Jeffrey Cook

Jeffrey Cook is Professor of Architecture at Arizona State University, Tempe.

Solar education is for everyone. It is a subject appropriate to all age groups and levels of interest. Children quickly understand how solar technology works—often before their parents do. And, because it is the common denominator of all human activity, the sun is adaptable to any educational situation. It is or can be a component of most specialized arts and sciences. The integration of solar education into every aspect of the public curriculum holds the key to a healthy energy future.

Basically, solar education is the application of day-to-day physics within the classroom of nature. Unlike science classes taught in labs, solar education is not restricted to a particular setting. Special fixed materials, test demonstrations and comparative routines can, of course, be useful and effective. But, in most cases the fundamentals of solar technology can be conveyed through outdoor exercises using the simplest thermometers and black boxes, mirrors, old beverage cans and maybe a spare window, a box of rocks, or a barrel of water. For the present, then, solar self-learning attached to a background in a more "traditional" field can be productive. Examples from among current solar professionals show this to be true.

Solar energy's leaders possess knowledge gained from years of experience in a variety of fields. My solar colleague John Yellott is a mechanical engineer with invaluable professional experience in the Manhattan Project, as well as in applied research to improve the efficiency of burning coal and gas fuels. My own experience as an architect creating total structures has been enriched by work in the history of architecture and in the evolution of patterns of human settlement. When such experiences are collectively focused to develop solar heated and cooled buildings it is a uniquely productive situation.

Among my Arizona neighbors with international reputations in solar energy, Drs. Aden and Marjorie Meinel come from astro-physics. Their professional responsibilities relate to astronomical observation and instrumentation. Dan Halacy is a former English teacher who has published successful books on many subjects in addition to his solar books and activities. Gene Zerlaut, President and Technical Director of Desert Sunshine Exposure Tests, began with a degree in Chemistry and courses in Chemical Engineering. His work with a paint chemist concerned with ultraviolet degradation led to doing research for NASA's Explorer I on coatings that would be stable in the hard radiation of outer space. That work in testing, measurement, and materials design became the basis for creating his solar testing business. None of these successful solar professionals began their careers in solar energy.

Today, however, we are in the second generation of solar activities and even high school graduates can seriously consider solar energy as a career. But, as solar becomes more pervasive, social and political pressures will require more formalized solar training. Proper "credentials" will be a prerequisite for a solar job. Where does this second solar generation begin?

Few undergraduate degrees are now available in solar. There are the beginnings in Junior Colleges and Community Colleges to develop curricula in solar applications, particularly in Texas and California. However, these programs are still a few years away and are aimed at training skilled technicians, contractors, entrepreneurs, and installers.

At the undergraduate level, solar energy might be the subject of only one course, emerging as a particular degree program at the graduate level. Yet, even here, the resulting degree generally comes with a traditional rather than some new title.

An example of a new specifically solar program is the graduate degree program in Solar Studies in the College of Architecture at Arizona State University. This consists of one intensive year of course work and thesis that leads to a Master of Environmental Planning Degree—a new title with a deliberately ambiguous meaning. It is structured around advanced courses in the application of solar energy to the design of buildings and communities. Because solar is a capital intensive investment, we have established energy conservation as a primary consideration. For our program, sensitivity to climate and its interaction with buildings is the beginning of any solar design.

The program is restricted to those with a demonstrated capacity for advanced work. The solar design concentration is open to those who hold architectural degrees and preferably have some professional experience. The solar technology concentration requires a degree in engineering or physics. Developed professional skills, as well as specific motivation, are a necessary prerequisite to such a short and intense graduate program. Approximately twenty-five students are taught primarily by two mechanical engineers and an architect. Active for only three years, the program has already received national and even international recognition.

Solar energy skills overlap and extend professional boundaries of architects, engineers, and planners. Our graduates are solar designers, both of whole buildings and systems, as well as experts in the component parts of those buildings and systems. But, equally important to the development of these professional skills is a commitment to an energy ethic.

Graduates seem to have almost limitless job opportunities. Their only difficulty is the decision of where to go. Professionals with a solar major are so much in demand that many of the A.S.U. students accept jobs after completing their course work, but without completing their thesis and qualifying for a degree. In a world of uncertain economics, it is nice to be in demand. These solar specialists bring to the world an understanding and a talent that bridges the separate disciplines of their instructors.

It could be argued that energy in general and solar energy in particular, are so timely and so valuable that they should be the subject of earlier specialization. A separate series of courses, instructors, sequences, text books, credits, certificates and degrees would be required. A few comprehensive solar courses could then tie it all together. But rather than isolating another subject that must somehow shoulder its way into already overloaded catalogs of specializations, the content of existing courses, curricula, and programs could be modified to cover specific solar questions. The sun's energy might become a repeated theme in many academic areas and not only in technical and mathematically based courses. Energy issues are as central to programming and management areas as they are to economics and resource use. Attitudes about energy are also tied to cultural value systems and the world of ideas. Philosophy and political and social history need to interact with solar technology and its implications. Ultimately, one's life work can include energy as a considered quality at every level of activity. The sun must not be left exclusively to specialists.

While the elements of solar energy are easily and quickly understood, they become only a curiosity in isolation from the value systems and practices of daily life. The young have the intellectual health and enthusiasm that turn solar education into an intense and exciting experience. But, energy as a concept is still subtle and elusive. It is fundamentally invisible and abstract. Energy is pervasive. The perception and use of energy as an objective quantity and quality of the human experience improves in practice with time and length of experience. And, involvement and commitment through continuing education, whether in the formal classroom, or in the curriculum of life itself is the goal. It is not an instant lesson plan but a lifetime learning goal.

Solar Calculations with the Scientific Hand Calculator

by Charles Michal

Charles Michal is the Design Group Vice President for Total Environmental Action, Inc., Harrisville, N.H.

The Scientific Pocket calculator is an extremely useful aid for the designer of solar energy systems. Inexpensive and widely available, this tool permits the user to work quickly and precisely through the numerical analysis and calculations common to most forms of solar engineering. With the right calculator, the repetitive calculations necessary for multiple-hour analysis of solar system performance can be handled easily. Some examples of how the calculator is used to solve typical solar engineering problems follow.

Calculators are precision instruments. They repeatedly provide identical answers to any given mathematical calculation, worked out to as many as ten significant digits. Common sense must be used to insure that this precision does not obscure the actual level of accuracy inherent in the numbers being used. Temperatures such as 101° F are much more sensible than 100.59° F; percentages expressed as whole numbers (23%, 48%) are more realistic than those that include fractional parts (23.3%, 47.6%). Fortunately, the better calculators let you select the number of significant digits that will be shown in the calculator display. Slide-rule accuracy is usually more than appropriate in solar engineering.

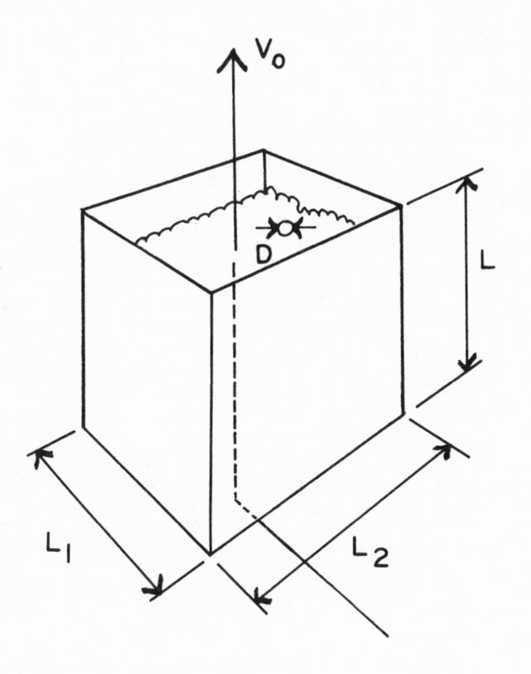

Fig. 1. A diagram of variables affecting rockbed pressure drop.

Solving Tedious Equations

Most engineers rely heavily on simplified equations, tables, nomographs and other design aids that are developed from the more complete equations describing various physical phenomena. Such time-tested shortcuts are not yet widely available to the solar designer, who must learn to use the longer equations found in the basic engineering literature. For example, pressure drop in the ductwork of an air-rock solar system is easily determined, as in any other duct system, using published tables and curves. Pressure drop due to air flow through a rock thermal storage bin is just as easily determined with the following equations and a calculator.

1) pressure drop in storage = P_d

2) $P_d = 1.096 \times 10^{-2}\ F$

3) $F = \dfrac{fV_o^2\ (1/D_e + 4/D_t)\ L}{32.2\ e^3}$

4) $f = 0.45 + (9.1 \times 10^{-4}/D_eV_o)$

5) $D_e = D/\ 6(1\text{-}e)$

6) $D_t = 2L_1L_2/L_1 + L_2$

These equations are directly derived from the formula for pressure drop due to flow of incompressible fluids in beds packed with particles of uniform size. The basic formula can be found in Sanders' *The Engineers Companion* and other engineering handbooks.

At the temperatures and low pressures developed in real systems, air can be treated as an incompressible fluid with a fixed viscosity. This assumption enables us to use the basic equations above. The variables used in the equations are described below and illustrated in Figure 1.

D_t =

hydraulic "diameter" of the rectilinear rock bed, where L_1 and L_2 are dimensions of the sides parallel to the air flow path.

```
Ø1 *LBL5 25 14 Ø5
Ø2 STO5      45 Ø5
Ø3 RCL3      55 Ø3
Ø4 ENT↑         21
Ø5 RCL4      55 Ø4
Ø6   X          51
Ø7 ST.5      45 .5
Ø8   6          Ø6
Ø9   Ø          ØØ
1Ø   X          51
11 RCL5      55 Ø5
12  PSE      16 64
13  PSE      16 64
14  X⇄Y         11
15   ÷          61
16 ST.3      45 .3
17 RCL4      55 Ø4
18 ENT↑         21
19 RCL3      55 Ø3
2Ø   +          41
21 RC.5      55 .5
22   ÷          61
23   2          02
24   ÷          61
25  1/X      25 64
26 ST.4      45 .4
27 RCL1      55 Ø1
28  CHS         22
29   1          Ø1
3Ø   +          41
31   7          Ø7
32   .          63
33   1          Ø1
34   8          Ø8
35  EEX         23
36  CHS         22
37   4          Ø4
38   X          51
39 RC.3      55 .3
4Ø ENT↑         21
41 RCL2      55 Ø2
42   X          51
```

```
43   ÷          61
44   4          Ø4
45   .          53
46   9          Ø9
47   3          Ø3
48  EEX         23
49  CHS         22
5Ø   3          Ø3
51   +          41
52 RC.3      55 .3
53 ENT↑         21
54   X          51
55   X          51
56   3          Ø3
57   2          Ø2
58   .          63
59   2          Ø2
6Ø   ÷          61
61  RCL1     55 Ø1
62  ENT↑        21
63   3          Ø3
64   Y^X     16 54
65   ÷          61
66  STO6     45 Ø6
67   4          Ø4
68  RC.4     55 .4
69   ÷          61
7Ø   7          Ø7
71   2          Ø2
72  ENT↑        21
73  RCL1     55 Ø1
74  CHS         22
75   1          Ø1
76   +          41
77   X          51
78  RCL2     55 Ø2
79   ÷          61
8Ø   +          41
81  RCL6     55 Ø6
82   X          51
83  R/S         64
```

Fig. 2. HP-19C listing for rockbed pressure drop. (To use store void fraction in Register 1, rock size in Register 2, bed height (L_2), and width (L_1) in Registers 3 and 4, enter air quantity in display and "65B5.").

D_e = ratio of the total volume of rock to total surface area of rock, where D is average rock diameter (in feet) and e is fraction of voids that result from packing.

f = a friction factor, related to the shape factor Ø (a shape factor of 1.5 is used for the irregular rock particulars), the ratio D_e and the superficial velocity, V_o, expressed in feet per second. (V_o is simple total air flow, cubic feet per *second*, divided by bed cross-sectional area, $L_1 \times L_2$).

F_f= energy lost as a result of air flow through the packed rock bin.

Pd = pressure drop given in inches of water, a way of expressing energy loss.

Using your calculator to solve equations such as those above is a straightforward application of this new tool. Conceptually, of course, the process is the same as you would use with the hand-held slide rule. Step-by-step explanations of such problem-solving are not discussed here. The hand-held calculator is really used to advantage when a programmable model is available. With a programmable calculator, a simple "computer" program can be written to solve equations automatically after only a few keyboard operations. Used this way, pressure drop through rockbeds under different conditions are calculated as fast as the descriptive data can be entered. The code, or program listing, in Figure 2 was written for the Hewlett-Packard Model 19-C, which can store and execute a program of 98 steps. Figure 3 shows a plot of pressure drop against face velocity for a typical rock bed design, obtained using this program.

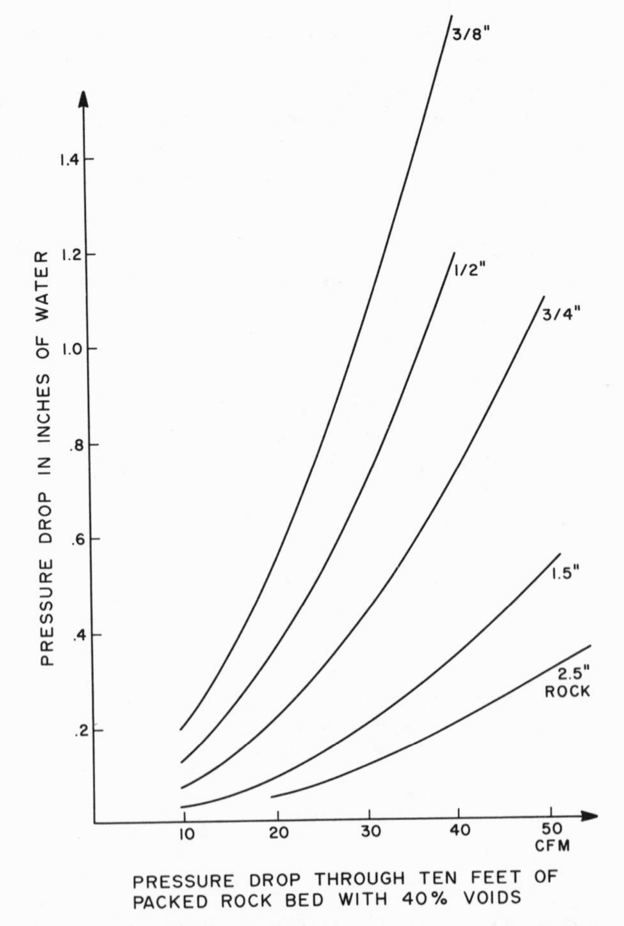

Fig. 3. Pressure drop through ten feet of packed rock-bed with 40 percent voids.

Active System Performance

The hand-held calculator can be a great aid in solving isolated sets of equations; such as those in the rock-bed pressure drop example just presented. For the designer who wants to go beyond rule-of-thumb estimates for the thermal performance of a given solar heated building, the programmable calculator becomes indispensable. Simulation of seasonal solar system performance is an iterative process: certain calculation sequences are carried out again and again using new information concerning the weather or other conditions each time. The sum or average of the product of these calculations is used to represent some aspect of a solar building's performance—usually the fraction of the heating load carried by "active" solar systems. The number of iterations performed depends upon the technique used; 8760 repetitions, one for every *hour* of the year, are common in the large computer simulations using codes such as "NBSLD" or "TRNSYS". Twelve sequences or even nine (one for each *month* of the period of interest) are more common to a multitude of simpler simulation techniques that various individuals and groups have devised. Certain of these, such as "F-Chart", are available programmed to fit Texas Instruments and Hewlett-Packard calculators. The listings of these codes can be purchased from the authors or sources such as Scotch Programs, P.O. Box 430734, Miami, Fla., 33143, and Solar Environment Engineering Company (SEBCO), P.O. Box 1914, Fort Collins, Colo., 80522. Other simulations have been outlined in the literature and can be freely used by interested individuals.

To illustrate the use of the programmable calculator for system simulation, I have selected a technique described by George Stickland, Jr. of Battelle-Columbus Laboratories, Columbus, OH. Using monthly averages of solar radiation and outdoor temperature, this method determines the long-term average daily energy available from an active solar collector system. A comparison with an average daily space heating load can then determine the "percent solar" of a given building and system combination.

The collector must be described using the following parameters:

a) heat removal efficiency factor, F_R

b) the product of cover transmissivity and collector absorbtivity evaluated at a 50 degree angle of incidence, $\bar{t}\bar{a}$

c) the monthly average absorber plate temperature

d) collector loss coefficient, U_L

```
Ø1 *LBL9 25 14 Ø9
Ø2   R/S      64
Ø3  STO5   45 Ø5
Ø4   R↓       12
Ø5  STO6   45 Ø6
Ø6   R↓       12
Ø7  STO7   45 Ø7
Ø8   R↓       12
Ø9    3       Ø3
1Ø    Ø       ØØ
11    X       51
12    9       Ø9
13    6       Ø6
14    -       31
15   SIN   16 42
16    2       Ø2
17    3       Ø3
18    .       63
19    4       Ø4
2Ø    5       Ø5
21    X       51
22  STO9   45 Ø9
23   SIN   16 42
24  RCL4   55 Ø4
25   SIN   16 42
26    X       51
27  RCL9   55 Ø3
28   COS   16 43
29  RCL4   55 Ø4
3Ø   COS   16 43
31    X       51
32    +       41
33  STO9   45 Ø9
34    9       Ø9
35    5       Ø5
36  RCL5   55 Ø5
37    -       31
38  RCL2   55 Ø2
39    X       51
4Ø  RCL1   55 Ø1
41    ÷       61
42  RCL9   55 Ø9
43    ÷       61
44  RCL7   55 Ø7
45    ÷       61
46    .       63
47    1       Ø1
48    9       Ø9
49    1       Ø1
```

```
5Ø    X       51
51   SIN   16 42
52  STO9   45 Ø9
53    1       Ø1
54  ENT↑      21
55    .       63
56    6       Ø6
57    Ø       ØØ
58    2       Ø2
59  ENT↑      21
6Ø    .       63
61    7       Ø7
62    2       Ø2
63  ENT↑      21
64  RCL7   55 Ø7
65    X       51
66    -       31
67  RCL9   55 Ø9
68   X2    25 53
69    X       51
7Ø  RCL9   55 Ø9
71    +       41
72    -       31
73  RCL6   55 Ø6
74    X       51
75  RCL1   55 Ø1
76    X       51
77  RCL3   55 Ø3
78    X       51
79   R/S      64
8Ø  STO9   45 Ø9
81    6       Ø6
82    5       Ø5
83  ENT↑      21
84  RCL5   55 Ø5
85    -       31
86  RC.2   55 .2
87    X       51
88   X>Y?  16 41
89  GTO6   14 Ø6
9Ø    1       Ø1
91  ENT↑      21
92   R/S      64
93  GTO9   14 Ø9
94 *LBL6 25 14 Ø6
95    ÷       61
96   R/S      64
97  GTO9   14 Ø9
98   R/S      64
```

Fig. 4. HP-19C listing for Strickland collector performance simulation.

e) tilt angle

f) area

The average absorber plate temperature is an indicator of the storage/distribution system efficiency. A system with two heat exchangers and a baseboard hot water distribution loop will demand higher collector temperatures than a simple air/rock-forced hot air system. An average plate temperature of 95° F, representing an air/rock system is

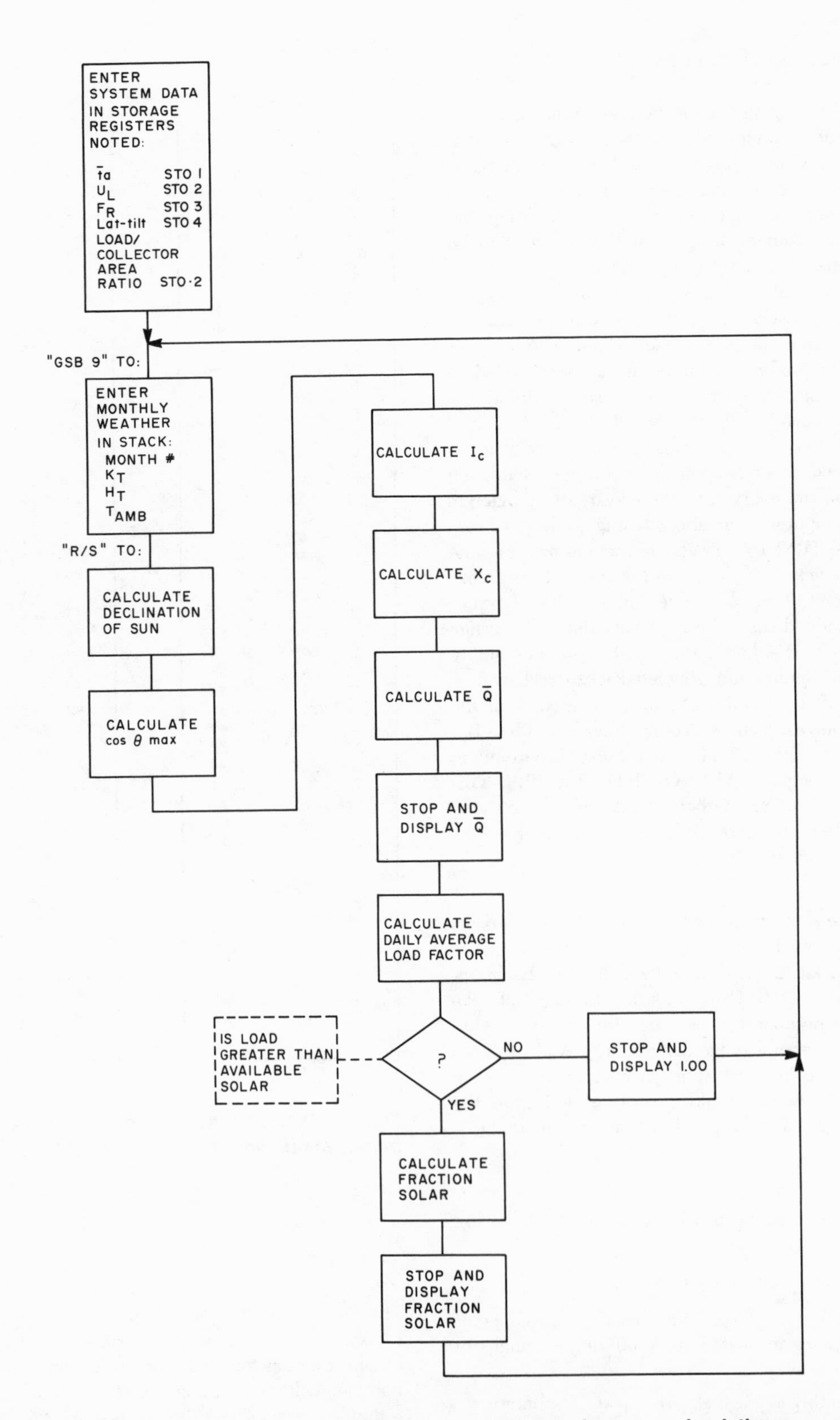

Fig. 5. Operational flow chart for the Strickland collector performance simulation program.

used in the following example. Weather and other related information needed are:

a) latitude

b) monthly average ambient temperature

c) monthly degree-day figures

d) the monthly atmospheric transmission factor, K_T

e) average daily insolation on the collector, by month, H_t

Using the concept of a "utilizability" factor (introduced by Liu and Jordan, 1963), Strickland calculates the average daily solar collection, $\overline{Q}$, as:

7) $\overline{Q} = F_R \bar{t}_a \overline{H}_t \overline{Ø}$

The utilizability factor $\overline{Ø}$ can be regarded as that fraction of the available solar energy that could be utilized by an imaginary collector with no optical losses ($\overline{ta}$ = 1.0) and perfect heat-transfer (F_R = 1.0). The factor, which is always less than 1.0, accounts for thermal losses from the collector at operating temperatures, and the "threshold" temperatures at which all the available solar energy is used to replace these losses. Strickland has devised a mathematical equation that he feels adequately relates this utilizability factor, Ø, to the major operating parameters. For latitudes greater than 30° north during the winter season the monthly utilizability is given as:

8) $Ø = 1 - X_c + (0.602 - 0.72\ K_T)\ X_c^2$ where

9) $X_c = \sin(0.191\ I_c / K_t \cos Ø \max)$

10) $Ic = U_L\ (Tplate - Tamb) / \overline{ta}$

11) $\cos Ø \max = \sin 5 \sin (L - B) + \cos 5 \cos (L - B)$

12) $5 = 23.45 \sin (30\ M - 96)$

(Note: L and B are site latitude and collector tilt, respectively. 5 is declination, a function of the month, M)

To use this method the code shown in Figure 4 was written using the flow-chart in Figure 5. Data are entered for each month and two outputs are obtained; the Btu per square foot available from the collector and the fraction of heating load this will carry. Where the fraction exceeds one, 100 percent is shown instead of the actual number. This is accomplished using a comparison function that permits the calculator to branch to one of two different program sequences.

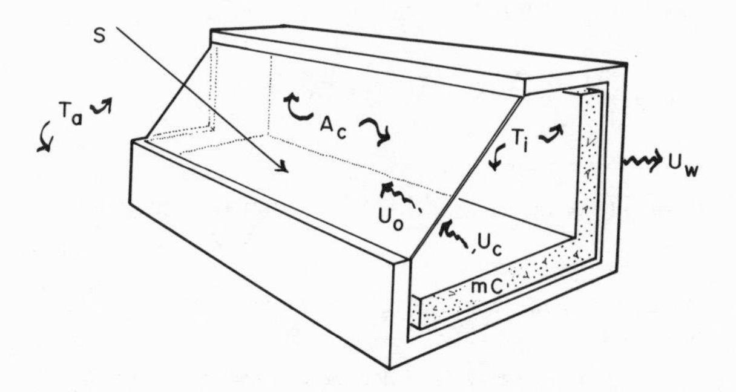

Fig. 6. Diagram of a simple passive solar structure.

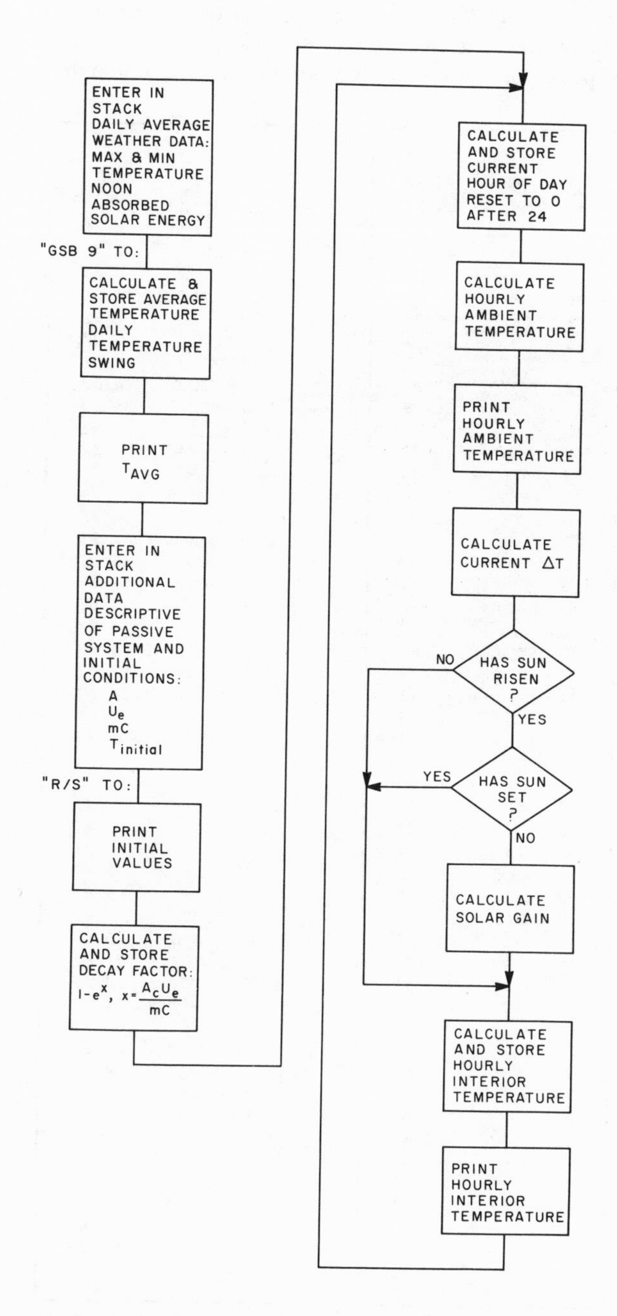

Fig. 7. Operational flow chart for greenhouse simulation program.

Usually, even what seems to be the most complicated simulation will result in only one or two "answers". The previous example showed daily energy outputs and percent solar heating. Seasonal energy output per square foot of collector and number of years to payback are other common answers sought by users of such simulations. Single answers can be stored and recalled one-by-one. If there are intermediate answers that you do not wish to or cannot store, a good programmable calculator will provide a "pause" or "stop" function which can be used to stop the program and display the number you are interested in long enough for you to record it with pencil and paper. This approach was used on the active system performance program above. In an identical fashion, the number of data entries needed for a given run may be large enough to require you to "baby sit" the calculator, feeding it additional information at prearranged points in the program sequence.

These techniques, although workable, can limit you in your application of the calculator. For true "computer" flexibility, both data input and output should be automated. Automatic *input* of stored data is possible only with the top-of-the line machines featuring magnetic cards or chips. *Recorded* automatic data *output* is more easily achieved in the *printing* programmable calculators that are usually one step below the magnetic card machines in a manufacturer's product line.

```
Ø1 *LBL9 25 14 Ø9
Ø2  STO9    45 Ø9
Ø3   R↓        12
Ø4  STO8    45 Ø8
Ø5   Σ+        35
Ø6   R↓        12
Ø7   Σ+        35
Ø8   x²     16 11
Ø9  PRTX       65
1Ø  STO7    45 Ø7
11  RCL8    55 Ø8
12   -         31
13  STO8    45 Ø8
14  R/S        64
15  PRST    16 65
16  STO1    45 Ø1
17   R↓        12
18  STO2    45 Ø2
19   R↓        12
2Ø  STO3    45 Ø3
21   R↓        12
22  STO4    45 Ø4
23  RCL3    55 Ø3
24   X         51
25  RCL2    55 Ø2
26   ÷         61
27  CHS        22
28   eˣ     25 32
29  CHS        22
3Ø   1         Ø1
31   +         41
32  STO5    45 Ø5
33 *LBL8 25 14 Ø8
34  GSB1    13 Ø1
35   1         Ø1
36   4         Ø4
37   -         31
38   1         Ø1
39   5         Ø5
4Ø   X         51
41  COS     16 43
42  RCL8    55 Ø8
43   X         51
44  RCL7    55 Ø7
45   +         41
46  PRTX       65
47  RCL1    55 Ø1
48   -         31
49  STO6    45 Ø6
```

```
5Ø   7         Ø7
51  RCLØ    55 ØØ
52  X>Y?    16 41
53  GTO5    14 Ø5
54  GTO3    14 Ø3
55 *LBL5 25 14 Ø5
56   1         Ø1
57   7         Ø7
58  X≤Y?    16 31
59  GTO3    14 Ø3
6Ø  RCLØ    55 ØØ
61   1         Ø1
62   2         Ø2
63   -         31
64   1         Ø1
65   8         Ø8
66   X         51
67  COS     16 43
68  RCL9    55 Ø9
69   X         51
7Ø  GTO4    14 Ø4
71 *LBL3 25 14 Ø3
72   Ø         ØØ
73 *LBL4 25 14 Ø4
74  PRTX       65
75  RCL3    55 Ø3
76   ÷         61
77  RCL6    55 Ø6
78   ÷         41
79  RCL5    55 Ø5
8Ø   X         51
81  RCL1    55 Ø1
82   +         41
83  PRTX       65
84  SPC     25 65
85  STO1    45 Ø1
86  GTO8    14 Ø8
87 *LBLØ 25 14 ØØ
88   Ø         ØØ
89  STOØ    45 ØØ
9Ø *LBL1 25 14 Ø1
91  ISZ     25 55
92   2         Ø2
93   4         Ø4
94  RCLØ    55 ØØ
95  X>Y?    16 41
96  GTOØ    14 ØØ
97  PRTX       65
98  RTN     25 13
```

Fig. 8. HP-19C listing for greenhouse simulation.

Passive System Performance

A programmed simulation that predicts hourly temperatures inside a simple direct passive structure provides a final example of the versatility of the new programmable calculators. Duffie and Beckman[1] have introduced some simplified equations dealing with heat balance on a flat-plate collector with finite heat capacity when no fluid flow is present. Their approach was adopted by the author[2] for the assemblage illustrated in Figure 6, which can be defined as a simple passive design characterized by large areas of south-facing glass or plastic and a well-insulated structural envelope all enclosing an interior space having a high thermal mass *positioned to absorb* incoming solar radiation.

Equation (13) below defines the energy stored in the "interior" of the structure as *the absorbed solar energy minus the heat lost through the enclosing insulated envelope and the glazing*. The absorbed energy is that which is transmitted through the glazing and absorbed by the interior surface materials.

This can be written as:

$$13)\ mC\,\frac{dTi}{dt} = AcS - \overline{AU}\,(Ti\text{-}Ta)$$

where $\overline{AU}$ is the overall heat loss from the structure per

1. *Duffie & Beckman*, Solar Energy Thermal Processes, *John Wiley & Sons, 1974.*

2. *C. Michal, "Glazed Area, Insulation and Thermal Mass in Passive Solar Design," proceedings of the First Annual Conference of the New England Solar Energy Association, June, 1976.*

unit time. The solution of this differential equation is:

$$14)\ AcS - \overline{AU}\,(T_i - T_a)$$

$$= e^{\frac{-\overline{AU}t}{mC}}\,[AcS - \overline{AU}\,(T_i,\ start - T_a)]$$

Manipulation of equation (14) gives:

$$15)\ \frac{-\overline{AU}t}{mc} = \ln\left[1 - \frac{\overline{AU}\,(T_i, - T_i,\ start}{A_cS - \overline{AU}\,(T_i,\ start - T_a)}\right]$$

which can be further simplified to

$$16)\ \frac{-A_cUet}{mc} = \ln\left[1 - \frac{Ue\,(T_i - T_i,\ start)}{S - U_e\,(T_i,\ start - T_a)}\right]$$

if U_e is defined as the overall heat loss normalized to a unit of glazed area.

Referring once more to Figure 6, we see that equation (16) relates the variation in interior temperature, T_i, of a "passive" design having given quantities of glazed area, A_c, insulation against heat loss, U_e, and thermal mass, or capacity, mC, to the time varying conditions of outdoor temperature, T_a, and *absorbed* solar energy, S. The analysis assumes that T_a and S are constant over the time frame in question. The time interval chosen should be the smallest consistent with the data on outdoor temperature and insolation. The degree of error introduced when the time frame is large (greater than one hour) has not been determined. Rewritten to give the interior temperature change over time when other variables are known, equation (16) has the form:

$$17)\ \triangle T_i = \left[1 - e^{\frac{-A_cUet}{mc}}\right]\left[\frac{S}{Ue} - (T_i,\ \ start - T_a)\right]$$

This form permits an easy simulation of a given passive structure's response to varying outdoor conditions.

A program can be written to produce hourly values of the interior space temperature. Since the method requires repeated solutions of the basic equation using hourly data on the previous interior temperature, the exterior (ambient) temperature and absorbed solar energy, it is convenient to include equations that define hourly

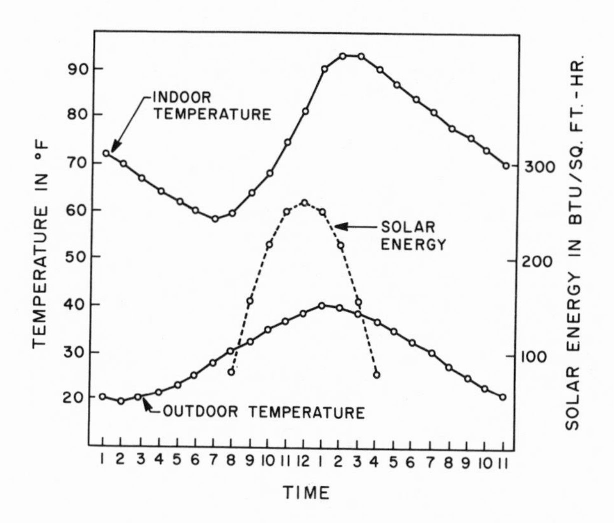

Fig. 9. Graphic plot of Greenhouse simulation results.

ambient temperatures and solar insolation rates as sinusoidal functions of daily average values. The flow chart shown in Figure 7 includes these additional computations which greatly increase the convenience with which the program is used.

The code for this passive simulation program is shown in Figure 8. In using the program, careful judgement is required to select reasonable value of the effective thermal mass and the transmitted solar energy.

This program illustrates the advantages of the printing calculators. Output data is in the form of hourly value for:

1) ambient air temperature
2) solar gain per square foot
3) interior average space temperature

Printed on the paper tape is a convenient, permanent record of the system simulation. From this tape, graphic plots of the data, illustrating the characteristic behavior of simple passive systems, can be prepared. Figure 9 is such a plot, taken from a typical run using the descriptive parameters noted.

These and similar uses of the programmable calculator save time when repetitive analyses must be carried out. For the solar system designer, the use of such a time-saving tool is important. Tools are not substitutes for knowledge, however, and a basic understanding of and skill in the applications of solar energy serves the designer best of all.

A Survey of Solar Simulation Programs

by Joseph Kohler

Joe Kohler is a member of the research staff at Total Environmental Action, Inc., Harrisville, N.H.

The computer is one of mankind's greatest labor saving tools. All fields of scientific and technical endeavor can benefit from their services, and solar energy, where many complex interactions must be considered at once, is no exception. The motion of the sun through the sky, the weather conditions of a given site, the performance capacity of a particular collector in relation to its storage tank (or mass), the angle of sun and collector, and the characteristics of the building to be heated or cooled must all be factored together to determine the best way to size and orient a solar system.

Computer simulations can be used to accomplish this as well as to compare the performance of several collector designs and to identify the least expensive design that will work well for a particular job. Simulations are quite useful for determining the effect of varying the tilt and orientation of the collector array and for evaluating the performance of more complex systems (e.g., heat pumps or solar absorption coolers).

For the design of passive systems, where the building, the collector, and the storage are all essentially the same structure, simulations are even more vital. Passive systems—including direct gain, greenhouses, convective loops, and mass storage walls and roofs—vary in performance depending on the particular building location and design. Weather conditions are considered in order to determine the system area and the thermal mass necessary to prevent uncomfortable variations in temperatures. Computer simulations also help the designer evaluate various glazing and energy conservation options.

Simulation of solar system performance is, however, complicated: incident solar radiation, cloud cover, ambient temperature, user demand, and other factors change continuously throughout the day and year and are difficult to predict. Furthermore, there are numerous interactions between the components of the system. In active systems, the solar collector efficiency depends on the storage temperature which is related to the building load, which in turn is related to the ambient temperature. Passive systems are even more complicated because performance depends on the geometry of the building in addition to the factors mentioned for active systems.

A variety of prediction techniques for analyzing solar system performance have been developed in the past few years, ranging from detailed, hour by hour computer simulations to simple rules-of-thumb. Six of these programs, which are widely used or frequently referenced in the technical literature, are described in the following paragraphs.

TRNSYS

TRNSYS is a versatile, generalized program developed at the University of Wisconsin Solar Energy Laboratory. Although it was originally designed to simulate the performance of active systems, the program is also useful for simulating direct gain passive buildings.

TRNSYS consists of approximately 30 subroutines, each of which simulates a specific component of a solar system. Typical components include a card reader that provides hour by hour weather data, a solar collector, a pump controller, storage, a room model, and a solar radiation processor. Each component model contains inputs, outputs and parameters. The user is required to develop an information flow diagram of the system, connecting the inputs and outputs of the component. For example, the radiation processor, the temperature and flow rate from the pump, and the ambient temperature from the card reader would be inputs to the collector. The temperature and flow rate out of the collector would be inputs to the storage tank. Building loads are simulated by combining room, wall, and roof models in which the construction type, insulation, window area and configuration, and other structural characteristics are user specified.

The program uses the ASHRAE transfer function approach to predict the dynamic thermal response of the building. Although the time-step is variable, most users perform hour by hour simulations, and hourly weather tapes are thus required. Also needed are parameters describing each separate component of the system. For example, the collector component needs nine constant parameters and seven time dependent inputs for operation. Models of rooms, storage, the hot water system pumps, heat exchangers, and other components have similar requirements. It takes several hours to "connect" the component models and specify the parameters in order to run TRNSYS. The outputs are user defined and can be as extensive as desired; useable energy from the collector, net change in storage and internal energy, room temperature and solar heat gain through windows are a few examples.

TRNSYS is often used as a research tool to evaluate the effect of various parameter changes on performance. It is generally too expensive to use to perform a "one-shot" analysis of a given design. The computer code and user manuals for TRNSYS can be purchased from the University of Wisconsin.

f-chart

F-chart, also developed at the University of Wisconsin Solar Energy Laboratory, provides a simple method for designing active solar domestic hot water and space heating systems when the detail provided by TRNSYS is unnecessary. The fraction of heating demand that is provided by a solar system, f, is correlated to two dimensionless ratios, the ratio of total energy absorbed on the collector to the building load, $F'_RAS(\tau\alpha)/L$, and the ratio of collector plate loss to building load, $F'_RAU_L(T_{REF}-T_O)/_L$. To establish the correlation, the design parameters were varied in over 300 TRNSYS simulations using "average weather" synthesized from eight years of meteorological data at Madison, Wisconsin.

To use f-chart, the user simply calculates the load ratios using the collector area (A), the monthly insolation (S), building load for the climate (L), and the intercept ($F'_R\tau\alpha$) and slope (F'_RU_L) of the standard collector efficiency curve. The fraction supplied by solar energy for each month is then determined. The user must do preliminary calculations to convert horizontal radiation to radiation on the appropriate tilted surface.

A computer version of f-chart is also available. Weather data for 172 cities is stored in the computer program. To run the program, the user inputs the values of fifteen parameters, including the location, the collector orientation, efficiency data, and the overall heat transfer coefficient for the building; default values are included for all inputs. The program can be run interactively from a remote computer terminal. A version of f-chart that can be run on hand-held programmable calculators is also available. The output data from f-chart includes monthly values for percent of load supplied by solar, incident solar energy, auxiliary and conventional building energy use, and an economic analysis if desired. f-chart can be purchased from the University of Wisconsin.

SOLCOST

SOLCOST was developed by Martin Marietta Aerospace for ERDA's Division of Solar Energy. SOLCOST, like f-chart, is a simplified design procedure useful for estimating the size and economics of solar space conditioning systems. SOLCOST is similar to TRNSYS in that the collector and thermal performance of the building are simulated using design equations. SOLCOST simulates the performance of the solar energy system on the fifteenth day of each month using average monthly weather data.

SOLCOST also has algorithms to model passive systems, heat pumps, air conditioning, and tracking and evacuated tube collectors. However, there is no provision for modeling the day to day dynamics of storage and load.

Input and output data for SOLCOST are similar to that in f-chart. Weather data for 124 cities is internalized. SOLCOST will do a building heating load analysis if the user supplies the detailed construction specifications, the fuel use, or the overall heat transfer coefficient. Default values for many parameters are also internalized.

SOLCOST is a public domain program available on the General Electric and Cybernet Timesharing systems. Access is by teletype over phone lines.

f-chart and SOLCOST Economics

F-chart and SOLCOST provide an economic analysis as well as collector area optimizing routines. The basic methodology used is to compare two thermally identical buildings, one with solar and one with a conventional energy system. The programs calculate the conventional building fuel cost, auxiliary fuel costs and the solar contribution to load. The amortized cost of the solar system plus backup fuel costs, less income tax savings, is compared to the energy costs of the conventional system. Fuel and maintenance costs are escalated over the period of analysis.

Both programs optimize collector area based on life cycle costs. The SOLCOST methodology sizes the collector to provide from 10 percent to 100 percent of the building load and identifies the size with the greatest life cycle savings. F-chart uses a numerical search method to find maximum, discounted life cycle savings.

CAL-ERDA

CAL-ERDA was developed to provide a detailed energy analysis of a large multizone building incorporating heating, ventilating, and air conditioning equipment as well as active and passive solar systems. There are four basic routines in CAL-ERDA. A *load* routine does a full, hour by hour building energy load analysis using a transfer function algorithm to simulate the dynamic response of the building. This load routine feeds a *systems* routine that determines the heating and cooling necessary to meet the building load based on user specified thermostat settings. A *plant* routine then sizes the boilers, chillers, and active solar systems. The solar simulation in this routine is based on the solar algorithm in TRNSYS but is faster and uses a simplified input. It models both air and liquid solar systems. Finally, an *economic* routine performs an analysis similar to f-chart and SOLCOST analysis. The program is driven by hour-by-hour weather tapes. A file of 60 test-reference-years of actual weather data, screened to eliminate typical conditions, has been developed. The program is designed to be user oriented. It requires input of building characteristics such as wall construction and parameters describing HVAC equipment. It is aimed at mechanical engineers and architects familiar with HVAC systems.

Although it is designed specifically to provide an analysis of multizone commercial buildings, CAL-ERDA may also be used to analyze single family residential buildings. It is presently limited to the analysis of active solar systems. However, a passive component that will include direct gain and water and masonry mass storage walls, is now being developed at Los Alamos Scientific Laboratories (LASL). CAL-ERDA is based on ASHRAE procedures but has not been verified experimentally. A detailed analysis of two buildings is being performed by LASL to validate the program.

CAL-ERDA was developed by Lawrence Berkeley Laboratories (LBL), LASL, and Argonne National Laboratories in cooperation with the Consultants Corporation Bureau of Oakland, California with support from the U. S. Department of Energy and the State of California Energy Research Conservation and Development Commission. The basic program has recently been completed and released. The computer code is available on tape, along with user and program manuals. The program is also available on the Cybernet Timesharing system. Information on availability and cost may be obtained from Argonne Code Center, Argonne National Laboratories, Argonne, Illinois.

PASOLE

PASOLE is a generalized computer algorithm developed and used at LASL to predict the thermal performance of passive building systems. The program does an hour by hour simulation of the performance of water and concrete mass storage walls and direct gain systems using a thermal network approach. The program is extremely versatile in that the user specifies the thermal network depending on the geometry of the system and the desired accuracy.

PASOLE has been verified by experiments at LASL. However, it is not user oriented and is not generally available. The results of extensive PASOLE simulations have been used to develop solar-load ratio correlations for Trombe and water walls with and without night insulation. These "passive f-charts" are available in a passive handbook put out by DOE. Similar correlations for direct gain systems are now being developed at LASL.

DEROB

DEROB is a system of programs developed at the University of Texas School of Architecture for the analysis of multi-room passive buildings. The program is designed to perform an hour by hour simulation of building performance. A transfer function approach is used to model the thermal response of walls and ceilings. The program calculates direct gain radiation and distributes it based on view factors and reflectivity of the interior spaces. The algorithm is geometry independent, permitting the user to model multi-room spaces of virtually any geometrical configuration. A peripheral graphics routine is available to input building characteristics and other required data. The program also requires hour by hour weather tapes.

DEROB is presently used as a design tool for studies at the University of Texas. A project funded by DOE is underway to combine parts of PASOLE and DEROB to provide a user oriented program for simulating the thermal performance of multi-room buildings with passive solar systems. This program will be available by early 1979.

Conclusion

The recent development of sophisticated simulation routines has provided the architect and engineer with a convenient means for predicting the performance of active and passive solar systems. It is now possible to evaluate various solar design options rapidly and at relatively small expense. This permits the designer to investigate new ideas and to best use existing systems.

Although modeling solar systems is inherently complicated, it has been recognized that it is essential that the simulation programs be easily accessible and relatively simple to use. Now that many of the basic computational algorithms have been written and verified, increased attention is being given to making programs more user oriented. The DOE sponsored project to combine the user oriented DEROB program and the well-verified PASOLE algorithm is an example.

It should be noted that the results of extensive simulations at universities, government laboratories, and architect-engineering offices are increasing our knowledge regarding the efficient application of solar energy. Simple correlations and rules-of-thumb are being developed to serve as guidelines for the design of cost-effective, energy conserving solar buildings.

Acknowledgements

The author wishes to thank Bruce Anderson and Paul Sullivan (TEA), Jack Breese (University of Michigan), Bob McFarland and Bruce Hunn (LASL), and Francisco Arumi (University of Texas, Austin) for providing the information on the simulation models which formed the basis of this article.

Doug Kelbaugh's house in Princeton, New Jersey, maximizes the use of south-facing glass. The wall behind the glass is composed of black concrete for heat storage and windows for light.

Glass—Windows—Sunlight

by John I. Yellott

John Yellott is Visiting Professor in Architecture at Arizona State University, Tempe.

You probably don't think about glass very often, unless there's a neighborhood baseball game going on in your back yard or you happen to be one of those oft maligned people who live in a house made of it. Glass is so common that we generally take it for granted and forget that it really has some remarkable qualities. It's breakability not withstanding, glass is extremely weather-resistant; it suffers neither from oxidation nor other forms of deterioration when exposed to summer's heat or winter's cold. And, although certain antique bottle collectors might be sorry to hear it, most window glass will not be transformed into some marketable shade of purple by "solarization," i.e., long exposure to ultraviolet radiation. About the only weaknesses that glass has are brittleness and breakability—deficiencies that can be diminished by various tempering processes. In short, glass is almost ideally suited to its most common task—the glazing of windows.

The clear window glass available today is made from sand that is as low in iron as possible and is tempered to prevent breakage from thermal stress and falling objects. "Soda-lime," the type of window glass used for most glazing purposes is particularly attractive because it is highly transparent to the sun's radiation between 0.3 and 3.0 micrometers (μm) wavelength (the radiation that tends to heat things up), while being completely opaque to the longwave infra-red radiation between 3.0 and 30 μm that is emitted by sun-warmed surfaces. This is why window glass is so useful as both the cover plates for solar collectors and as the world's foremost fenestration glazing material. All of the services glass offers—its ability to admit or reject solar radiation, its protection from the winds—are valuable, but here we are most interested in the first of these and in the methods currently available for estimating the rate at which heat will be admitted through vertical windows under most normal weather conditions.

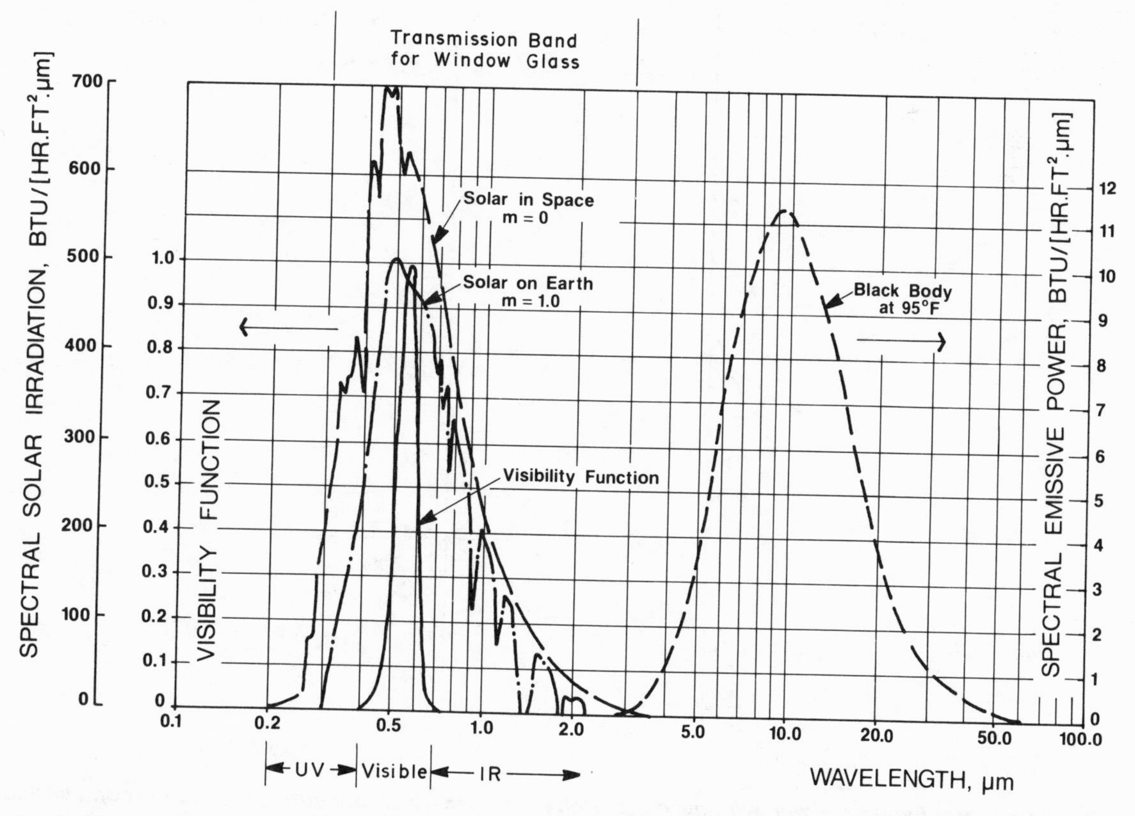

Fig. 1. ***The solar spectrum in space and on earth; sensitivity curve for the human eye; spectral emittance for a black body at 95° F; transmission band for window glass.***

Glass and Its Unique Place in the Electromagnetic Spectrum

Because the spectrum of electromagnetic radiation covers such a wide range, it can best be shown on the logarithmic scale, used in Fig. 1. Above the earth's atmosphere, solar radiation can first be detected in the spectrum's very short ultraviolet region at about 0.2 micrometers (μm); and 99.9 percent of the sun's radiant output is shorter in wavelength than 5.0 μm. Peak radiation intensity is encountered near 0.48 μm, in the blue-green region of the visible spectrum, not far from the wavelength at which the human eye is most sensitive.

After passing through the earth's atmosphere, solar rays are reduced in intensity by at least 30 percent on even the clearest day because of the scattering action of atmospheric molecules of oxygen and nitrogen, the reflection by dust particles and the absorption by certain atmospheric components such as ozone (O_3), carbon dioxide (CO_2) and water vapor (H_2O). The absorption bands shown in Fig. 1 reflect the condition when the sun is directly overhead and the observer is at sea level (air mass $m = 1.0$).

Despite its invisibility, the 5 percent of the total energy that arrives in the ultraviolet portion of the spectrum at sea level is extremely important. It can be either beneficial or deleterious to nature and people, but what matters here is the quantity of heat that it can produce indoors, particularly on winter days, when transmitted solar radiation is absorbed by floors, walls, furniture and draperies or other indoor shading devices.

Glass and Its Solar-Optical Properties

When a beam of solar radiation falls on a sheet of window glass, Fig. 2, some of the radiant energy is reflected from the front and rear surfaces of the glass, some is absorbed as the radiation passes through, and some is transmitted to the glazed space. The quantitative values of these solar-optical properties—reflectance, ρ, transmittance, τ, and absorptance, α —vary with the incident angle θ between the incoming ray and the line, OP, normal (perpendicular) to the surface. Fig. 3 shows

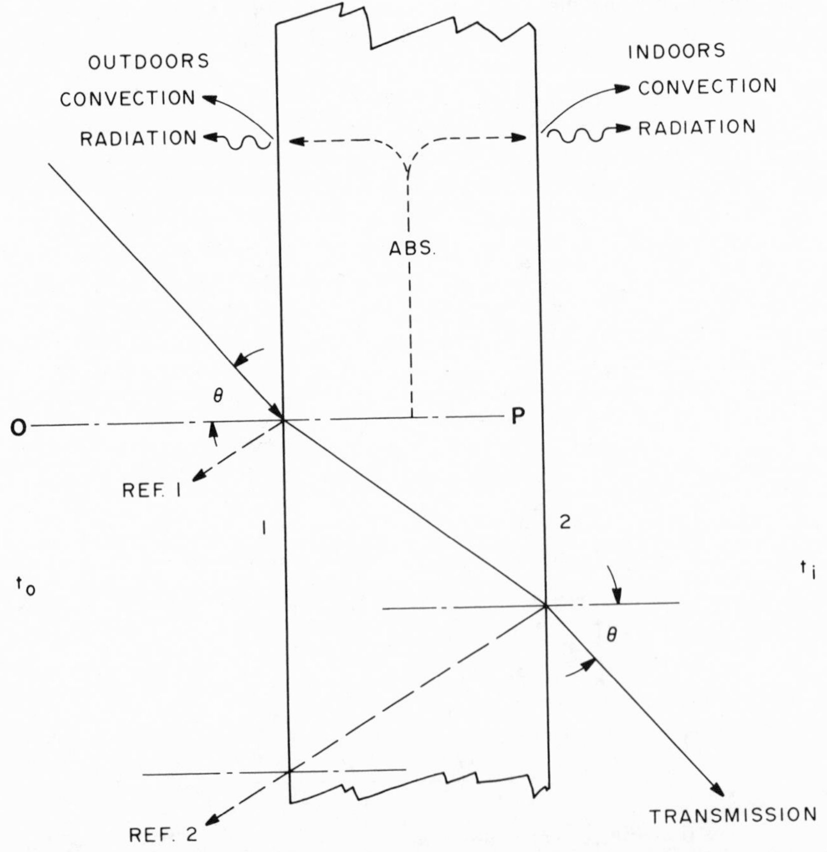

Fig. 2. Interaction of solar ray with window glass. Reflection occurs at first surface, 1, and second surface, 2; refraction towards the normal occurs at first surface, back to original direction at 2.

Table 1. Solar angles, transmittance for ¼ in clear glass, direct normal irradiation and amount of radiation transmitted on Dec. 21 and June 21 for south-facing windows at 40 deg north latitude.

Date		**December 21**					**June 21**				
Solar Time	a.m.	8	9	10	11	12	8	9	10	11	12
	p.m.	4	3	2	1		4	3	2	1	
Solar Altitude, B		6	14	21	25	27	37	49	60	69	74
Solar Azimuth, ϕ		53	42	29	15	0	91	80	66	42	0
Incident Angle, θ		53	44	35	29	27	91*	84	78	75	73
Cosine 0		.60	.72	.82	.87	.90	0	.11	.21	.26	.29
I_{DN}, Btu/(ft^2.hr)		176	257	288	301	304	245	264	274	279	280
I_{DN}, × Cosine θ		107	185	236	262	274	0	28	58	73	81
Transmittance,		.68	.72	.76	.77	.77	0	.20	.38	.45	.46
Radiation Transmitted, Btu/(ft^2.hr)		73	133	179	202	211	0	6	22	33	37

*When θ exceeds 90 degrees, no direct radiation strikes the glass.

how these properties vary as the incident angle rises from 0 to 30, 60 and finally 90 degrees. Little change occurs until θ exceeds 30 degrees and then transmittance begins to decrease as reflectance begins to rise and a minor increase occurs in absorptance due to the increasing length of the path through the glass. Beyond 60 degrees, reflectance begins its rapid increase toward 1.00 when θ reaches 90 degrees. Both τ and α decrease rapidly at high values of θ, a fact that is very beneficial in summer when there might otherwise be an excessive admission of unwelcome solar radiation.

For south-facing vertical windows, the incident angle θ is relatively low in winter and very high in summer. This produces the two results shown in Table 1, which gives the solar angles for the hours from 8:00 A.M. to 4:00 P.M., solar time, at 40 degrees north latitude on December 21 and June 21. The direct component of incoming solar radiation is the product of the direct radiation intensity (I_{DN}) on a surface perpendicular (normal) to the solar beam, and the cosine of the angle of incidence θ, between the solar rays and the line (OP in Fig. 2) normal to the surface. In equation form,

$$I_{D\theta} = I_{DN} \times \text{Cosine } \theta \quad (1)$$

The "bottom line" of Table 1 shows the difference between the performance of ¼ inch clear glass in south-facing windows on December 21 compared with the performance of the same glass on June 21. Even without any exterior shading, very little direct radiation is transmitted through the glass in mid-summer because of the combined effects of the high incident angles, with their low cosines, and the low transmittance of the glass at those high angles.

In mid-winter, by contrast, transmitted radiant energy is high during all of the sunlit hours of the day. This explains why properly designed south-facing windows can be significant sources of winter heat. The simplified estimate reported in Table 1 does not take into account diffuse radiation or the beneficial effects of radiation reflected from the foreground to the glass. The calculation procedure outlined in the next section takes all three sources of radiation—direct, diffuse and reflected—into account and allows for speedy and reliable estimates of clear day solar heat gains for windows with any orientation.

The ASHRAE Method of Estimating Heat Flow Through Windows

In the preceding section, a step-by-step procedure was used to find the solar heat gain through south-facing windows in winter, when the gains are high, and in summer, when the gains are low. The following procedure adopted by ASHRAE's Technical Committee on Fenestration can be used to make estimates for any month for vertical and horizontal surfaces with any orientation, using any of the common types of glazing materials.

Prior to 1963, the ASHRAE Guide contained solar data for only one day in the year (August 1) and for one location (Pittsburgh). In 1963, the chairman of the Fenestration Committee, Donald Vild of Libbey-Owens-Ford Glass Co., proposed a new computation procedure that could be used anywhere in the United States (and ultimately in most of the world!) at any time of the year. Vild found that the *ratio* between the *solar heat gain* through any kind of

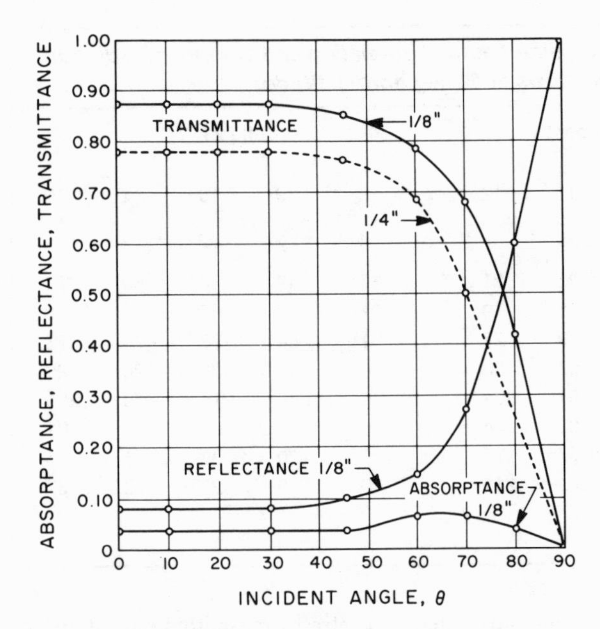

Fig. 3. Transmittance, reflectance and absorptance for 1/8 inch clear glass; transmittance for 1/4 inch clear glass.

fenestration under a given set of conditions and the *solar heat gain* through ordinary ⅛ inch clear window glass under the same conditions is essentially constant, regardless of incident angle. This ratio is called the *Shading Coefficient* (SC). The next step was to show that the solar heat gain through the reference glass (clear, Grade A, double-strength, Transmittance 0.86 at near-normal incidence) could be calculated easily for clear days when the latitude, date and time were specified. Vild called this quantity the Solar Heat Gain Factor (SHGF) and he produced tabulated values of SHGF for the summer months for latitudes from 24 to 56 degrees north.

Vild's original tables were based on Parry Moon's 1940 calculations of the variation of solar radiation intensity with solar altitude. It was discovered, however, that these values were far too low for winter conditions. Therefore, the 1967 ASHRAE Handbook of Fundamentals contained a new set of Solar Heat Gain Factors developed with the aid of computers, and a new set of clear day solar irradiation data formulated by Professors Jordan, Liu and Threlkeld at the University of Minnesota. The actual computation of the tables was done at the Canadian Bureau of Building Research under the direction of Donald Stephenson. SHGF values published in the ASHRAE Handbooks of 1967, 1972 and 1977 contain an allowance for 20 percent foreground reflection. Additionally, the 1977 edition gives data for orientations around the compass by 22.5 degrees increments, and the latitude range is extended to cover the entire northern hemisphere from the equator to 64 degrees north, by 8 degree increments.

Values of the Shading Coefficient for all of the commonly-used glazing systems, including shaded and reflective glasses and non-glass materials, are given in the ASHRAE Handbooks of Fundamentals and the data summarized below are extracted, with permission, from Chapter 26 of the 1977 edition and those in earlier editions. For ⅛ inch clear glass, since it is the reference glass SC = 1.00; for ¼ inch clear float glass, with 0.77 transmittance at near-normal incidence, SC = 0.94, and for ¼ inch heat-absorbing glass, with 0.46 transmittance, SC = 0.69. For a sheet of opaque black glass with zero transmittance and 4 percent reflectance, SC would be about 0.30, but no one would use this for a window!

For double-glazing, SC = 0.88 for two lights of ⅛ inch clear glass and 0.81 for two lights of ¼ inch clear glass. For the combination of an outer light of ¼ inch heat-absorbing float glass and an inner light of ¼ inch clear float, the transmittance is 0.36 and SC = 0.55. The latter glazing system is used where exclusion of summer sunshine is required, while the first two combinations are selected where admission of winter sunshine is desired. Fig. 4 shows Shading Coefficients for a number of unshaded single and double glazings. Beyond this, the 1977 ASHRAE Handbook of Fundamentals gives reliable data on Shading Coefficients for virtually any kind of glazing combination available.

Windows, regardless of what the sun may be doing, conduct heat inwardly or outwardly, in response to temperature differences between indoor and outdoor air. The complete equation for heat flow through fenestration in winter is:

$$Q = \frac{\text{Btu}}{\text{Hour}} = \text{Area} \times [SC \times SHGF - U \times (t_i - t_o)] \qquad (2)$$

where U = overall coefficient of heat transfer, Btu/ (hr.ft^{2}F)

t_i, t_o = indoor tempatures, F

A = area of glazed surface, ft^2

In summer, when t_o is higher than t_i the second term in Eq. 2 becomes $+U\,(t_o - t_i)$.

For all uncoated single glasses, U = 1.10 in winter, when the wind speed is assumed to be 15 mph. In summer, when the wind speed is assumed to be 7.5 mph, U = 1.04. For double glazing, the width of the air space is

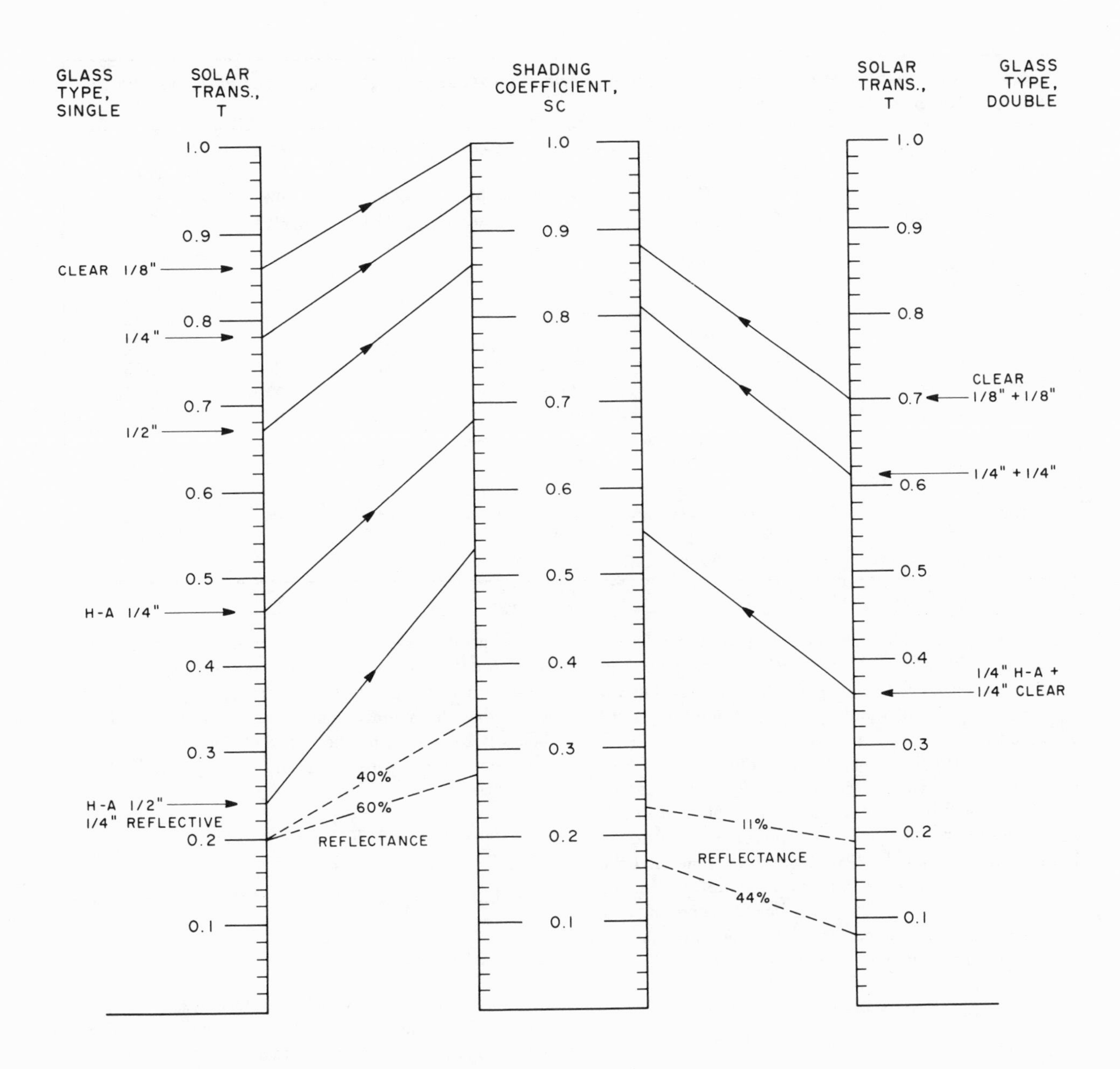

Fig. 4. Shading coefficients vs solar transmittance for single and double glazing (data from 1977 ASHRAE Handbook of Fundamentals).

important; winter values of U are 0.58 for ¼ inch air space and 0.49 for ½ inch air space. For summer, values are 0.61 for ¼ inch and 0.56 for ½ inch. U values for other combinations and for coated glasses, are given on page 26.10, Table 13, of the 1977 ASHRAE Handbook.

Because of widely differing amounts of moisture in the atmosphere, summer SHGF values are consistently lower than winter values for comparable solar altitudes. At any specified time of day (solar time) the solar altitude is determined by a combination of local latitude and the sun's declination on that day (see 1977 ASHRAE Handbook for definition of declination). Table 2 gives selected values of SHGF in Btu/(hr ft^2) for south-facing surfaces at 40 degrees north latitude on the 21st day of each month, with the months paired according to the declination that exists on the specified day.

Table 2. Solar heat gain factors for south-facing vertical surfaces throughout the year at 40 deg north latitude; 21st day of each month.

Months		Dec.	Jan. / Nov.	Feb. / Oct.	Mar. / Sept.	Apr. / Aug.	May / July	June
Declination, deg		−23.5	−19.9	−10.6	0.0	+12.0	+20.4	+23.5
Time								
a.m.	p.m.							
5	7	*	*	*	*	*	*	1
6	6	*	*	*	*	6	11	13
7	5	*	*	13	21	17	21	22
8	4	50	74	92	73	41	30	29
9	3	151	158	154	126	82	52	45
10	2	210	211	200	168	118	82	69
11	1	242	242	227	194	144	104	88
	12	253	252	237	203	152	111	95
Daylong			1626	1642	1388	976	716	
Total		1550	1596	1582	1344	948	704	630

*denotes hours for which sun is not above the horizon

In Table 2, the SHGF values for the paired months are the averages of the two values for each date and time. The maximum error involved in this simplification is about 1.5 percent, which is not significant. The daylong totals are given separately for each month.

For any particular daylight hour, the net heat gain through a south-facing window at 40 degrees north latitude is found by using the data from Table 2 in Eq. 2, as shown by the following example. Let the date be January 21, the hour be 12:00 noon, solar time, and the outdoor air temperature be 35° F while the indoor temperature remains at 70° F. Let the glazing be single ¼ inch float glass, unshaded. The heat gain through 100 square feet of glazed area under these conditions is:

$$Q = 100 \times [0.94 \times 252 - 1.10 \times (70 - 35)]$$
$$= 100 \times (236.9 - 38.5) = 19{,}838 \text{ Btu/hr}$$

AT 8:00 A.M., when the SHGF is only 74 Btu/(ft^2 hr) and the air temperature would probably be closer to 25 than to 35, the answer is:

$$Q = 100 \times (0.94 \times 74 - 1.10 \times 45) = 2{,}006\text{-Btu/hr}.$$

For a day-long estimate, the average outdoor temperature should be used and for the example cited above, a reasonable value would be about 30° F. Here the net day-long heat gain would be:

$$Q = 100 \times [0.94 \times 1626 - 24 \times (70 - 30) \times 1.1] = 47{,}244 \text{ Btu/day}.$$

When moveable insulation is used, the window has a nighttime U-value of 0.20 instead of the bare glass value, 1.10, and maximum Btu/day increases significantly. Assume that the average temperature during the 16 dark hours is 25° F; during the 8 daylight hours, the average temperature would be 40° F and the day-long heat gain would be:

$$Q = 100 \times [0.94 \times 1626 - 8 \times (70 - 40) \times 1.1 - 16 \times (70 - 25) \times 0.2]$$
$$= 100 \times [1528.4 - 264 \text{ (day loss)} - 144 \text{ (night ins. loss)}]$$
$$= 112{,}044 \text{ Btu/day}.$$

For clear days in the contiguous 48 states, south-facing clear glass is always a net energy saver, since more heat is admitted through the glass than is lost by outward conduction. The use of moveable insulation is strongly recommended, even if it is only a heavy double drape. Moreover, double glazing will give more heat gain than single, as the following example shows. Use clear ¼ inch insulating glass with a ½ inch air space to obtain a Shading Coefficient of 0.81 and a U-value of 0.49. For the day-long conditions cited above:

$$Q = 100 \times [0.81 \times 1{,}626 - 24 \times (70 - 30) \times 0.49]$$
$$= 84{,}666 \text{ Btu/day}.$$

The improvement due to the lower U-value of the double glazing is 37,422 Btu/day, without any other insulation.

Values of U for combinations of double glazing and internal shading devices are given in Chapter 26 of the 1977 ASHRAE Handbook of Fundamentals.

Conclusions

South-facing glass is always a source of heat gain on clear days in winter because of the fortunate combination of high rates of irradiation and low incident angles, as shown in Table 1. Double glass, with a winter U-value less than half as great as that of single glazing, is more effective than single glazing. The use of moveable insulation is essential to give adequate protection on very cold winter nights and on sunless days. The ASHRAE procedure for estimating solar heat gains and the data now available on Solar Heat Gain Factors and U-values make it possible to obtain reliable estimates of clear day heat gains. The methods used to find *average day* heat balances are similar to those used with solar collectors because, in the final analysis, a south-facing window is a first-rate collector with a very high efficiency!

A Solar Installer's Checklist For Domestic Hot Water Systems

by Rick Schwolsky

Rick Schwolsky is a partner in Sunrise Solar Services, Suffield, Conn., and chairman of the National Association of Solar Contractors, 139 Washington St., Weymouth, Mass. 02188.

Paying attention to detail is one of the most important considerations when installing domestic hot water systems. Materials and procedures should be carefully planned ahead of time and care should be taken throughout the entire construction process.

The skills required to successfully complete the installation of active solar heating equipment for domestic hot water should be complemented by persistence, and the maintenance of high trade standards. Installation consciousness must extend to all decisions made in the course of a project. This includes choice of materials, equipment, and theoretical design of a system, as well as an awareness of the impact the installation will have on a structure. The trade-offs must be dealt with, the decisions made, and total familiarization with the project should begin.

Possibly the most important factor of a successful project is the open flow of information between the manufacturer, supplier, installer, and owner. The manufacturer's installation manual should be reviewed before any work begins. Any questions not covered in the literature should be answered by either the supplier or the manufacturer. In the case of site fabricated systems, the designer is the source of information concerning installation, operation and troubleshooting. By installing systems as specified by the manufacturer, engineer, or designer, the installer then has recourse if a problem arises. Look for confident answers to any technical questions. If you are not satisfied by the information received, seek the next highest level in the manufacturer-supplier-installer-owner network. If the manufacturer cannot provide satisfaction, you've been dealing with the wrong company.

The following general guidelines illustrate many of the details to be aware of during the system layout, installation, testing and troubleshooting. Each of the manufacturers has a different approach, and specific details must be obtained from them. Plan ahead, take your time, double check every step, and enjoy your role in helping to reduce fossil fuel consumption. There will always be on-site decisions to make. Use good trade practices and common sense in finding solutions.

Installation: Ground mount

SITE LAYOUT

Collector location should always:

- Face within 15 degrees of true south.
- Receive six hours of sunlight per day without shading.
- Be installed at pitch equal to site latitude.

Layout of pipe runs with pipes to be trenched always:

- Use annealed copper tubing of appropriate size (no joints underground).
- Choose suitable insulation material for use underground.
- Determine entry method through wall or foundation.

With pipes above ground, always:

- Choose proper insulation and protect it from UV and weather.
- Pitch to drain.
- Support outside runs properly.

With collector stands always remember to:

- Determine material of choice (speed rail pipe stand, welded angle iron, cedar, redwood, or pressure-treated wooden stand).
- Construct to provide proper pitch, means of fastening collectors, aesthetically pleasing appearance.
- Install to provide long term stability to wind and frost, proper orientation, protection from shading.
- Install to present as little of a safety hazard as possible.
- Avoid long runs of pipe either above or below ground.

Installation: Roof mount

SITE LAYOUT

Collector location should always:

- Face within 15 degrees of true south.
- Receive at least six hours of sunlight per day without shading.
- Be installed within five degrees plus or minus of site latitude. (Example, Hartford, Conn.—42° N. lat.—collectors can be installed between 37°–45° angle to the horizontal.)

With layout of pipe runs always remember to:

- Be aware of chase locations, pipe support and pitch, and insulation of lines important while doing layout.
- Use pipe chases as provided in new construction.
- Use accessible pipe chases or closets, utility rooms, etc. in retrofits.
- Run outside pipe along side of structure to enter through siding.

With collector mounting always remember to:

- Use common sense in handling collectors for safety to installers, homeowners, and equipment.
- Assure proper pitch (within guidelines).
- Install frame or stand to adjust pitch, or
- Mount collectors to finished roof if within guidelines in retrofits.
- Provide proper pitch on roof or section of roof for collectors in new construction.

Retrofit to finished roof surface, always remember to:

- Follow manufacturers installation manuals to determine necessary fastening hardware* and other supplies for securing collectors.
- Carefully seal all fastening penetrations.
- Be conscious of water run-off, air circulation, build-up of leaves, snow and ice damming.
- Be conscious of wind loads and possible problems at each site.
- Mount according to manufacturer's details.
- Follow manufacturer's suggestions.
- Follow good construction practices.
- Adopt mounting practices to room construction and site influence.

New construction mounting to unfinished roof surface

Collectors mounted to sheathing or rolled roofing: always remember to:

---

* *Fasteners include lag bolts, hex-head bolts, speed rail hardware, threaded rod and others.*

- Be aware of flashing requirements (i.e., spacing of collectors from other roof penetrations, spacing between collectors).
- Be aware of material compatability, avoid dissimilar metals.

Plumbing connections

Plumbing details vary with manufacturers, so remember to:

- Follow guidelines in installation manuals to determine necessary fittings and other supplies.
- Make proper connections and run lines to storage tank and/or heat exchanger.
- Insulate all lines—assure protection from UV degradation, and weathering.
- Install control sensor and run wire with pipe run.
- Seal feed and return penetrations with suggested material (roof flashing, silicone caulking, other roofing sealants).
- Provide any components necessary for air elimination or entry to the system.
- Try for the most direct route available with as few turns as possible.
- Pitch to drain—to assure proper drainage and to help air elimination.
- Support piping or ducts as per trade practices, and code requirements.
- Pipe components as directed by manufacturer's guidelines and system schematic.
- Support all components.
- Be aware of sensible installation layout to avoid interference of space.
- Be aware of ease for future maintenance.

When locating storage tank and heat exchanger always aim for closest:

- Proximity to feed and return of heat transfer fluid.
- Proximity to cold feed (water).
- Proximity to any existing water heating equipment to be placed in series with solar water heater.

Electrical

When determining need for electrical hook-up for storage tank, pump, fan, or controls (some building codes require separate circuits for powering solar system components) remember to:

- Check out controls *before* installation.
- Install controller in proximity to pump.
- Wire pump and controller.
- Wire sensors to controller (after sensors are in place).
- Wire any solenoid or control valves, or heating elements required.
- Install and wire any monitoring devices.

Installation pressure test

Install all components except those unsuitable for high pressure (air vents, expansion tanks, temperature and pressure relief valves).

- Pressure test system at 80–100 psi for one hour.

Install all components

- Pressure test system at 30 psi for 24 hours.

Check with manufacturers for test and working pressures for all components.

Discharge pressure test after satisfying requirements

- Install all missing components.
- Flush system prior to final charge, and drain to remove dirt and flux.
- Charge system with heat transfer fluid (if antifreeze, assure proper ratio for designed protection).

Pressurize to manufacturers suggested operating pressure and then remember to:

- Operate circulator (either automatically or by directly plugging pump into electrical source).
- Bleed air from venting locations as installed.
- Assure final working pressure.
- Reset all controls for automatic operation.
- Label all relevant valves with simple explanation of function and operating position (cold feed to collectors: leave open).

TROUBLESHOOTING CHART

The following guidelines illustrate troubleshooting procedures in general. Contact component manufacturers for specific information concerning their products. These steps are described by symptom, component, problem, and corrective action, and can save some of the guesswork involved in troubleshooting faulty operation.

SYMPTOM	COMPONENT	PROBLEM	CORRECTIVE ACTION
Not enough hot water	Collectors	Improperly faced	Check direction. Face collectors due south.
		Improperly sloped	Check slope. Dom. water, make lat. – or + 5°
		Partially shaded	Remove shadowing material or move collectors.
		Insufficient area	Increase collector area.
		Unequal flow	If collectors are unequally warm, repurge system to equalize flow, assure reverse return piping layout.
	Differential Thermostatic Controller	Loose or incorrect electrical connections	Check wiring schematic for correct connections, tighten loose wiring.
		Sensors not insulated from surrounding air	Check and insulate.
		Faulty sensors	Dip alternately in hot and cold water to test whether switches start and stop pump.
		Faulty thermostat	Check to see if thermostat contacts close. CAUTION: DO NOT JUMP COMMON AND LOAD TERMINALS TO TEST.
	Tank	Too small	Install second tank.
		Improper electrical connection	Check power source wiring.
		High storage losses	Check insulation and location of tank.
	Piping	Night convection losses	Check piping. Install check-valve if necessary.
		Excessive heat loss	Check insulation.

TROUBLE SHOOTING CHART (cont.)

SYMPTOM	COMPONENT	PROBLEM	CORRECTIVE ACTION
	Mixing valve	Improperly adjusted	Check adjustment temperature indicator and set higher if necessary.
		Faulty	Replace
Water leak	Collectors	Leak at connec-tions	Check and tighten fittings.
		Internal leak	Repair leak or contact manufacturer.
	Relief valves	Set to low	Adjust pressure setting or replace.
		Do not re-seat (close)	Clean seat or replace unit.
		High system pressures	Install pressure reducing valve in water supply line.
Drop in system pressure	Collectors	Slow leak	Inspect and repair or replace.
	Relief valve	Spitting fluid	Inspect and repair or replace adjustable relief valve.
	Air vents	Spitting fluid	Inspect and repair or replace
	Expansion Tank	Loss of pressure	Recharge to 12 psi.
	Piping	Leak in heat ex-changer	Replace tank.
		Leaky joints	Locate and repair.
Noisy System	Piping	Entrapped air	Purge system and install vents as needed at high points.
		Air purge installed backwards	Reverse. Face arrow in direc-tion of flow.
		Pipe vibration	Prevent vibration. Isolate tube from hard surface.
		Air hammer	Install shock suppressor.
	Air vents	Not working	Check for tight cap. Operate plunger manually. Replace if necessary.

TROUBLE SHOOTING CHART (cont.)

SYMPTOM	COMPONENT	PROBLEM	CORRECTIVE ACTION
		Insufficient number	Install additional vents at intermediate high points.
	Pump	Air locked	Loosen venting screw in body. Vent air and tighten.
No flow	Collectors	Air locked	Purge system and install vents as necessary.
No flow	Pump	Too small	Check system pumping head and change pump if pump is not adequately large.
		Air locked	Loosen venting screw in body. Vent air and tighten.
		Impeller bound	Loosen impeller screw.
		Installed backwards	Reverse. Check flow direction arrow and reverse if necessary.
		Closed off	Open throttling lever on pump head.
		Installed incorrectly	If installed horizonally, place motor to side. Check pump installation book.
		Low speed on pump	Increase speed to high and check pump size.
	Piping	Air locked	Force purge; install air vents if necessary.
		Too small	Increase piping size or add second pump.
	Vents	Insufficient number	Install additional vents.
		Faulty	At high points not already vented operate vent plunger manually. Replace vent if necessary.
	Shut-off valves	Shut off	Open
	Flow Regulator	Clogged	Open and clean out venturi.

TROUBLE SHOOTING CHART (cont.)

SYMPTOM	COMPONENT	PROBLEM	CORRECTIVE ACTION
		Installed backwards	Check arrow for flow. Reverse if necessary.
Decreasing performance	Collectors	Increased shading	Remove shade or move collectors.
		Dirty	Clean periodically.
		Deterioration of absorber coating	Contact Reverse.
Decreasing performance	Piping	Night convection losses.	Install solenoid shut-off valve or thermal loop.
	Tank	Sludge in bottom	Drain water from tank periodically.
System stays off	Differential thermostat controller	Sensors improperly placed.	Inspect, fasten securely to surface, cover with insulation.
		Sensors faulty	Inspect, replace.
		Faulty on-off auto switch setting	Inspect, replace. Check and reset if necessary.
		Wired incorrectly	Check wiring diagram for correct connections.
System stays on	Differential thermostat controller	Sensors improperly placed	Inspect, fasten securely to surface, cover with insulation.
		Sensors faulty	Inspect, replace.
		Faulty on-off-auto switch setting	Inspect, replace. Check and reset if necessary.
		Wired incorrectly	Check wiring diagram for correct connections.

Introduction to Small-Scale Wind Power:

by Joseph Carter

Joseph Carter is Associate Editor with Wind Power Digest and a freelance writer/photographer in the alternative energy field.

For centuries, people around the world have mustered the spunk to actualize their ideas about harnessing the wind. Like all solar-renewable energy technologies, wind power is ideally suited to small-scale independent power production.

To the novice, wind power might seem a subject better left to the professional engineer. Even the smallest systems employ some fairly sophisticated gadgetry. The design and installation of Wind Energy Conversion Systems (WECS) is a topic that fills volumes with "technese"—the jargon of technology. Yet, wind energy can be distilled and condensed to a point well within the grasp of any person who gives it proper consideration.

Electric bills come every month and are small in relation to the initial cost of an independent power system. A wind system requires a sizeable initial cash outlay ($15,000 to $20,000 for the equipment, plus additional costs to make it operative). After that the "fuel" is free and the only continuing expense is occasional maintenance. Wind energy converted to useful power creates *savings* by not using electric or fossil fuel power. You are ahead of the game when these savings (you save more as energy prices rise) surpass the cost of the WECS. If a WECS site is subject to frequent and intense winds (those averaging over 13 mph or 5.8 meters per second, m/s), the break-even day can come in as few as eight years. For marginal wind sites "pay back" might not be realized for as long as twenty years; don't worry—the machines *can* last that long.

Another way to look at WECS economics is to make an immediate cost comparison between wind and conventional energy. On a cost per kilowatt/hour basis, recent studies show that the power produced by WECS installed today will run somewhere around 10 to 20¢ per kwh. The national average for utility electricity is about 5¢ per kwh. It goes without saying that WECS will not be cost competitive in many suburban, urban and rural areas served by utility power. But there are many places where energy costs far exceed the national average.

A perfect example of this situation is the growing settlement of remote, backwoods regions. Areas like these are often far from existing power service and the thought of paying thousands of dollars to string power lines is, in many cases, unthinkable. As fuel prices mount, even the cost of running an alternative system such as a gas generator will surpass that of a WECS. The generating facilities of many island communities are fueled by coal, oil or diesel fuel that must be shipped from the mainland. Residents of these localities (which are often windy) pay up to 25¢ per kwh for electrical service.

A group of urban homesteaders in New York City, where electricity is sold for 8 to 10¢ per kwh, have installed a WECS to light the hallways and to run the circulating pumps of the solar domestic hot water system in their restored walkup. This particular system is connected directly to the standard utility grid. The need for battery storage and other expensive power conditioning components is thereby eliminated, and the result is a much cheaper system with good back up, compliments of ConEd. Given a monthly wind average of 8–13 mph (3.5 to 5.8 m/s) and a consistant annual rise in utility costs of 10 percent the thirty year savings from this system should be over $32,000. Where there's a wind there's a way.

The prerequisite to putting together any size wind system is planning. Begin with what is known as a *load analysis*, a measure of how much electrical or mechanical (for water pumpers) power you need and when you need it. If you're presently using utility electricity, natural gas or propane (or any combination of the three), your load analysis has already been done. The readout is printed on your monthly energy bill. If you don't have the bills, call the utility or distributor, they'll have a record of your energy consumption. To total all your energy use in terms of kilowatt/hours, use these conversion factors:

one therm (100 cubic feet) of natural gas = 29.3 kwh, and

one gallon of propane = 28 kwh.

It should be noted that at today's regulated prices natural gas and propane are both cheaper than wind for space and water heating and for cooking loads.

Table 1 will also give you a good ball park estimate of your total load. Remember that tables of this kind are developed from average United States residential figures. This would make them overestimated for a conservation-oriented user. It's important to think in terms of *minimizing* the total load; conserving energy is less costly than

Table 1. Appliance power consumption.

	Average Wattage	Average hours used per year	Approx. kWh used per year
Air cleaner	50	4,320	216
Air-conditioner (room)	860	1,000	860
Blanket	177	831	147
Blender	386	39	15
Broiler	1,436	70	100
Carving knife	92	87	8
Clock	2	8,760	17
Clothes dryer	4,856	204	993
Coffee maker	894	119	106
Deep fryer	1,448	57	83
Dehumidifier	257	1,467	377
Dishwasher	1,201	302	363
Egg cooker	516	27	14
Fan (attic)	370	786	291
Fan (circulating)	88	489	43
Fan (rollaway)	171	807	138
Fan (window)	200	850	170
Floor polisher	305	49	15
Freezer (15 cu. ft.)	341	3,504	1,195
Freezer (frostless 15 cu. ft.)	440	4,002	1761
Frying pan	1,196	156	186
Hair dryer	381	37	14
Heat lamp (infrared)	250	52	13
Heater (portable)	1,322	133	176
Heating pad	65	154	10
Hot plate	1,257	72	90
Humidifier	177	290	163
Iron (hand)	1,008	143	144
Mixer	127	102	13
Microwave oven	1,450	131	190
Range with oven	12,200	96	1,175
Range with self-cleaning oven	12,200	99	1,205
Radio	71	1,211	86
Radio/record player	109	1,000	109
Refrigerator (12 cu. ft.)	241	3,021	728
Refrigerator (frostless 12 cu. ft.)	321	3,791	1,217
Refrigerator/freezer (14 cu. ft.)	326	3,488	1,137
Refrigerator/freezer (frostless 14 cu. ft.)	615	2,974	1,829
Roaster	1,333	154	205
Sewing machine	75	147	11
Shaver	14	129	1.8
Sunlamp	279	57	16
Television (b&w, solid state)	55	2,182	120
Television (color, tube)	300	2,200	660
Toaster	1,146	34	39
Toothbrush	7	71	0.5
Vacuum cleaner	630	73	46
Waffle iron	1,126	20	22
Washing machine (automatic)	512	201	103
Washing machine (non-automatic)	286	266	76
Waste disposer	446	67	30
Water heater	2,475	1,705	4,219
Water heater (quick recovery)	4,474	1,075	4,811

*Courtesy of **Wind Energy and You,** Robert C. Griffin, copyright 1976.*

installing a larger-than-necessary WECS to meet an unnecessarily high load.

Peak load is another factor to be determined from your load analysis. This is the maximum amount of power that might be used at any time during the day. In electrical terms, power is expressed by watts (w). An electrical device draws so many watts or kilowatts (one kw = 1000w) and the peak load factor is determined by adding the wattages of whatever devices might be used simultaneously. Before you consider buying any wind-electric device be sure that its peak rated output is close to your peak load factor. It's a good idea, whenever possible, to stagger appliance and tool use; it will minimize your peak load and help reduce initial costs while boosting overall efficiency.

If you're thinking about including a water pump in your wind system, start by determining your average daily water use. Table 2 shows the average water needs of a number of things, animals and people. Average human use runs anywhere from 75 to 125 gallons per day. Here again, simple conservation techniques will increase the effectiveness of the windmill pumping system.

Having established your energy needs, you must then consider the energy supply—where it's coming from, when and how much is available. Wind energy is a capricious commodity, but by using the right data it is possible to develop averages and to make a reasonable prediction about the local abundance of wind.

High and low pressure air systems move over the planet in fairly regular patterns. According to the laws of equilibrium, air flows from a high pressure zone to an adjoining one of low pressure. This is wind. Great differences in pressure between two zones create a high *pressure gradient* that results in a rapid flow of air, i.e., strong winds. The regularity of atmospheric circulation gives wind its generally consistent patterns of what are commonly called *prevailing winds*. If more than one wind site is available, pick one that is open to the prevailing wind direction both up and downwind. Any downwind obstruction of the air stream will impede incoming winds. A primary rule of thumb in siting is to allow at least thirty feet above any obstruction within a 300 foot radius of the machine. A higher support tower is a sound investment in cases where obstructions exist. The greater initial expense will pay for itself many times over through increased wind harvest.

Natural terrain can serve to accelerate the wind coming to the WECS. Uphill slopes that face prevailing winds compress the air stream and force it to move faster, up and over the top of the rise. The air stream will tend to become turbulent at the very top of such a rise, so the ideal wind machine location would be just below the top. But, if the top is the only possible site, a higher tower (90 feet is not inconceivable) will help avoid the turbulence. To reduce tower costs, rooftop siting has been used successfully in some residential installations, but only for the smallest machines. Aside from structural stress, framed buildings often resonate with the windmill, tower and guywires, causing a good bit of racket inside. In general, it's better to use an earth-anchor.

Other land forms such as linear valleys, river gorges, and glacier cuts that receive prevailing winds act like

Courtesy of Aermotor Corp. Div. Valley Industries, Conway, Ark.

Table 2. Average water needs.

Type	Gallons
Milking cow, per day	35
Dry cow or steer, per day	15
Horse, per day	12
Hog, per day	4
Sheep, per day	2
Chickens, per 100, per day	6
Bath tub, each filling	35
Shower, each time used	25-60
Lavatory, each time used	1-2
Flush toilet, each filling	2-7
Kitchen sink, per day	20
Automatic washer, each filling	30-50
Dishwasher	10-20
Water Softener	up to 150
¾-inch hose, per hour	300
Other uses, per person per day	25

funnels and make the wind flow faster through a smaller space.

After choosing one or two potential sites, make a careful study of the local winds. The amount of time you spend at this depends on how accurate you want to be. There are many wind measuring and averaging devices available (see Wind Product Listings). Some come fully equipped with automated data collection systems, while others that are cheaper and simpler require more personal attention.

One device combines a cup-type anemometer with a counter that records miles of wind "run." Using simple mathematical averaging techniques, this instrument will help you to determine representative windspeeds over predetermined time periods—an hour, a day, whatever. The body of averages obtained should be compared to those recorded at the nearest weather station. If a proportional relationship exists between the two, you're in luck! You can go on to extrapolate years of data and come up with a useful wind history. If such a comparison is not possible, site evaluation should continue for *at least* three months. Average weekly blows of 14 mph (6 m/s) indicate a healthy site, but even in these winds a machine might produce only half what it would with a 20 mph average. Strip chart recorders will also yield basic windspeed averages. However you derive them, they can be used with Table 3 to determine the monthly electrical output of a wind machine based on its peak output rating.

Windspeed can be measured in pre-selected "velocity bins" (i.e., groupings of 0 to 7, 7 to 9, 9 to 11 mph and so on) by another instrument. What this tells you is the total number of hours the wind blew at each velocity level within a given period. By using the power output curve of a particular wind machine (from manufacturer's data) and again, a little math, you can determine the total number of kwh that could have been produced by the machine(s) you're modelling. The same data can be produced automatically by means of a device that can be programmed to simulate a mill's power output. One thing that you should always remember about these periodic kwh averages is that they refer only to power produced *at* the wind machine. With all its components, the overall efficiency of a WECS is never 100 percent and reductions of 50 percent or more in a system's *useable* power are not uncommon. This may seem unacceptable, but compare it to the typical 30 percent efficiency of fossil-fueled electrical generating facilities. It's not that bad.

To my knowledge, no independent studies of the performance of water pumping wind systems exist. There is a need to know more about them. You can get some idea of pump performance from Table 4, but use it with care. Like any WECS, the variables affecting final water output are many and the figures here are suggestions at best.

Back to wind-electrics. If the results of your preliminary investigations are promising, it's time to think about hardware. At the moment there are about two dozen commercially available wind-electric machines. Some are good enough and others are better, and not surprisingly, you generally get what you pay for. But, different machines, regardless of the price, are suited to different applications. Selecting the right machine is primarily a matter of choosing from the one to four models that fit your output range. A further understanding of some of the fundamentals of WECS design will help you make the choice that best fits your needs and pocketbook.

*Courtesy of **Wind Energy and You,** Robert C. Griffin, copyright 1976.*

Table 3. Wind speed vs output in kilowatt-hours.

Nominal Output Rating of Generator in Watts	Average Monthly Wind Speed in mph					
	6	8	10	12	14	16
50	1.5	3	5	7	9	10
100	3	5	8	11	13	15
250	6	12	18	24	29	32
500	12	24	35	46	55	62
1000	22	45	65	86	104	120
2000	40	80	120	160	200	235
4000	75	150	230	310	390	460
6000	115	230	350	470	590	710
8000	150	300	450	600	750	900
10,000	185	370	550	730	910	1090
12,000	215	430	650	870	1090	1310

AVERAGE MONTHLY OUTPUT IN KILOWATT-HOURS

Wind-electric machine watt and kilowatt ratings are a function of the unit's rotor size. The *rotor* is the prime mover, the wind catcher, that converts kinetic wind energy into the rotation of a vertical or horizontal main shaft. Modern rotor blades are made of wood, steel, aluminum or plastic. Wood is the most resilient to the bending forces exerted by wind, but blades made from other materials are strengthened with spars and struts to prevent stress fatigue.

Safety features are an important part of any wind system, and the most important safety feature is over-speed protection. Strong winds can easily accelerate a rotor to speeds that could damage or destroy the blades or the whole machine. To prevent this, governing devices are standard. One method uses centrifugally activated weights, springs, and gears located in the hub of the machine that twist or *feather* the blades during the high winds, causing them to catch less wind. In another method, centrifugal force is used again to project air spoilers from the blades. These make the blades aerodynamically "dirty" and reduce rotor speed. A simple way to control speed is to turn the rotor out of the wind. In upwind, horizontal axis machines, the rear mounted tail-vane performs this task. Water pumpers and some wind-electrics use this method. In machines where the rotor is downwind of the generating unit, a small motor takes the place of the tail-vane, but the basic technique is the same.

The connection between the wind-electric and its generator or alternator drive shaft can be direct or geared (regardless of whether the machine has a horizontal or vertical axis). Gears absorb more energy, but also produce more power by multiplying rotor rpms into the generating unit. Direct drive machines use slower-speed generating units to produce full power. Gears can wear out; direct drive bearings shouldn't. Because of the lack of suitable slow speed generating units (they're mostly found in older machines), most new designs use gear or belt-and-pulley drives.

Support towers are of two types: freestanding or guyed. The absolute minimum rotor height is 40 feet (a tower of 60 feet will of course find the stronger winds). Because available wind energy increases by the *cube* of the windspeed, it makes economic sense to maximize tower height before going to a larger machine.[1] For example, if winds average 11 mph at 40 feet and 13 mph at 90 feet, the energy available at the higher elevation is about 200 percent greater and useable power increases by as much as 40 percent. Remember, however, that local zoning ordinances may include tower height limitations. Consult these before doing anything else.

Other WECS components depend on how electrical power is to be used. If your load analysis shows that everything can operate on direct current (DC) you will simply need storage batteries. Lead acid batteries will both even out the variable power output of the wind machine and

1. The best plain-speaking discussion of tower height can be found in "Planning a Wind Powered Generating System", a guide published by the Enertech Corp. of Norwich, Vt.

Courtesy of Dempster Industries, Inc., Beatrice, Neb.

Table 4. Pumping capacity of back geared windmills.

Cylinder Size	6 Ft. 5″ Stroke Elev.	G.P.H.	8Ft. "A" 7½″ Stroke Elev.	G.P.H.	10 Ft. 7½″ Stroke Elev.	G.P.H.	12 Ft. 12″ Stroke Elev.	G.P.H.	14 Ft. 12″ Stroke Elev.	G.P.H.
1⅞	120	115	172	173	256	140	388	180	580	159
2	95	130	135	195	210	159	304	206	455	176
2¼	75	165	107	248	165	202	240	260	360	222
2½	62	206	89	304	137	248	200	322	300	276
2¾	54	248	77	370	119	300	173	390	260	334
3	45	294	65	440	102	357	147	463	220	396
3¼	39	346	55	565	86	418	125	544	187	465
3½	34	400	48	600	75	487	108	630	162	540
3¾	29	457	42	688	65	558	94	724	142	620
4	26	522	37	780	57	635	83	822	124	706

These capacities are based on a 15-mile per hour wind for small mills and 18 to 20 miles per hour wind for larger mills. Capacities are based on longest stroke of Dempster mills. If short stroke used, capacities will be reduced in proportion to length stroke used.

If the wind velocity be increased or decreased, the pumping capacity of the windmill will also be increased or decreased. Capacities will be reduced approximately as follows, if wind velocity is less than 15 miles per hour: 12 mile per hour wind, capicity reduced approximately 20%; 10 mile per hour wind, capacity reduced approximately 38%

store whatever power is not immediately used. Storage and discharge capacity is measured in ampere/hours (AH), which essentially means how much power can be drawn from a battery and for how long. In designing storage capacity, consideration must be given to matching the battery voltage with the rated voltage output of the wind machine. Also important is knowing the total daily load on the batteries and the frequency with which strong energy winds will replenish them. Generally, storage capacity should be sized to provide adequate power over three to five windless days.

It may not be economical to size either the wind machine or the storage capacity to meet 100 percent of the electrical load unless the winds are exceptionally strong and frequent (averaging at least 14 mph, 6.2 m/s). Therefore a gasoline, diesel or propane-powered back up generator will be needed, and it should have a rated output at least ⅓ to ½ that of the wind machine rating.

Your load analysis may show that some devices (TV, stereo, clock) will operate only with alternating current. Since battery discharge is always DC, an *inverter* must be used to convert the DC to AC power. Because of the cost of inverters, it is most economical to keep the AC load at a minimum and to match closely the peak AC load with the smallest possible inverter rating. (Inverters are rated by their peak AC output.) Inverters are most efficient when operated at or near their peak capacity.

As mentioned earlier, a WECS can be connected or *interfaced* with a utility power line that would serve as the back up for the system. The hardware required is a *synchronous inverter,* and the implications of its use are rather interesting. On the face of it, a WECS-synchronous inverter system which eliminates the need for storage, back up and power conditioning components is lower in cost and can compete with utility power in certain regions—as in the New York City example. Equally interesting is that with the synchronous inverter, any WECS output that exceeds the load at any given time will flow into the utility grid. How then is the "reverse flow" measured and how is the WECS owner reimbursed for supplying power to the utility? A few utilities in the United States do have at least one synchronous inverted in their service territory and agreements between parties have been reached, though not always to the complete satisfaction of either.

As you look at the wind power products in this catalog, you'll run across WECS components other than the primary ones discussed here. They have been variously developed to increase versatility, automation and/or efficiency, and they may be useful for your application. However you choose to harness the wind, there are wind machines and components available that will provide years of reliable service. When you do go out looking to buy, consult with more than one manufacturer or seller; find out about actual working systems similar to what you have in mind and talk with some of the owners. There's a lot to be learned from folks who've been doing it for a while. With careful planning you won't need good luck when you actually have your mill flying, just good winds.

Woodstove Shopping: What to Look For

by William Hauk

William Hauk is an inventor and innovator who lives in Maine and is a freelance writer in the alternative energy field.

There's something magical and mesmerizing about a wood fire, just as the fire on the hearth remains a symbol of the home.

In recent years, millions of people have taken up heating with wood. The most apparent reason for this wholesale return to mankind's oldest fuel is economics: the geometric increase in the price of fuel oil, gas and electricity; the increasing return to natural life styles during the past decade is certainly another factor. In most of rural America, firewood is readily available and relatively cheap, and as the makers of woodstoves are fond of pointing out, wood is a renewable fuel and a natural repository of solar energy.

A million woodburners were sold in the U. S. last year. A market that grossed about $60 million in 1974 swelled to a quarter-billion in 1978. Although such an exponential rise cannot continue for long, this heavy and unprecedented demand for woodburners has resulted in an ample and variegated supply, and in some new and ingenious designs. There are now literally hundreds of different woodburners on the market, woodstoves, furnaces, boilers, and cookstoves, all of them claiming superior efficiency, quality or value. Faced with such a staggering array of choices, the buyer may become confused and buy the wrong stove for the intended purpose, or the right stove for the wrong purpose, caught in the welter of manufacturers' claims, sometimes contradictory and frequently nonsensical. In Northern New England the choices are even more awesome, since it seems as though every village welder there has become a stove-maker, and such locally-made or even homemade heaters must account for a significant percentage of the market. I shall confine my discussion here to woodstoves alone, that is, to woodburning space heaters, since the limitations of space forbid writing about other woodburners as well.

The prospective buyer must ask some basic questions: How much space has to be heated? Is that space well-insulated or drafty? It is good to match the heater to the space, although most stoves are capable of a range of heat output. Even though an oil-fired furnace may be rated at, say, 100,000 Btu per hour, it (hopefully) isn't running all the time, and a woodstove rated at 50,000 Btu per hour, burning steadily, may be adequate for the job. Woodstove manufacturers may rate their products in terms of Btu per hour, in watts, or more frequently, in the less precise terms of the number of cubic feet of space that they will heat adequately. In any case, manufacturers' ratings of power and efficiency are to be taken as one drinks tequila, with many grains of salt. A heater that's too small for the job will need frequent tending, and will still leave the buyer shivering, and a heater that's too large will be hard to regulate, and if left to smolder at low power for months on end will load up the chimney with creosote, a certain precursor to chimney fires.

Is cost a real consideration? Good stoves range in price from $200 to $800. My advice is to buy the best you can possibly afford, perhaps even a bit more than you can afford. No other appliance holds its value so well as a good stove, and the steady increases in the price of stove materials, iron and steel, and labor, neatly balances resale depreciation, so that a used stove in good condition can frequently be sold for its original purchase price. Cheaper stoves, of course, depreciate faster. A high quality woodburner is one of the best investments you can make in these perilous times, and some banks will now make woodstove loans as part of a home-energy-saving package.

Antiques

When the Arabs tried to bring us to our senses a few years ago, people rushed out and bought up all the old parlor stoves available, and within a year those that had long been rusting in a barn, for sale for $5, were fetching fancy prices. Ornate old stoves continue to bring high prices. My advice is to avoid antiques unless you really know what you're doing. The vast majority are inefficient, overpriced, and were built to burn coal, not wood. Many will be cracked or warped beyond repair, leaking at the seams, or with missing plates that cannot be replaced. If you're serious about heating with wood, buy a modern airtight stove. If you can't stomach the stark, utilitarian appearance of many of the modern airtights, buy one of the many antique-looking ones. On the other hand, some antiques are fine heaters, as good as most stoves built today, but these finds are *exceedingly* rare, and to evaluate them requires an expert. And finding a disinterested expert is as difficult as finding a good antique.

Combustion and Efficiency

In order for wood to burn, there must be enough heat and enough oxygen. A fire progresses in three stages: first, heat drives water vapor out of he wood, and then various volatile gasses burn at temperatures ranging from 725° F for methanol to 1128° F for carbon monoxide (we see these as flames of various colors). Finally, the volatiles consumed or driven off, the remaining charcoal burns down to ashes. In a roaring fire, of course, all three stages are going on simultaneously. About half the energy in a dried piece of wood lies in the various volatiles and half in the charcoal, so that a flameless smoldering fire may waste almost half the energy potential in wood.

The problem of woodstove design thus folds itself almost in half, to the considerations of combustion—burning the wood completely—and heat transfer—getting the heat from the fire into the living space. The best stoves have tried to increase both by various means, some of them quite ingenious. There is a theoretical limit to the efficiency of a woodburning appliance, since a sizable percentage of the heat from a fire must be expended to keep the chimney warm and drawing well. Any woodburner is only as good as its chimney; even the best stove will be a balky, smoky horror if hooked up to a marginal chimney.

Incomplete combustion is almost always the result of inadequate oxygen supply in airtight stoves, or of wet or green wood dampening firebox temperatures. Green wood causes a smoky, smoldering fire that in turn increases the flow of unburned volatiles into the chimney, where they may condense to form creosote (once again, the dangers of creosote and chimney fires are well-known). Non-airtights, like the standard Franklin stove, may not generate much creosote since they can't choke off the oxygen supply. But because it gives the fire more air than it needs the Franklin stove is notoriously hard to regulate, and can seldom hold a fire for more than a couple of hours. Most people like to burn their wood slowly, to keep a low, smoldering, smoky fire going throughout the winter months, taking their chances with creosote. Curiously enough, this is also the most efficient way to burn wood in airtight stoves, even though as much as 35 percent of the energy may be lost. The trade-off here is one of time. There is more opportunity for heat transfer in a low fire than there is in a roaring blaze; this increased heat transfer makes up for the less efficient combustion. (Most of the heat in a high fire is lost up the chimney.) In addition, there is the convenience of adding wood just a few times a day.

Another approach to woodburning, but one that few people take, is that of using a small stove at high power with a stack-type heat-exchanger, to capture most of the heat normally lost up the chimney. This is one of the most efficient ways of burning wood, because it maximizes both combustion and heat transfer, and virtually eliminates the creosote problem. This approach involves frequent tending, however, and most people are too lazy. It also doesn't allow for overnight fires, and there is something to be said for awakening to a warm stove with a glowing bed of coals ready for a new charge of wood. There is only one stack heat-exchanger on the market worthy of serious consideration, the Sturges Thriftchanger, and it's known mainly to those in the vanguard of the woodstove industry.

Some preliminary study of woodstove efficiency has been done in the last four years by Jay Shelton, formerly an assistant professor of physics at Williams College. Shelton tested various woodburners in a small calorimetric chamber where heat output from a weighed amount of uniform wood could be measured accurately. His preliminary findings indicated that stoves varied only slightly in efficiency, from about 40 percent for an open Franklin stove to 65 percent for a modern airtight, with most stoves clustered around the 50 percent median. His results, not too surprising in themselves, show that the prospective stove buyer should pay more attention to other factors than claims of efficiency, such as quality of construction and materials, ease of operation, appearance and cost.

The Great Iron vs. Steel Controversy

Neither iron nor steel is clearly superior as a stove material. Iron is the traditional favorite, largely because in the previous century steel was expensive, welding unknown, and steel stoves were flimsy affairs that burned out quickly. There were many iron foundries in the east (over 100 in Troy, N.Y. *alone!*), all but a few of which either died out or were closed by the Occupational Safety and Health Administration. If you've ever seen any of the amazing parlor stoves of the last century, you know the aesthetic potential of cast iron, but iron has its drawbacks. It's an astonishingly fragile material; it can crack if castings are poor or if the stove is abused. Hot-rolled mild steel plate, the material of the modern steel stoves, can't crack or break, but has an inherent molecular stress, which may cause it to warp under intense heat. Thin iron plates can also warp, but you'll see this only in the very cheapest of iron stoves. Plates of iron or steel of the same thickness will have identical thermal properties and an equal resistance to corrosion. Whether of iron or steel, look for heavy plates, $^{3}/_{16}$ inches or better. Get the most metal for your money.

Iron stoves will need rebuilding from time to time, since they're assembled from separate plates and sealed at the joints with refractory cement, which invariably fatigues under repeated thermocycling (cold to hot to cold) and "throws out" of the joint, leaving a leak. Flexible plastic sealants have emerged recently, but have just begun to penetrate the market. Steel stoves are welded permanently airtight. The ideal stove material would be cast steel, without the built-in stress of rolled steel plate, but able to be welded airtight. No cast steel stoves exist, to my knowledge, and such stoves would be expensive indeed.

Liners

Many stoves have liners to protect their side plates from warping or cracking, liners that serve to *insulate* the plates (and the buyer) from hot coals. Liners thus partially defeat the purpose of a wood heater, which is ostensibly to get heat into the living space. Some manufacturers of the new steel stoves contend that their firebrick liners boost combustion temperatures (and therefore combustion efficiency) by insulating the combustion zone, but they do so at the expense of heat transfer. If you're considering a stove with liners, whether of iron, steel or firebrick, make sure that there's enough uninsulated surface above liner level to take some heat out of the stove. Some stoves have curved plates to resist thermal stress and avoid the need for liners, and some steel stoves use internal reinforcement, such as welded-in baffles or angles, to keep plates from warping.

Thermal Mass

Thermal mass is the heat-holding capacity of a stove, its ability to store heat from a fire, which it gives up to the living space. For our purposes thermal mass refers to the weight of the stove's hull, or exterior shell. When the buyer is comparing the relative weights of stoves, the gross weight should not be considered. Gross weight frequently includes the weight of insulating liners, which shed most of their stored heat not to the exterior but to the interior of the firebox, whence it may be swept up the flue and lost. Consider only the weight of exterior plates, which a knowledgable dealer should be able to estimate accurately by subtracting the weight of liners from the gross weight.

A stove with great thermal mass takes a while to warm up from a cold start, but by the same token will cool off slowly once the fire is spent. A massive stove will moderate hot spots or flare-ups in the fire itself, distributing the heat throughout its mass and releasing it more gradually than a lightweight stove. For garages or workshops or other places where quick or occasional heating is required, light-gauge stoves are recommended. Where steady, constant heat is needed, heavy stoves are preferable. Heavy stoves last longer, too.

Baffles

Many of the best stoves now feature baffles of various configurations to increase the efficiencies of combustion and heat transfer. Baffles are a good idea in most stoves because they allow only part of the fuel load to burn at a time, rather than permitting the entire charge to smolder,

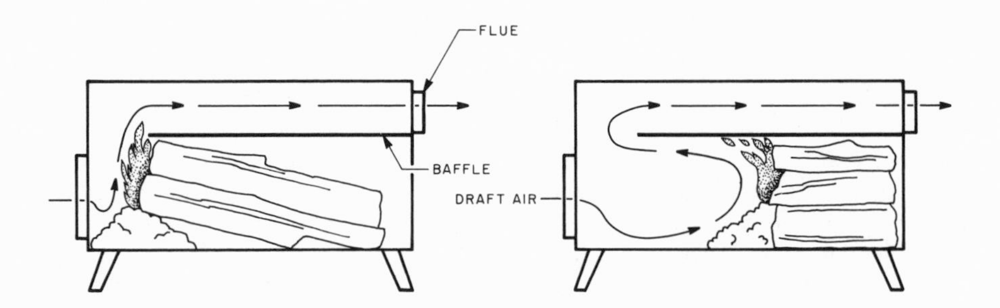

Fig. 1. Horizontal or Scandanavian-style baffle system

as an unbaffled box stove will. A smaller, hotter fire means more complete combustion, and better control of the stove's heat output, for longer, steadier fires.

By far, the most common baffle system is the horizontal or Scandanavian-style baffle that forces the draft flow into an S-shaped pattern so that logs burn from front to back. Only the forward part burns, until a pile of coals remains at the rear of the firebox. The coals are then raked forward to the front of the firebox and a new charge of wood is placed on top. This elegantly simple design assures a long and steady power output without the need for thermostatic draft controls, and for this reason the horizontal baffle is found on a great number of the newer brands of stoves offered today. There are even some established manufacturers who have incorporated this baffle design into what was previously an unbaffled box stove.

The horizontal baffle design does have some built-in flaws, however slight. These stoves are relatively unforgiving of wet or green wood, and of poor chimneys or chimney hookups. Some buyers may not relish the chore of raking the coals forward before adding each new load of fuel, but this procedure is quick and easy. Furthermore, it is not always possible to add more wood at just any time. One must sometimes wait until one load is fully consumed before more fuel may be added, and that may sometimes be inconvenient.

Probably the best-known of the Scandanavian baffled boxes is the Jøtul 118 from Norway, which has made a heavy impact on the American market in the last four years. It has spawned a host of imitations in steel, and some shamelessly close copies in iron. In an industry rife with rip-off, the 118 has more imitators than any stove made, bar none. Having lived with a 118, I can testify that it's an excellent heater, but its baffle plate dislodges easily, and the side plate liners can then fall into the firebox. Like many imports, its door is rather small, and its firebox will accept only a 24 inch stick, no longer. But if imitation is the sincerest form of flattery, then the 118 has been flattered indeed. Other Norwegian imports are Trolla and Ulefos, both with product lines similar to Jøtul's, but nowhere near as successful.

From Denmark come Lange and Morso, also with a comparable range of products. Personally, I prefer the Danish stoves to the Norwegians—quality and workmanship are visibly better. Once again the doors are too small, but in the Danish stoves the doors and doorframes are machined and hand-filed to fit airtight without using gaskets. All the Scandanavians make a "combi-fire" model, a fireplace/stove with a large door or doors that open for viewing the fire but close airtight for holding the fire overnight.

The baffled boxes that impressed me most are American-made: the cast-iron Cawley-Lemay, the Ram, fabricated of ¼ inch steel plate, and the Independence, of lighter $^{3}/_{16}$ inch plate. The Upland 207 is an attractive iron stove that can be operated as an open Franklin or, closed, as a Scandanavian-style heater. The Portland Foundry recently introduced their Atlantic 228, which also may be operated as a Franklin, with a sliding horizontal baffle like the Upland 207.

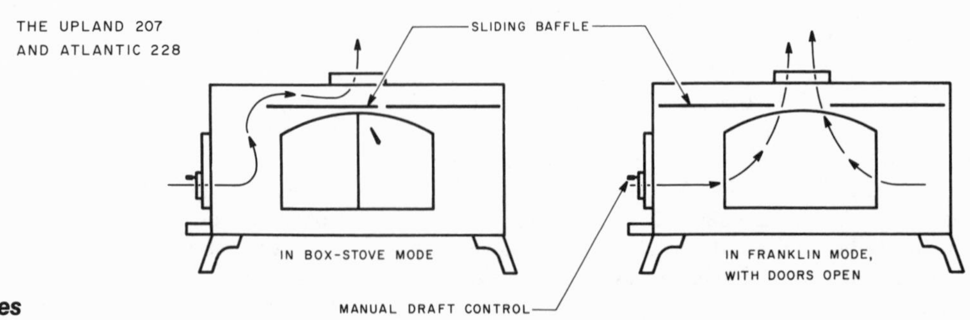

Fig. 2. Fireplace stoves

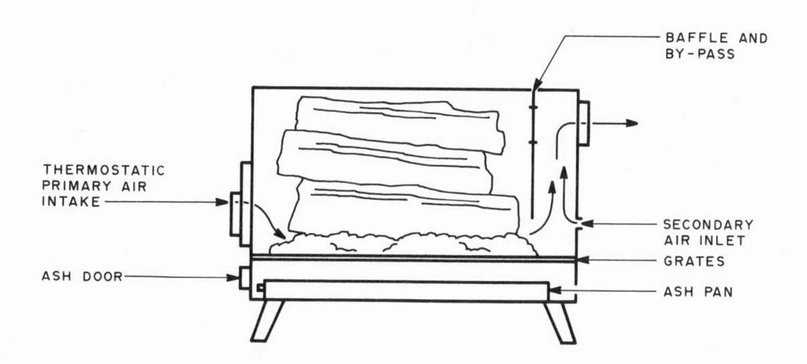

Fig. 3. Base-burner system

The Base-Burners

A less common but equally advantageous baffle system is shared by the base-burner stoves: a vertical baffle that allows only the base of the fuel charge to burn.

Wood feeds down from above, as in a hopper, being dried as it sinks, so by the time it reaches the fire zone it burns with a relatively hot flame. Volatiles baked out of the wood must pass over the coals in order to escape under the baffle, and as they round the baffle they are mixed with a stream of secondary air to aid combustion. All base-burners must have a smoke by-pass to dump smoke from the firebox during fuel loadings, and all have a thermostatically-controlled primary draft to maintain a steady power output.

The base-burner design is not new (the basic design is well over a century old), but one of the best of the new stoves is a base-burner. This stove is the Defiant, which opens like a conventional Franklin for viewing the fire, but close the doors and flip the by-pass lever, and *voila*, a very well engineered and efficient base-burner. For my money, the Defiant is one of the best stoves made, and it too has been ripped off by a West Coast foundry. The Riteway 2000 and the larger 37 are well-known base-burners, as are Perley Bell's more expensive Bellway stoves.

Base-burners are probably the easiest of baffled stoves to use. Wood may be added at any time, fed into the top of the firebox as there is room to fit it in. On the whole, they tend to be more forgiving of wet or green wood than are horizontal-baffled stoves, but even here moisture will dampen combustion temperatures (and combustion efficiency) and cause smoky fires and creosote deposits.

All of the base-burners have secondary air inlets near the exhaust side of the baffle, to mix oxygen with the oxygen-poor exhaust gasses in an effort to burn more of them before they're lost forever up the stack. In spite of manufacturers' claims, I remain unconvinced that secondary combustion occurs in most stoves, except in high fires. Riteway and Bellway stoves make no provision for preheating the secondary air, so that it must cool the resulting mixture and fail to ignite many of the gasses. The only base-burner I know of that preheats its secondary air to any extent is the Defiant, which might have a prayer of achieving a degree of secondary combustion. In most other stoves, base-burners, Scandanavians, or box stoves, secondary air inlets strike me as so much useless window-dressing, harmless enough to be sure, but not to be taken seriously. It certainly shouldn't be the basis for choosing one stove over another.

Thermostatic Controls

Here I must also confess to a certain prejudice against thermostatic draft controls. All employ a bimetallic spring coil, that closes the draft as it gets hot and expands, to restrict the fire. Some work well, and others do not. When they aren't sensitive or aren't working properly, they will let a fire get quite hot and then choke off its oxygen, resulting in a great flow of unburned volatiles into the chimney, and the persistant creosote problem. Ashleys are notorious for this failing, but they are by no means unique in this. Indeed, any airtight stove will produce creosote, but I feel that the use of cheap bimetallic coils is asking for trouble. Riteway stoves employ a magnetic catch on the intake flap that closes the draft completely, more or less halting primary combustion. Riteways are thus either "on" or "off," not choked little by little, and their secondary air inlets provide oxygen to burn some of the volatiles that escape when the primary draft closes. To provide a steady heat output, base-burners and updraft box stoves have no recourse except to use thermostatic controls.

The Downdrafters

Among the most interesting of the new stoves are the downdrafters, so called because their primary draft is

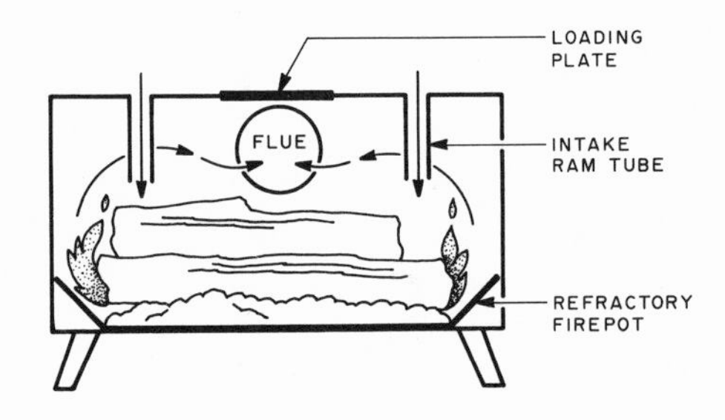

Fig. 4. Downdrafter - Tempwood

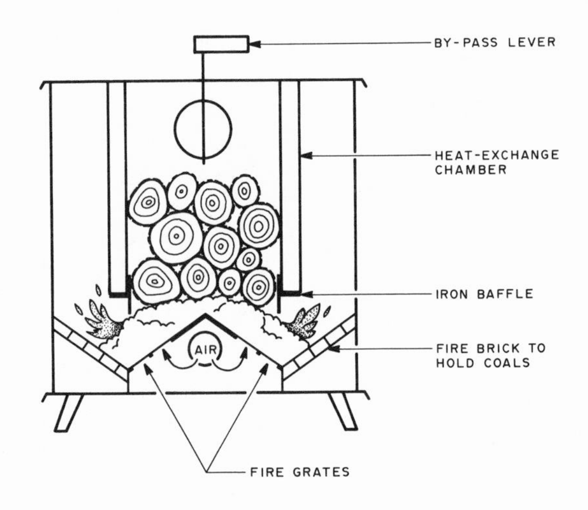

Fig. 5. The Vermont Downdrafter →

actually directed downwards, into the coals. The draft blasts volatiles into the coals, where they are consumed. Once again, the basic design concept is not new. Ben Franklin designed a downdraft coal burner two centuries ago, but the best-known downdrafters, the Tempwood and the Vermont Downdrafter, have appeared in the last four years. In the Tempwood, two 8 inch steel draft tubes ram the intake air at considerable velocity down into the bed of coals, which lie in a refractory-lined firepot at the bottom of the firebox. The stove is simplicity itself. Instead of a conventional loading door, the Tempwood has an 11 inch iron plate on top, where wood is loaded in and ashes shovelled out. Much of the labor and materials cost in any stove is the door, in fabricating, fitting and sealing. By eliminating the door, the Tempwood can sell for much less than most stoves its size. The Tempwood is fabricated of light-gauge ⅛ inch steel, but since the coals are contained in its refractory firepot, its manufacturer can probably get away with using these light plates. Unloading ashes might seem like a hassle, but users say it really isn't, and a bent shovel is available for the purpose. There is plenty of uninsulated surface above the firepot for heat recovery, but since that surface is thin plate, the Tempwood is a bit short on thermal mass. Its combustion is by no means complete either, since the intake air can't carry *all* the volatiles into the coals, but for its simplicity and low cost the Tempwood remains one of the best buys in the industry.

The Vermont Downdrafter is much more sophisticated, complex, massive and expensive. Wood feeds down from the top, as in a base-burner, into the fire zone at the apexes formed by the grates and the baffles. There incoming draft air, admitted by a thermostatic control at the rear, is mixed with the volatiles from the wood above, *inside* the two coal beds, the hottest part of the fire (1800° F or better). Inside the coal bed there should be enough heat to warm both volatiles and draft air hot enough to burn. I have nothing but praise for the new Downdrafter. (The first year's models had flimsy grates and baffles, which have been considerably beefed up.) It is a heavy-duty, powerful and steady heater, and relatively easy to operate. Its looks won't win any prizes, though. It is available in two sizes, the larger with twin heat-exchange ducts and a thermostatically-controlled blower.

The Double-Tops

The best rags-to-riches story of the last two years in the woodstove biz is that of Fisher stoves. Conceived a few years back by Bob Fisher, an unemployed welder in Oregon, the Fisher line now has over 30 submanufacturers throughout North America, and within the past two years has spawned as many imitators as Jøtul did four years ago. It seems as though every time I open a magazine with woodstove ads I see yet another Fisher rip-off. These copycats are taking their chances, since Fisher has a patent on their unique shape, and at this writing Fisher is at lawsuits with the most prominent imitator, asking for $3 million in damages. If Fisher wins, the lesser copycats may be forced to change their shapes as well.

Fisher's secret is its massive two-level top, bent from a single plate of $^5/_{16}$ inch steel. The upper chamber in the rear helps keep smoke from billowing out during fuel loading and tending, and also provides two cookng surfaces with different temperatures. More importantly, Fisher claims that it provides turbulence of gasses, in conjunction with the flue that protrudes into the upper chamber, for better combustion and heat transfer. The sides are ¼ inch plate, and are lined in the burning zone with firebrick. Fishers might put out more heat if the

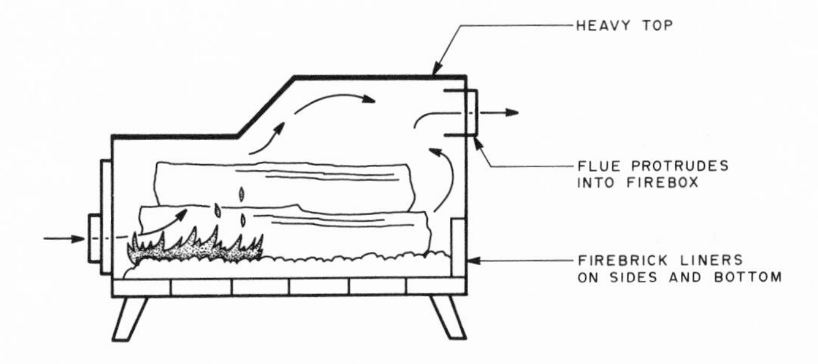

Fig. 6. The Fisher woodstove

liners were removed, but since there is no reinforcement of the side plates, some warpage would surely result, and so the insulation is necessary. There is plenty of uninsulated surface above these liners, however, particularly in the massive top. With or without turbulence, Fisher's top would put out more heat than a conventional flat top if only because of greater surface area. The doors are heavy cast iron, with screw-type draft controls. Three sizes of box stoves are offered, and two fireplace models. Fisher has won many loyal fans, and has done more than any other manufacturer to make steel stoves respectable.

Circulating Heaters

Circulating heaters are stoves intended to provide primarily convected heat rather than radiant heat. Most circulators are surrounded by a light steel cabinet, and radiant energy from the stove heats the air between it and the cabinet, causing the air to rise rapidly. Most of the energy from an unenclosed stove escapes as radiant heat, but in circulators there is little radiance, rather, rapid streams of warm convected air. Circulators are recommended for well-insulated homes only. Drafty homes do better with radiant heaters. For the most part, circulators are thermostatically-controlled updraft box stoves, such as those made by Ashley, Shenandoah, Wonderwood, and Warm Morning. All resemble a large humidifier or a kerosene heater more than a woodstove, but all are effective heaters, and relatively easy to use. And since the hot surfaces are enclosed, small children are less likely to burn themselves. Circulating tile stoves from Europe are now available in the U. S., which use decorative tiles as an exterior shell. Some are quite beautiful.

The ingenious Larry Gay offers soapstone slabs with his Independence, a large baffled box stove. The slabs turn it into a circulator. But the most ingenious circulator I've

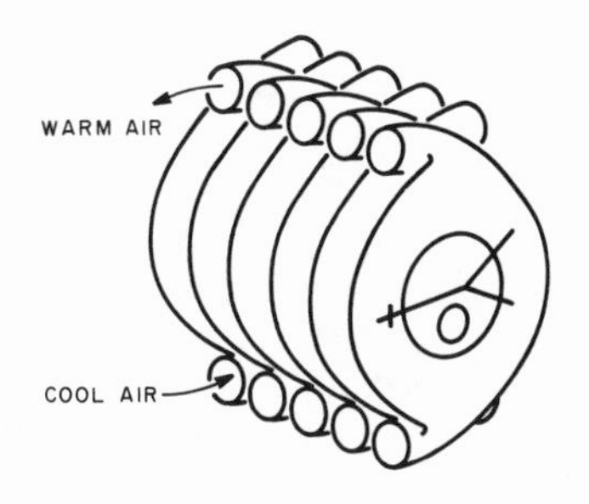

Fig. 7. Circulating heater

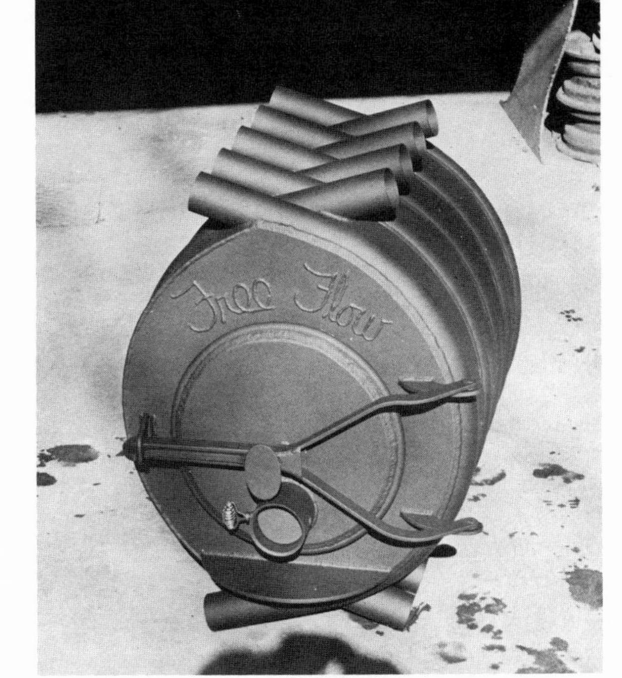

seen is Eric Darnell's Free Flow stove, made from semicircular sections of heavy steel pipe, that emit streams of convected air when hot. The firebox has a horizontal baffle. The Free Flow has to be the most bizarre-looking heater on the market.

Fireplace Stoves

Many stoves are available that enable the buyer to enjoy the warmth and romance of an open fire, but that will close airtight to hold an overnight fire. For serious heating, the so-called Franklin stove should be avoided, except for those few that will close airtight. (Poor ol' Ben had nothing to do with the stove that bears his name, and probably rolls over in his grave every time it's mentioned.) The buyer should be warned that an open fire is inherently wasteful. When the doors and damper are open most of those precious Btu race up the chimney. Many fireplace stoves feature tempered glass panes in their doors, so that you can watch the fire even when the doors are closed. It's not the same as an open fire, but those who prefer TV to reality won't appreciate the loss. These panes may break from thermal shock or physical abuse, but are relatively cheap to replace. They may smoke up during a low fire, but are easily cleaned with an oven cleaner. (Mr. Muscle has been highly recommended.) The buyer should be sure that the doors fit well and close tightly.

Automatic Boxes

The thermostatically-controlled updraft box stoves are probably still the best-sellers in the industry, dominated by the Ashley line. Martin Industries, the giant that makes Ashley and the similar King line, had half the American market until a few years ago, when foreign and domestic competition began in earnest. Ashley remains America's favorite woodburner. The best-selling Ashleys or Kings may be loaded from the front or top. Buyers should check to see that the iron loading top is not warped and that it fits its frame, that there is an adequate seal between the thin-gauge steel firebox and the iron doorframe, and that the doorlatch fits securely and can't be knocked open by a log falling inside. Ashley's automatic controls are not the most sensitive, and may allow wide swings in room temperature. Even for its few faults, Ashley is one of the best buys. They are easy to use and inexpensive, and besides, can hundreds of thousands of Americans be wrong? Other popular automatics are made by Shenandoah, Atlanta, and Thermo-Control.

A Caveat

In the past couple of years there has been a veritable flood of cast iron woodstoves from Taiwan on the American market, most of them look-alike copies of American and Scandanavian models, and some of them even have Scandanavian-sounding names. All of them are remarkably inexpensive, costing a fraction of those stoves they seek to imitate. At the risk of sounding xenophobic, I would advise the buyer to beware of Taiwanese heaters. The ones I've seen (there are dozens) are merely passable at best and utter trash at worst, with visibly flawed castings, poorly sealed joints, and ill-fitting doors. Even in the best of them was there little evidence of good workmanship. The Taiwanese foundries, buoyed up by preferred trade arrangements with the U. S., cheap materials, and coolie labor, can frequently manufacture and ship a stove to the U. S. for what an American manufacturer pays for materials alone. Ask the dealer where the stove was made, and if it's Taiwan, take a closer look, or better yet, save your money. There are few real bargains in the competitive world of woodstoves, and if one stove costs a *lot* less than a similar product there's probably a good reason for it. You get what you pay for.

The Future of Wood Heat

Even though I've owned perhaps a dozen woodstoves in my life, and used dozens more, I regard the woodburning space heater as an obsolete artifact of the last century, an inefficient waste of our precious forest resource. They're the best we can do, I suppose, to retrofit wood heat to a house not designed for wood heat, or to a drafty old barn not designed for *any* sort of heat. For those who want to build a house *around* wood heat, I would advise consideration of the so-called Russian Fireplace, or the sort of massive masonry tile stoves long in use in Germany, Switzerland and Central Europe. These have just started to appear in Northern New England, chiefly in Maine. They have relatively small, kiln-like fireboxes set in a mass of masonry, with convoluted flue passages in a massive central chimney. The owner builds one or two small, hot fires a day. These heaters provide maximum combustion efficiency, heat transfer, and tons of thermal mass for heat *storage*, something no ordinary stove can supply. The chief flaw of such systems is obvious at once: it takes many hours to get such a mass warm (and many more hours to cool it off). One or two small, hot fires a day may be all that is necessary, because almost all the heat may be captured and released slowly into the living space. Such heaters are obviously expensive, costing thousands of dollars, but aren't really that much more costly than the usual unwise suburban fireplace. And with tons of masonry involved, you can't buy one in a store and bring it home, a real problem for package-oriented American shoppers. The ideal system, in my mind, would be a heater of high thermal mass, such as a Russian Fireplace, coupled with some provision for quick space heating, like a Sturges Thriftchanger heat-exchanger. So far as I know, no such system exists, but at this writing several people are working on it, myself included.

Further Reading

Readers who wish to do some more homework are encouraged to read the following:

The Woodburners Encyclopedia by Jay Shelton. The best, most complete, and most rigorously scientific work. Includes a products catalogue. 155 pp. $6.95 from Vermont Crossroads Press; Waitsfield, VT 05673.

Modern and Classic Woodburning Stoves by Bob and Carol Ross. Well-thought-out and profusely illustrated. 143 pp. $10 from the Overlook Press; Woodstock, NY 12498.

Heating With Wood by Larry Gay. Still the best basic book of the genre. Start with this one and work your way up to Shelton. 128 pp. $3.50 from Garden Way Publishing; Charlotte, VT 05445.

Wood Heat by John Vivian. Complete and nicely illustrated. 320 pp. $4.95 from Rodale Press; Emmaus, PA.

Wood Stoves: How to Make and Use Them by Ole Wik. The best how-to book. Build your own. 144 pp. $5.95 from Alaska Northwest Pub. Co., Box 4-EEE; Anchorage, Alaska 99509.

Stove Book by Jo Reid and John Peck. A lovely color picture book of European enamelled stoves, a must for your coffee table. 110 pp. $5.95 from St. Martin's Press, 175 Fifth Ave., New York 10010.

Antique Woodstoves by Will and Jane Curtis. Mind-boggling American antiques in black and white. 64 pp. $2.95 from Cobblesmith; Ashville, ME 04607.

Woodstove Directory. A good shopper's guide, with fixed-format ads, and manufacturers' outrageous claims. Let your fingers do the walking. 102 pp. $2 from Mitchell Associates, P.O. Box 4474; Manchester, N.H. 03108.

Wood Burning Quarterly and Home Energy Digest. The organ of the industry. Look here for new developments and products. $6 per year from WBQ, 8009 34th Ave. So.; Minneapolis, MN 55420.

Articles on Russian Fireplaces

"Masonry Stoves" by Albert A. Barden III in the Feb. 1978 issue of *Country Journal*.

"What's So Hot about a Russian Fireplace?" by M. R. Allen in the Feb. 1978 issue of *Yankee*.

"Stability and Control in Wood Heating Systems" by Jay Shelton, in the Summer 1977 number of *Wood Burning Quarterly*.

WHAT
is
HOME
without a
MOTHER
DEFIANT

The Wood Stove Buyer's Guide

Factors ranging from saftey and performance to convenience, aesthetics and cost will influence the wood stove buyer's final decision, and no buyer's guide can substitute for a personal inspection of the various models on the market. Yet the following charts should prove valuable by providing fingertip information for comparision of the ever-growing number of woodburners on the market, and hopefully by assisting our readers in choosing a wood stove that will still prove satisfactory in the chill grip of winter. Every effort has been made to produce here the most comprehensive and complete woodheater buyer's guide ever published.

This Guide has been reprinted in its entirety with permission from the weekly newspaper, ***New Hampshire Times****. For even more current stove information, write for the fall edition of their Annual Wood Stove Buyer's Guide, available the last week in September. Address your inquiries to: The Annual Wood Stove Buyer's Guide,* ***New Hampshire Times,*** *P.O. Box 35, Concord, NH 03301.*

Wood Stoves

Considerations of safety having first been taken into account by the potential wood stove buyer, and personal evaluations of convenience, aesthetics, value and cost aside, the most frequent question regarding a woodheater's performance is its **efficiency.** Unfortunately, this factor is exceedingly difficult to measure, much less compare among various stove models, because its computation involves variables ranging from combustion chamber designs to species and condition of wood used as fuel. There are definite stove characteristics that affect heating efficiency, however, and the most important are the following:

✓ Airtightness: Stoves that have tight-fitting joints and doors burn wood much more efficiently than stoves that allow unregulated air to seep into the firebox.

✓ Complete combustion: Among the stoves designed to be airtight, those that have effective baffle systems and/or smoke chambers will burn a higher percentage of volatile gases emitted from wood and extract more heat from the same charge of wood than those that do not.

✓ Surface area: The larger the number of square inches of total surface area a stove presents to the air around it, the more heat the stove will radiate into the living space.

✓ Conductivity and mass: Certain stove materials, such as cast iron and steel, will readily conduct heat from the fire to the outside of the stove. Firebrick, glass and other insulating materials are poor conductors of heat, and although they may be desirable for reasons of safety or aesthetics, interfere with the stove's heat transfer. The mass of a stove's construction material also affects its performance. A stove made of sheet steel will be lighter than one constructed of heavy-gauge cast iron. It will heat up faster, but will also lose heat faster after the fire goes out. Cast iron warms more slowly, radiates heat longer than steel, and is also considered more durable.

Cookstoves

The woodburning cookstove is a creation unto itself. Producer of the finest of breads, pies and roasts under the guidance of an expert master or mistress, the kitchen stove can also turn ornery, sullen and well nigh impossible to light. Its primary purpose is to cook food, not warm people, and though the midwinter cook is sure to appreciate its cozy by-product, the cookstove should be judged primarily on the performance of its intended chore.

The largest trade in kitchen wood stoves is still the resale of refurbished 19th century models —the most practical of antiques. Some hardware stores carry the white-enameled and 20th century improvements on these models, and traditional cookstoves are now being cast by several American manufacturers today. In addition, some Scandinavian manufacturers are producing highly efficient kitchen stoves for the American market, and combination wood-electric and wood-gas cookstoves are also becoming available in response to the rising costs of fossil fuels.

Fireplace Units

These are the latest generation of devices designed to render the traditional fireplace more efficient. Unfettered by such improvements, the fireplace enriches the spirit and warms the heart, but the heat that it generates in the home is largely illusory. In fact, the functioning fireplace is usually a net **extractor** of heat from the living space.

This situation can be remedied to varying degrees by the addition of circulating fireplaces or fireplace stoves. Some units fit into an existing fireplace where the naked fire once burned, while others are designed to replace the fireplace in new home construction. Retrofitted fireplace units are **never** as efficient as airtight wood stoves, but individual aesthetic considerations may make their purchase worthwhile.

Wood Boilers and Hot Air Furnaces

It may well be that efficient woodburning boilers and furnaces will be the home heating appliances of the future. Heating a home with wood stoves allows for greater flexibility in the choice of which areas are to be warmed and which are to be closed off and abandoned for the winter, but basement woodburners provide the comfort of central heating throughout the house. Heating an entire home with boiler or hot air furnace requires a substantial initial investment for the heating plant and necessary ductwork, and also consumes a large volume of cordwood each winter. In addition to the advantages of comfort, however, are those of convenience. A central woodheating system means there is only one heater to feed—usually no more than two or three times in 24 hours—and that all the bark chips, wood splinters, ashes and associated mess that goes along with heating with wood can be confined to the basement. Woodfired furnaces can easily be adapted to provide domestic hot water too.

Wood Stove Kits

One of the newest developments to rise out of the wood stove market is the do-it-yourself stove kit. The kits range from an enameled, Scandinavian-type wood stove to barrel heaters made from recycled materials, and generally result in efficient, reliable woodburning units if care is taken during construction.

Besides the satisfaction of building your own wood stove, do-it-yourself stove kits have the advantage of lower retail prices than their ready-made counterparts.

Combination Stoves

The Franklin or combination stove as it is known today bears little resemblance to the "Pennsylvania Fire-Place" first invented by the great American statesman in 1740. Benjamin Franklin set out to improve upon the efficiency of the traditional fireplace, but being culturally English, insisted on being able to view the fire burning inside his creation. Today's Franklin stoves achieve both of their namesake's aims, with varying degrees of success. Open, the combination stoves provide the cheery warmth of an open fire. With doors closed, some become efficient woodheaters, although few models approach the efficiency of the standard woodburning stove. Traditional Franklin stoves are a compromise between raw efficiency and pure aesthetics, and function as well as a marriage between these two virtues might be expected to perform.

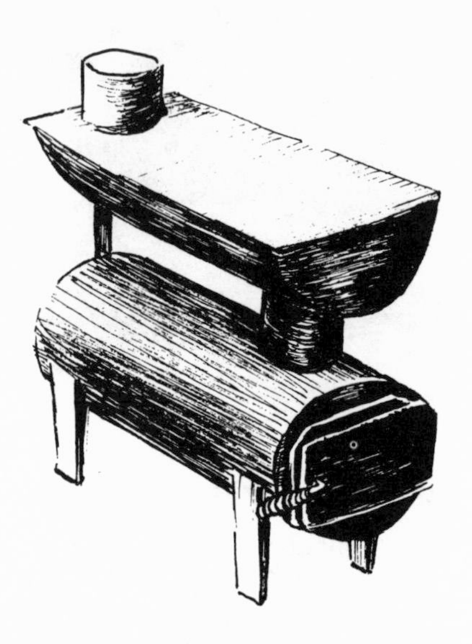

Footnotes

A word about heating capacity — These figures, provided by the wood stove manufacturers, are the most general estimates of a wood stove's capabilities. The ability of a stove to heat a given area depends not only on the cubic feet of space, but also on the configuration and contents of that space.

For the convenience of our readers, we have also included distilled statements of stove manufacturers' claims for their products. These should be viewed as manufacturers' claims and nothing more. With the glut of wood stoves on the market, the hard sell is on. Take your time examining the woodheaters that interest you in the dealer's showroom so that you can make an informed decision about the type of woodburner that's best for you.

Every effort has been made to assure that this guide is accurate and complete. However, we cannot be responsible for omissions, typographical or factual errors, manufacturers' changes in specifications provided for the compilation of this chart, or variations from the suggested retail prices listed here.

compiled by Michael Harris
illustrations by Rose Cravens

THE STOVES	Construction	Guar-antee	Mfr's Claims	The Models	Weight & Size (HxWxDepth)	Heating Capacity	Log Length	Price
All Nighter	Steel, firebrick lined, cast iron door; airtight type.	30 days	UL approved for safety; patented air system provides 24-hr. burning.	Big Mo Mid Mo Little Mo Tiny Mo	524 lbs./31½x23½x41½ 436 lbs./31½x21½x36 333 lbs./28x19½x32 267 lbs./26¾x17½x28½	7-10 rooms 5- 7 rooms 3- 5 rooms 1- 3 rooms	28" 23" 18" 15"	$515 $445 $375 $335
Alpiner	Firebrick lined steel and cast iron stepstove; airtight type.	25 yrs.	Consistently offer more stove for the money.	Chamonix Mont Blanc Matterhorn	325 lbs./28x16½x27 410 lbs./29¼x16½x34 485 lbs./29¼x18½x39	8,000 cu. ft. 11,000 cu. ft. 14,000 cu. ft.	19" 27" 30"	$325 $375 $425
Arctic	Cast iron, airtight type, baffeled box heater.	1 yr.	An extremely efficient, simply terrific stove.	30	140 lbs./31x16x36	None stated	22"	$350
Ardenne	Cast iron, firebrick lined box stove; airtight type.	1 yr.	Solid, exceptional effectiveness with distinguished French design.	Ardenne	259 lbs./28½x9½x25¾	10,000 cu. ft.	24"	$440
Ashley	Thermostatically con-controlled steel and cast iron circulating heaters.	None stated	Burns anything but rocks.	C-60 C-62 23-HF 23-HF	242 lbs./36x35¼x21¼ 223 lbs./35x28¼x20¾ 125 lbs./34x20x30 105 lbs./30x18x23½	4-5 rooms 3-4 rooms 4-5 rooms 1-3 rooms	27" 19½" 24" 22¼"	$360 $310 $189 $175
Atlanta	Thermostatically controlled steel circulating heater.	None stated	Efficiency, economy, durability.	Homesteader 24 WGE Model 2502	240 lbs./33½x32¼x19¼ 317 lbs./36½x35x20 147 lbs./35½x17½x22	4-5 rooms 4-5 rooms 3 rooms	24" 24" 20"	$289 $497 $149
Atlantic	Cast iron box stove, airtight type.	None stated	First original American cast iron, airtight box stove.	224	180 lbs./24¼x20¾x35½	7,000 cu. ft.	22"	$300
Autocrat	Thermostatically controlled steel circulating heater.	None stated	8-12 hrs. between fuelings.	Model FF76	245 lbs./34¼x32½x21¾	5 rooms	25"	$399
Birmingham	Thermostatically controlled steel circulating heater.	None stated	Efficient. One fueling lasts all night.	Knight 224 Knight 124 Majik 122-A	317 lbs./36½x24x35 230 lbs./33½x32¼x18 156 lbs./35½x20x26	4-5 rooms 3-4 rooms 3-4 rooms	24" 24" 22"	$438 $328 $196
Bullard	Steel and firebrick, airtight type.	Lifetime	Energy efficient for fuel conservation.	The Eagle The Hawk The Falcon	550 lbs./36x24x30 450 lbs./25x24x30 310 lbs./20½x18½x31	3,000 sq. ft. 8 rooms 3 rooms	30" 22" 18"	$539 $490 $339
Canadian Stepstove	Boilerplate steel, cast iron door, baffled.	Lifetime	Scandinavian design and efficiency at a domestic price.	Stepstove	360 lbs./30x17x33	11,000 cu. ft.	24"	$395
Cawley/LeMay	All cast iron, airtight type box stoves with baffles and recessed cooking surfaces.	25 yrs.	Uncompromising quality and efficiency.	400 600	300 lbs./35½x18x36 385 lbs./35½x18x44	6,500 cu. ft. 10,000 cu. ft.	16" 24"	$545 $675
Chappee	Refractory lined steel with enamel finish and cast iron grate.	None stated	Cooktop surface; capable of efficiently heating small areas.	8008	141 lbs./20¼x21x12½	4,240 cu. ft.	14"	$300

Comforter	All cast iron, airtight type.	5 yrs.	Not only beautiful, but functions with highest efficiency and serves as excellent cooker.	The Parlor Stove	270 lbs./26¾x24¼x21½	10,000 cu. ft.	21"	$549
Culvert Queen	Black steel culvert pipe with welded seams.	3 yrs.	55% efficiency.	The Culvert Queen	140 lbs./31x33 dia.	10,000 cu. ft.	18"	$245
De Dietrich	All cast iron, decoratively sculptured heater with oven and food warming area.	1 yr.	True replica of 1684 French traditional model; efficient.	AL-77	240 lbs./39¾x28¾x18¼	19,840 BTUs/hr.	20½"	$635
Dover	Steel plate, lined with 3,000° refractory cement; heat exchanger included.	None stated	No other stove is as efficient.	Super Box	225 lbs./36x22x44	Still unknown	28"	$495
Dynamite	Boilerplate steel, airtight stove. Water heater available.	Unconditionally guaranteed	Rugged. Burns long and slowly, uses less wood.	Dynamite Greenwood Dynamite	140 lbs./27x18x34 240 lbs./38x18x34	5,000 cu. ft. 12,000 cu. ft.	24" 24"	$150 $225
Elm	Cast iron, steel and fire-brick baffled stoves with Pyrex viewing windows and cooking surface.	1 yr.	Durable, steady fires for up to 14 hrs., and romantic view of the fire.	The Elm The Short Elm	275 lbs./26x23x33 240 lbs./26x23x27	7,000-9,000 cu. ft. 5,000-7,000 cu. ft.	24" 18"	$405 $350
Energy Harvester	All cast iron box stove; airtight type.	5 yrs.	High-quality casting, baffled with dual controls for more efficient burning.	Chocorua	220 lbs./26¾x18¼x34½	55,000 BTUs/hr.	20"	$395
Fatsco	Stainless steel body with optional cooking surface; can be used to barbecue.	None stated	Ideal for small spaces; efficient little cookers and heaters.	Tiny Tot Pet Chummy Buddy	15 lbs./14x10½ dia. 13 lbs./11¾x10½ dia. 25 lbs./15x10½ dia. 31 lbs./21½x10½ dia.	None stated None stated None stated None stated	8" 8" 8" 8"	$36 $32 $51 $92
Fisher	Welded steel stepstoves with cast iron doors; fire-brick lined, two level cooking surface.	25 yrs.	All day or night burning on one loading.	Baby Bear Mama Bear Papa Bear	245 lbs./26¼x15½x28 345 lbs./30½x18x34 410 lbs./30½x20x39½	7,000 cu. ft. 10,500 cu. ft. 14,000 cu. ft.	18" 24" 29"	$335 $410 $455
Fjord	All cast iron replica of Scandinavian airtight stoves.	None stated	Efficient, long-burning box stove.	Fjord	231 lbs./30x14x31	8,000 BTUs/hr.	24"	$200
Free Flow	Steel tubing body forms with baffles; airtight type.	1 yr.	Far-reaching heat distribution and uniform temperatures...without electricity.	The Circulator The Wonder Furnace	200 lbs./33x23x34 260 lbs./35x25x36 325 lbs./37x28x37	8,000 cu. ft. 12,000 cu. ft. 25,000 cu. ft.	22½" 27½" 30"	$480 $600 $720
Frontier	Steel box stove.	As long as you own the stove.	Quality hand-crafted, efficient; holds fire overnight.	Bx-24-6	250 lbs./23½x18x27	1,200 sq. ft.	24"	$280

Hede	Fiberglass, insulated, enameled steel.	1 yr.	Combines the best features of other wood stoves, with glass doors for viewing fire.	Hede	240 lbs./42x31x20	8,000 cu. ft.	24"	$499
Hinckley Shaker	Cast iron, steel baffles, airtight type.	None stated	40% more heating surface area than other 2 ft. log burners.	Basic Shaker Heat Exchanger	225 lbs./25½x26x39½ 265 lbs./32x26x39½	8,000-10,000 cu. ft. 8,000-10,000 cu. ft.	26" 26"	$425 $550
Home Warmer	Thermostatically controlled steel and cast iron; airtight type. Extensive baffling.	Lifetime	UNH tested at 68.03% efficiency.	Home Warmer I Home Warmer II	320 lbs./33x18x30 270 lbs./28x16x24	60,000 BTUs/hr. 39,000 BTUs/hr.	29" 23"	$449 $399
Huntsman	Welded steel, firebrick lined stepstove; airtight type.	Standard Sears home trial.	Efficient wood-burner; burns combustion gases.	42P84151N	390 lbs./30x16x20	None stated	24"	$299
Independence	Welded steel.	3 yrs.	60% efficiency.	Independence Independence Junior	280 lbs./31x16x36 190 lbs./27x12x30	10,000 cu. ft. 6,000 cu. ft.	28" 22"	$370 $275
Jotul	All cast iron, airtight type. Elaborate baffling. Optional enamel coatings.	2 yrs.	Exceptionally long burning time; impressive fuel economy.	118 602 606	231 lbs./30x14x31 117 lbs./25x13x21 175 lbs./41x12x22	8,980 BTUs/hr. 4,007 BTUs/hr. 5,565 BTUs/hr.	24" 16" 16"	$510 $280 $515
Kachelofen	Custom designed ceramic tile stove.	1 yr.	Efficient work of art, offers "healthiest type of heat available today."	Variety of styles	1,000-2,000 lbs./ dimensions vary	38,000 BTUs/hr.	24"-30"	$2,500
Kickapoo	Boilerplate steel with cement-lined firebox and cast iron door and frame; airtight type. Boilerplate steel with fire-brick lining; airtight type.	5 yrs. 5 yrs.	Economical; model of quality design and craftsmanship. High quality, efficient heater.	BBRS BBRA BBRB BBRC Boxer	370 lbs./34¾x20¼x28 364 lbs./34x20x27 258 lbs./34x20x27 235 lbs./34x20x20 330 lbs./34¼x21¼x26½	2,400 sq. ft. 6-8 rooms 1,000 sq. ft. 3-5 rooms 1,600 sq. ft.	24" 24" 24" 18" 22"	$500 $695 $435 $424 $345
King	Thermostatically controlled steel circulating heater.	None stated	Up to 12 hrs. of constant, even heating.	8801-B 9900-B	240 lbs./32x33x21 267 lbs./34x33x24	3-4 rooms 4-5 rooms	25½" 25½"	$361 $389
Koppe	Airtight, firebrick and cast iron ceramic tile stoves. Some models have windows for viewing fire.	None stated	Artistic masterpiece, well designed and well made.	KH 77 KH 56 KK 100/s KK 150/s KK 400	313 lbs./31x28x16 375 lbs./37x28x16 452 lbs./32x35x16 562 lbs./41x35x16 440 lbs./36x36x20	9,600 cu. ft. 10,700 cu. ft. 10,700 cu. ft. 14,400 cu. ft. 9,600 cu. ft.	24" 24" 30" 30" 32"	$ 720 $ 788 $ 969 $1,011 $ 990
Lakewood	A-shaped steel and cast iron, firebrick lined stove; airtight type.	5 yrs.	Full Scandinavian baffling system gives long burn.	Cottager Workhorse	250 lbs./25½x16½x29 500 lbs./37½x21x39½	1,000 sq. ft. 2,000 sq. ft.	21" 32"	$319 $530
Lakewood Stepstove	Steel and cast iron, fire-brick lined; airtight type.	5 yrs.	Baffled for long burn, with two cooking surfaces.	The Stepstove	350 lbs./30¼x17¾x33¼	1,500 sq. ft.	26"	$390
Lange	All cast iron, elaborate baffling; airtight type. Optional enamel exterior.	None stated	Durable, safe, tightly made, producing fuel efficiency and long burning time.	6303A 6303 6302A 6302K 6203BR 6204BR	145 lbs./23½x16x25 220 lbs./37½x16x25 272 lbs./34x16x34 370 lbs./50½x16x34 213 lbs./41x13¼x20 250 lbs./41x13¼x25	3,000-5,000 cu. ft. 4,000-6,000 cu. ft. 7,000-9,000 cu. ft. 8,000-10,000 cu. ft. 4,000-6,000 cu. ft. 5,500-7,500 cu. ft.	18" 18" 24" 24" 14" 18"	From $320 From $430 From $550 From $770 From $430 From $550

Locke Stove	Steel, brick, cast iron grates.	None stated	Heavy-duty construction outclasses all similar stoves.	Warm Morning	290 lbs./33x18x36	4-5 rooms	24"	$520
Montgomery Ward	Thermostatically controlled steel circulating heater.	Standard Ward's home trial.	Sensitive thermostat, circulates heat evenly.	68A5718R 68A5722R	212 lbs./33x32x21 285 lbs./34x32x21	3-4 rooms 4-5 rooms	24" 25"	$259 $289
Morso	All cast iron, airtight type. Optional enamel exterior. Some models with baffles.	None stated	Very efficient, economical. Elegant design, hand workmanship. Lasts a lifetime.	1B 2B 1BO 2BO 6B	254 lbs./34x14x30 124 lbs./28x13x27 353 lbs./51x14x30 164 lbs./40x13x27 146 lbs./24x14x24	6,000 cu. ft. 4,800 cu. ft. 9,000 cu. ft. 6,400 cu. ft. 4,300 cu. ft.	22" 20" 22" 20" 18"	$450 $379 $749 $449 $519
Nashua	Boilerplate steel, firebrick lined. Radiant plus circulatory heat design.	Lifetime guarante	Alone in its field. Heats an entire house in five minutes flat.	Nashua 18 Nashua 24 Nashua 30	400 lbs./29¾x20¼x38¾ 600 lbs./32¾x25x45½ 1,045 lbs./40¾x31½x61½	8,000-12,000 cu. ft. 14,000-20,000 cu. ft. 21,000-35,000 cu. ft.	18" 24" 30"	$ 525 $ 750 $1,295
Norflame	All cast iron, double-walled and baffled, airtight type.	5 yrs.	Produced in one of Europe's oldest foundries.	EX75	230 lbs./29½x14½x28	7,000 cu. ft.	24"	$459
Norman	All cast iron circulating heater; airtight type.	1 yr.	Recirculation chamber makes this stove outperform others twice its size.	Norman 400	111 lbs./20x14x17	13,500 cu. ft.	16"	$329
Norwester	Thermostatically controlled steel and cast iron heater.	1 yr.	Makes for a hotter and more efficient fire.	C-76	140 lbs./32¾x19x25	None stated	24"	$335
Old Timer	Welded boilerplate steel with cast iron doors. Firebrick lined, baffled.	5 yrs.	Extremely airtight, rugged and efficient. Cooking area.	Old Timer Stove	455 lbs./35¾x19x37	An average home	26"	$479
Pillsbury	Steel and firebrick, cast iron door. Optional glass viewing door.	25 yrs.	Very attractive, efficient baffled stove. Can fit into existing fireplace.	The Dove The Falcon	300 lbs./25x17¾x28¾ 380 lbs./27x19¾x33¼	8,000 cu. ft. 11,000 cu. ft.	20" 24"	$375 $425
Quaker	Steel and cast iron, firebrick lined.	Lifetime	Special baffling system provides longer burn and excellent cooking efficiency.	Buck Doe Fawn	510 lbs./34x17x35 480 lbs./34x15x32 380 lbs./34x14x21	17,000 cu. ft. 13,000 cu. ft. 9,000 cu. ft.	28" 25" 15"	$499 $449 $399
Ram	Welded steel stove, airtight.	25 yrs.	Most efficient wood stove manufactured today.	Ram Tile Stove	250 lbs./30x14x36 300 lbs./30x14x36	15,000 cu. ft. 15,000 cu. ft.	28½" 28½"	$449 $649
Riteway	Thermostatically controlled steel and cast iron. Firebrick or aluminized steel linings.	1 yr.	Heavy-duty construction for efficient burning and long life.	2000 37	215 lbs./33x21x33 375 lbs./40x24x33	50,000 BTUs/hr. 75,000 BTUs/hr.	24" 24"	$308 $448
Scandia	All cast iron box stove; airtight type.	5 yrs.	A rugged, economical, efficient Scandinavian stove.	100 400	253 lbs./30x15x31 165 lbs./26x13x23	3-6 rooms 3-5 rooms	26" 17"	$225 $144

Sears Roebuck	Thermostatically controlled steel circulatir heater.	Standard Sears home trial	Holds fire up to 12 hrs. without refueling.	42G84123N 42G84065N	147 lbs./35x20x27 235 lbs./33x32x19	None stated None stated	24" 22"	$170 $230
Sevca	Recycled steel gas tanks.	None stated	Highest wood stove energy efficiency.	Double Barrel Stove Baby Sevca	255 lbs./36x16¾x36 175 lbs./36x16¾x24	50,000 Btus/hr. 30,000 BTUs/hr.	30" 18"	$340 $315
Shenandoah	Thermostatically controlled steel and firebrick. R-76 circulating heater.	None stated	Economy, efficiency, durability, integrity.	R-76 R-77 R-65	260 lbs./36x24x35½ 164 lbs./35x11x24 dia. 158 lbs./37x11x18 dia.	4-6 rooms 1-3 rooms 1-3 rooms	24" 21" 18"	$399 $299 $259
Sierra	Steel and firebrick, cast iron door.	None stated	Very rugged, very efficient.	Sierra 300 Sierra 500	335 lbs./24x16x24 408 lbs./26½x18x32	None stated None stated	24" 30"	$387 $438
Styria	Firebrick lined, enameled steel with cast iron doors.	None stated	Superior stoves; handmade, efficient and safe.	Excelsior Reliable Imperial	390 lbs./39x18x15 450 lbs./42½x18x15 640 lbs./48½x21¾x18½	1-2 rooms 2-3 rooms 3-5 rooms	16" 16" 19¾"	$955 $800 $990
Suburban	Thermostatically controlled steel circulating heater.	None stated	Economical, efficient.	Woodmaster	225 lbs./32¼x32x19	None stated	23"	$260
Sunshine	Steel and firebrick, airtight type.	1 yr.	High efficiency.	1	300 lbs./31x14x34	6,000-8,000 cu. ft.	24"	$300
Tempwood	Welded steel, airtight type, top-loading stove.	15 yrs.	Efficiency, immediate heat and long burnability.	Tempwood II Tempwood V	200 lbs./28x28x18 100 lbs./24¼x24x14	55,000 BTUs/hr. 35,000 BTUs/hr.	18" 16"	$289 $239
Thermo-Control	Steel and firebrick. Features built-in water heating system. Airtight type, may be connected to existing oil furnace or adapted for use in fireplace.	10 yrs.	Thermostatically controlled downdraft burning produces high efficiency.	100 200 400 500 TM2 TM4 TM5	200 lbs./27x18x33 210 lbs./27x18x33 375 lbs./27x24x34 425 lbs./33x24x40 210 lbs./29¼x22x36 325 lbs./29½x28x36 430 lbs./33x28x40	1-3 rooms 800-1,200 sq. ft. 1,000-1,600 sq. ft. 1,400-3,000 sq. ft. 1-3 rooms 2-4 rooms 2-6 rooms	18" 18" 18" 24" 18" 18" 24"	$339 $369 $499 $639 $369 $499 $639
Timberline	Steel step-type box stoves; firebrick lined, airtight type, baffled.	5 yrs.	Rugged and efficient.	T-18 T-24 T-33	285 lbs./30x14x21 418 lbs./33x16x27 502 lbs./33x18x33	1,100 sq. ft. 1,600 sq. ft. 2,500 sq. ft.	18" 24" 30"	$365 $430 $475
Trolla	All cast iron, inside baffling and linings. Airtight type.	None stated	High fuel efficiency. Small stove does the work of larger conventional stoves.	102 105 107	76 lbs./24½x11x17½ 178 lbs./25½x13x21 253 lbs./28½x13x31	1,800-3,200 cu. ft. 3,000-5,000 cu. ft. 4,000-7,000 cu. ft.	14" 21" 28"	$260 $325 $479
Ulefos	Airtight, baffled cast iron Scandinavian stoves.	1 yr.	Precision-built, rugged, safe and efficient.	868 864 865 172	115 lbs./24½x11x19 143 lbs./26x13x20½ 253 lbs./31x14x33 374 lbs./68x13½x31	3,000 cu. ft. 4,000 cu. ft. 7,000-9,000 cu. ft. 9,000+ cu. ft.	13½" 15½" 25½" 23"	$260 $290 $490 $880
Vally Comfort	Steel and stainless steel. Airtight type.	5 yrs.	Durable and highly efficient.	C-26 C-31	170 lbs./35x28x22 195 lbs./35x34x22	11,000 cu. ft. 15,000 cu. ft.	18" 24"	$445 $465
Vermont Downdrafter	Thermostatically controlled steel and firebrick heater with cast iron doors. Blower and hot water coils available.	1 yr.	Only true downdraft stove on the market. Rated 60% efficiency.	DD1	500 lbs./32x26x34	60,000 BTUs/hr.	24"	$695

Warner	Boilerplate steel with cast iron door; baffled, airtight type.	Lifetime	Designed to be the most economical, efficient stove you can buy.	W118 W124 W130	285 lbs./25x15½x28¾ 385 lbs./29½x18½x33½ 510 lbs./29½x21½x42½	8,000 cu. ft. 12,000 cu. ft. 18,000 cu. ft.	18" 24" 32"	$300 $375 $410
Waverly	Steel circulating heater.	None stated	Provides circulating warm air.	109	60 lbs./24x16x26	3,000 cu. ft.	20"	$160
Weso	Cast iron/ceramic tile stove.	1 yr.	Efficient, beautiful, offers healthiest type of heat available today.	Weso	500 lbs./32x34½x17½	6,400 cu. ft. min.	18"	$985
Wonderwood	Thermostatically controlled steel circulating heater.	None stated	Amazingly efficient. One fueling burns 10 hrs.	Model 2600	210 lbs./33x19x32½	None stated	24"	$282
Wood King	Boilerplate steel with cast iron door and frame; airtight type. Top loading.	None stated	Extracts maximum heat from every fueling.	2600	147 lbs./35x19x25	None stated	24"	$216
Woodsman	Firebrick lined steel stepstove; nickle-plated cast iron doors. Airtight type.	25 yrs.	A radiant heater and cookstove in one elegant unit.	A B C D	275 lbs./28x18x28 320 lbs./28x24x26 415 lbs./31x30x26 450 lbs./31x36x26	1,000 sq. ft. 1,500 sq. ft. 2,000 sq. ft. 3,000 sq. ft.	22" 20" 21" 26"	$419 $519 $599 $627
Yankee	Steel barrel-type stove with cast iron fittings. Stainless steel construction available.	120 days	Economical, efficient and reliable.	Model 20 Model 30	75 lbs./36x18½x21 90 lbs./29½x18 dia.	7,000 cu. ft. 11,000 cu. ft.	18" 28"	$139.95 $149.95
	Steel box stove; airtight type.			Model K	225 lbs./28x20x36	15,000 cu. ft.	30"	$479.95

COOK-STOVES	Construction	Guarantee	Mfr's Claims	The Models	Weight & Size (HxWxDepth)	Cooking Area	Log Length	Price
Atlanta	Solid cast iron.	None stated	Steady, even temperatures. Designed to last a lifetime.	15-36 Range 8316 Cookstove	254 lbs./29x35x21 155 lbs./28x30x21	6 covers, oven 4 covers, oven	15" 15"	$522 $310
Birmingham	Solid cast iron.	None stated	Long-lasting, even heat for cooking.	Red Mountain "T" Bonanza	300 lbs./29x35x21 165 lbs./28x30x21	6 covers, oven 4 covers, oven	15" 15"	$400 $274
De Dietrich	Cast iron; water jacket included.	None stated	Even heat oven, outer door stays cool; can be used to feed heat through domestic radiators.	636 (other models available without water jacket)	440 lbs./33½x31½x23¾	Entire surface area	12"	$998
Dynamite	Boilerplate steel, airtight stepstove. Water heater available.	Unconditionally guaranteed.	Can be loaded to burn through the night, or to hold a small cooking fire.	The Kitchen Cooker	360 lbs./38x36x26	Entire surface area	24"	$350

Findlay Oval	All cat iron, black porcelain finish with nickel plating with water tank.	Lifetime	The Cadillac, a top-quality stove.	The Oval	400 lbs./58x48x24	Entire surface	16"	$1,595
Jotul	All cast iron.	2 yrs.	Airtight, efficient heating stove.	404 (with oven) 380 (without oven)	223 lbs./31½x25x17¾ 131 lbs./27½x23½x16¼	2 large burners, small oven 1 large burner, 1 small burner	12" 18"	$630 $340
King	Steel and cast iron.	None stated	Dependable and economical.	Perfection	202 lbs./25¼x22x15	4 covers, oven	16"	$300
Lange	Cast iron with brass rail surrounding cooking surface.	None stated	Large firebox. Will efficiently heat large areas.	911W	375 lbs./33¼x36¼x24	Entire surface area	10"	$725
Monarch	Cast iron woodburner with 36" elec. or gas range and oven.	None stated	Unbreakable, quick heating, safe.	6LEH (elec.) HG36HW (gas)	325 lbs./47x36x26 325 lbs./47x36x26	4 burners, 2 covers, oven	16" 16"	$850 $850
Olympic	Steel with cast iron grates and cooking tops.	1 yr.	Good design, excellent craftsmanship, painstaking attention to detail.	B-18-1 18-W 8-15	490 lbs./31½x35¼x26½ 320 lbs./32x32x22½ 245 lbs./30¾x34x23½	2 covers, end plate, oven	16" 14" 15"	$1,085 $ 700 $ 400
Pioneer	All cast iron; firebrick lined oven. Various size ovens, warmers and griddles available.	None stated	Classic 1911 reproductions with unequaled performance, durability and price.	Range	325 lbs./30x30x28	4 covers, oven	12"	$685
Stanley	Steel. Airtight construction.	None stated	Fuel economy, excellent performance.	Mark I	210 lbs./35x35x22	4 covers, oven	12"	$795
Styria	Steel, cast iron door, firebrick lined. Hot water tank and brick bread oven available.	None stated	Super efficient, holds heat overnight.	130 119 106	1,100 lbs./34x51x40 800 lbs./34x47x36 611 lbs./34x42x34	Entire surface area, with separate warming shelf and oven	18" 18" 16½"	$2,300 $2,100 $1,900
Tiba	Steel, firebrick, stainless steel or enamel finish.	None stated	Very efficient, stable cooking temperatures.	Tiba	725 lbs./36x35x24	35x24 oven	14"	$1,164 stainless $1,098 enamel
Tirolia	Cast iron and steel enamel finish.	1 yr.	Completely insulated, completely airtight. Efficient space heating.	SD4 D6 D5N D7N D9N	245 lbs./28x26x20 310 lbs./31x33x21 355 lbs./33x29x23 420 lbs./33x35x23 540 lbs./33x43x23	18x20 oven 25x21 oven 21x23 oven 27x23 oven 35x23 oven	10" 12" 14" 14" 15"	$429 $625 $699 $799 $899
Wamsler	Cast iron and steel; firebrick lined.	30 days	Maximum efficiency and economy; airtight construction.	LSCK 90 (models with other options available)	465 lbs./34x36x24	Entire surface area	14"	$650
Wonderwood	Sheet steel body, cast iron top.	None stated	Rapid, even heating.	Ranch Stove	105 lbs./29x27x20	4 covers, oven	12"	$170

FIREPLACE UNITS	Construction	Guarantee	Mfr's Claims	The Models	Weight & Size (HxWcDepth)	Heating Capacity	Log Length	Price
Free Heat Machine	Steel heat exchanger with tempered glass doors.	1 yr.	Five times more efficient than other systems.	23-20 26-20 26-24 28-20 28-24	150 lbs./27½x24½x19¾ 165 lbs./30½x24½x19¾ 185 lbs./30½x24½x23¾ 175 lbs./32½x24½x19¾ 200 lbs./32½x24½x23¾	12,000 cu. ft. 12,000 cu. ft. 12,000 cu. ft. 12,000 cu. ft. 12,000 cu. ft.	22" 22" 22" 22" 22"	$466 $466 $466 $507 $507
Charmaster	Steel with glass doors; water heating coil and blower available.	1 yr.	Elegant, functional, economical, safe.	Charmaster II	747 lbs./54x38x55	Entire house	30"	$1,923
Dover	Boilerplate steel lined with refractory cement.	None stated	We'll custom-build you a beautiful, heat-efficient fireplace stove."	Custom-built	Varies with installation 27x18x15 min.	Varies with size	Varies with size	Starts at $470
El Fuego	Steel with tempered glass doors.	None stated	Firebox units for new or existing homes. Increase fireplace efficiency.	III (module) IV (pre-built) V (free-standing)	3 sizes available to fit fireplace 325 lbs./48x44x24 250 lbs./48x44x28	av. 42,000 BTUs/hr.	18"-32" depending on model	$450 $750 $580
Firemagic	Steel fireplace with cooking surface.	5-yr. guarantee against warping.	Designed to produce the ultimate in heat and air circulation.	Independent	300 lbs./32x28x26	1-3 rooms	20"	$349
Hearth Mate	Steel box stove with fireplace cover panel.	None stated	Converts the fireplace into an efficient home heater.	Better 'n Ben's	150 lbs./24x18x24	10,000 cu. ft.	18"	$319
Heatilator	Steel circulating unit.	90 days	A comforting supplemental heating system.	Mark 123C	34½x46x21 (variable)	Depends on fire.	30"	$300-$1,000
Heatscreen Plus	Steel heat exchanger with tempered glass doors.	Standard Sears home trial	Reduced heat lost up chimney; channels heated air back into room.	34P9702N2	109 lbs./ 24½-32¼ x 33¼-45 x 23	None stated	20"	$199
Hydroplace	Double-walled water heating steel unit. Steel plate.	25 yrs.	Can be used as sole source of home heat.	40" Hydroplace	450 lbs./56½x42x25	50,000 BTUs/hr.	Up to 40"	$925
Leyden Hearth	Steel with tempered glass doors.	2 yrs.	Distributes more heat over a wider area. Economical.	Leyden Hearth	200 lbs./25x26x20	3 rooms	22"	$495
Majestic	Zero clearance prefabricated built-in fireplace system.	1 yr.	35% efficiency; UL approved.	ESF-II	263 lbs./varies according to installation	None stated	20"	$470
Martin	Sheet steel with glass doors.	Limited warranty with 25-yr. protection plan.	Can be built anywhere you can dream of.	BW-28-45 BW-36-45A BW-42-45A	235 lbs./47x25x38 287 lbs./49½x25x45 391 lbs./54x25x51	42,600 BTUs/hr.	28" 36" 42"	$283 $339 $465

	Construction	Guarantee	Mfr's Claims	The Models	Weight & Size (HxWxDepth)	Heating Capacity	Log Length	Price
Thermo-Control	Thermostatically controlled steel box stove with fireplace cover panel.	None stated	Airtight, efficient heating system.	Model 300	240 lbs./28x18x24	3-4 rooms	22"	$399
Thriftchanger	Fireplace is refractory brick, with tempered glass folding doors. Steel heat exchangers.	None stated	Highly efficient passive heating system.	Thriftchanger heat recovery fireplace	Variable/adaptable	Entire house.	24"-32"	Less than $2,000
Timberline	Airtight steel and firebrick stove that fits standard fireplace. Blower optional.	5 yrs.	Rugged and efficient. Holds fire overnight.	TFI	32x23x22	1,800 sq. ft.	20"	$595
Yankee	Steel with cast iron doors; chimney baffle not included.	120 days	Designed exclusively to fit existing fireplaces.	F	180 lbs./22x30½x22	None stated	24"	$299

BOILERS AND FURNACES	Construction	Guarantee	Mfr's Claims	The Models	Weight & Size (HxWxDepth)	Heating Capacity	Log Length	Price
Bellway	Welded steel hot air furnace. Hot water coils, automatic non-electric humidifiers and non-electric thermostats available.	None stated	Ten hours of heat from one fueling.	Four sizes available	Weight varies 48-60 x 30-46 x 44-72	10 rooms	24"-48"	$600-$3,500
Charmaster	Thermostatically controlled, welded steel hot air furnace.	1 yr.	Converts wood to charcoal, burning gases produced at high temperatures.	Charmaster Furnace	646 lbs/54x28x55	Entire house	30"	$1,278
Combo	Steel, wood-only and oil-wood boilers and hot air furnaces with optional domestic water heating. Lined with serra felt instead of firebrick.	1-20 yrs. on various parts	Efficient. Special quality and safety features.	WO 12 23 1 WO 12 23 3 W 12 22 6 WO B 22 11 WO B 22 13	525 lbs./53x28½x54 585 lbs./53x28½x57 620 lbs./52x28½x67 820 lbs./49½x28½x30½ 870 lbs./49½x28½x35½	95,000 BTUs/hr. 126,000 BTUs/hr. 140,000 BTUs/hr. 126,000 BTUs/hr. 156,000 BTUs/hr.	22" 24" 28" 22" 28"	$1,250 $1,300 $1,070 $1,595 $1,695
Controlled Combustion Systems	Welded steel water boiler with baffle system.	5 yrs.	Patented combustion system burns up to 48 hrs. on one loading.	A B	1,250 lbs./54x30x60 950 lbs./54x30x30	Entire house	48" 24"	$2,950 $2,450
Daniels	Welded steel and cast iron, insulated wood and wood-oil furnaces.	None stated	The original chunk wood furnace. Built right to work right.	R30W R30WO	450 lbs./41x36x41 450 lbs./41x36x41	10,000 cu. ft. 10,000 cu. ft.	28" 28"	$1,220 (wood) $1,990 (wood-oil)
Dover	Boilerplate steel lined with refractory cement. Blower and thermostat available.	None stated	Operates independently or attached to existing oil or gas furnace.	Super Box	600 lbs./55x41¾x34¼	Entire house	24"	$729

Dynamite	Boilerplate steel, airtight wood furnace. Water heater available.	Unconditionally guaranteed.	Rugged, reliable, recommended by owners to others.	Greenwood Furnace	300 lbs./38x18x34	Up to 30,000 cu. ft.	20"	$225
Hoval Boiler	Steel-plated boiler, Swiss manufactured.	5 yrs. (1 yr. on controls)	Efficiency: 87% with oil and gas; 74% with wood incl. domestic hot water.	Six models	Average: 1,400 lbs./70x33x28 with hot water tank	100,000 BTUs	14"-21"	$1,600-$2,700
Hunter	Stainless steel firebox, steel plate.	1 yr.	Economical multi-fuel heating system.	HWO-100	790 lbs./53x48x56	140,000 BTUs/hr.	28"	$1,495
Independence	Welded sheet steel hot air furnace with manual draft; automatic draft optional.	3 yrs.	55% efficiency.	A B C D E F	300 lbs./30x22 dia. 325 lbs./30x24 dia. 350 lbs./30x26 dia. 325 lbs./36x22 dia. 350 lbs./36x24 dia. 375 lbs./36x26 dia.	All models 90,000-150,000 BTUs/hr.	24" 24" 24" 34" 34" 34"	$400 $430 $460 $430 $460 $495
Kickapoo	Steel hot air furnace. Blower optional.	5 yrs.	Effective alone or combined with gas, oil, electric or solar heat.	BBR-D	334 lbs./35¼x26x30	2,400 sq. ft.	24"	$573
Logwood	Welded steel hot air furnace. Optional oil burner.	5 yrs.	Heaviest gauge burner on the market.	36 24	1,200 lbs./61x32x60 1,000 lbs./61x32x48	200,000 BTUs 140,000 BTUs	36" 24"	$1,833 $1,503
Longwood	Reinforced steel hot air furnace.	10 yrs. pro-rated	Most economical furnace on the market.	Wood-Oil-or Gas Combination Furnace	700 lbs./40x24x66	150,000 BTUs	5'	$1,400-$1,500
Lynndale	Steel and firebrick; cast iron grates.	None stated	Central heating system, maximum heat utilization.	810	1,200 lbs./51x34x68	125,000 BTUs/hr.	36"	$2,000
Monarch	Steel and cast iron, thermostatically controlled, firebrick lined hot air wood furnace. Blower optional.	1 yr.	Connects to existing gas or oil furnace. Maximum utilization of fuel load.	AF324	300 lbs./42x22x32	75,000 BTUs/hr.	24"	$644
Newmac	Welded steel hot air furnace.	10 yrs.	Top woodburner in the USA.	Wood-Oil Combination	1,100 lbs./50x48x54	167,000 BTUs	18"	$1,600
Northeaster	Cold rolled steel.	None stated	Super-efficient hot air furnace.	224B 101-B	550 lbs./48x36x35 460 lbs./48x36x29	125,000 BTUs/hr. 100,000 BTUs/hr.	24" 18"	$1,241 $1,106
Oneida Royal	Steel and firebrick, thermostatically controlled wood and wood-coal-oil-gas boilers and hot air furnaces.	10 yrs.	Provides economical, trouble-free home heating.	WOB112 (boiler) Woodcraft 120 (attaches to existing furnace) "All Fuel" AF 160 "Two in One" WGO 160	1,483 lbs./47¼x21x42½ 475 lbs./54x24x35 1,200 lbs./65½x30¾x67 1,360 lbs./59½x30¾x67	Entire house Entire house Entire house Entire house	20" 24" 22" 22"	$4,210 $1,058 $2,397 $2,600

Passat	Welded steel combination hot air furnaces and boilers. Optional blowers.	10 yrs.	Will burn large chunks of unsplit wood, also paper, cardboard and anything else.	HO20 (boiler) HOL20 (hot air)	442 lbs./35x29x48 600 lbs./55x29x72	72,000 BTUs/hr. 88,000 BTUs/hr.	36" 36"	$1,150 $1,600
Ram	Welded steel hot air furnace and boiler.	25 yrs.	Combines with existing oil furnace or boiler.	Ram (furnace) Ram (boiler)	350 lbs./48x27x42 350 lbs./30x16x38	Entire house Entire house	28" 28"	$595 $725
Riteway	Steel, firebrick lined wood-oil boilers and hot air furnaces.	1 yr.	Well constructed for long service and efficiency.	4 hot air furnaces 4 water boilers	Vary according to model Vary according to model	125,000 BTUs/hr. 350,000 BTUs/hr.	Varies Varies	$1,400-$2,700 $2,400-$3,800
Spaulding	Welded steel hot air furnace.	None stated	75% of the energy out of nearly anything.	The Spaulding	350 lbs./72x24x38	100,000 BTUs	200 lbs. of anything	$2,295
Tarm	Steel oil-wood boilers with optional domestic water heating. Steel wet-base burner.	5 yrs.	Energy-conserving, economical, strong.	OT 35 OT 50 OT 70 OT 28 MBS-55X	1,100 lbs./38x39x27 1,500 lbs./37x46x27 1,850 lbs./37x46x36 900 lbs./38x39½x19 1,100 lbs./44x15½x41	112,000 BTUs 140,000 BTUs 200,000 BTUs 71,430 BTUs (wood) 140,000 BTUs	18" 18" 28" 10" 27½"	$2,000-$3,000 (varies with installation) $1,552
Tasso	Cast iron wood boiler.	None stated	Well insulated; lasts a lifetime.	Series with 8-11 sections available	787-1,083 lbs. 40 x 18 x 12-27	126,000-185,000 BTUs/hr.	20"-30"	$1,250-$1,550
Thermo Pride	Welded steel, firebrick lined hot air furnace.	10 yrs.	Burns wood or coal with thermostatically controlled efficiency.	W/C 20 W/C 27	635 lbs./43¼x25x50¼ 850 lbs./46¾x27x58½	90,000 BTUs/hr. 130,000 BTUs/hr.	20" 27"	$1,586 $1,730
Valley Comfort	Welded steel hot air furnaces.	None stated	Better heat at lower cost.	RB-3D RB-4D	485 lbs./46x25x36 580 lbs./46x25x48	90,000 BTUs 120,000 BTUs	33" 45"	$895 $950
Volcano	Thermostatically controlled, welded steel.	5 yrs.	True secondary combustion visible through an inspection port.	Volcano II (hot air) Volcano III (boiler)	585 lbs./42x21x29 585 lbs./38x21x28	120,000 BTUs/hr. 120,000 BTUs/hr.	24" 24"	$695 $795
Yankee	Firebrick lined steel barrel; cast iron door.	120 days	Connects to existing hot water or forced hot air systems.	B (water system) R (hot air system)	642 lbs./36½x30x39½ 496 lbs./36½x30x39½	125,000 BTUs/hr. 125,000 BTUs/hr.	36" 36"	$995 $600
Yukon	Steel cast iron and firebrick hot air furnace.	None stated	Automatic firing oil-wood furnace.	LW085 LW0100 LW0112	905 lbs./50x20x30 905 lbs./50x24x30 905 lbs./50x30x30	106,000 BTUs/hr. 125,000 BTUs/hr. 140,000 BTUs/hr.	22" 22" 22"	$1,300-$1,700

THE KITS	Construction	Guarantee	Mfr's Claims	The Models	Contents	Heating Capacity	Log Length	Price
Country Craftsmen	Cast iron door assembly and flue; steel legs.	Lifetime	Ugly yet efficient; the lowest priced cast iron kit on the market today.	Model 15/55	Everything needed but drum.	1-3 rooms	26"-34", depending on size of drum used.	$37.50
Fatsco	Steel barrel and grease drum assemblies.	None stated	Neat, serviceable, long-life stove that assembles easily.	Woodsman	Everything needed but drum.	None stated	Depends on size of drum used.	$63
Fisher's	Steel barrel and grease drum assembly with Pyrex glass door for viewing fire.	None stated	Functional, eye-appealing, easy to assemble, inexpensive.	Kits to fit 15-, 30- or 55-gallon drum	Glass door, legs, finishing materials.	None stated	Depends on size of drum used.	$35
Reginald	All cast iron box stove; airtight type.	None stated	Elegant, durable, efficient.	Reginald 101	Entire stove.	4,700 cu. ft.	16"	$270
Washington	Steel barrel and grease drum assemblies.	1 yr.	You can have a sturdy, rustic stove that will last several seasons.	Oil drum	Door, legs and flue collar.	None stated	Depends on size of drum used.	$90
Yankee	Steel and firebrick box stove; airtight type. Cast iron and steel barrel assembly.	120 days	Rugged and dependable; perfect for do-it-yourselfer.	Model K Model 20 Model 30	Entire stove. Entire stove. Entire stove.	15,000 cu. ft. 7,000 cu. ft. 11,000 cu. ft.	30" 18" 28"	$249.95 $90 $100

COMBINATION STOVES	Construction	Guarantee	Mfr's Claims	The Models	Weight & Size (HxWxDepth)	Heating Capacity	Log Length	Price
AFS	Thermostatically controlled, steel and firebrick; cast iron doors.	25 yrs.	Baffle design allows more heat from less fuel.	THE AFS Fireplace	400 lbs./28x26x36	2,400 sq. ft.	21"	$529
Atlanta	Solid cast iron.	None stated	Lifetime use for heating or cooking.	32 26 22	400 lbs./32¾x44x30 310 lbs./32x38x25 248 lbs./30x34x23	3 rooms 2-3 rooms 2 rooms	27" 23" 18"	$530 $390 $315
Atlantic	Cast iron, airtight.	None stated	Precision-fitted castings. Horizontal baffling system combined with time-tested draft control means efficient overnight burning.	228	280 lbs./31x28¾x21¼	12,000-14,000 cu. ft.	24"	$525

Autocrat	Steel cabinet, cast iron linings. Thermostatically controlled.	None stated	Welded seams, seals tightly.	Americana	400 lbs./36¼x42¾x29	5-6 rooms	24"	$749
Cherokee	Steel double walls with baffle system.	Lifetime	80% efficient — 60% with doors open.	Cherokee Chief Cherokee Princess	360 lbs./25x33¾x17 345 lbs./22x29½x16	3,000 sq. ft. 2,700 sq. ft.	None stated	$675 $650
Comforter	All cast iron, airtight type.	5 yrs.	Air preheat system, baffle with Venturi for secondary air combustion means efficient burning.	Fireplace	270 lbs./26¾x24¼x21½	10,000 cu. ft.	21"	$549
Dover	Welded steel and cast iron.	None stated	Solidly built. Heating pipes increase efficiency.	The Dover	325 lbs./32½x33¼x32¼	6-7 rooms	24"	$350
Dynamite	Boilerplate steel, airtight stove. Water heater available.	Unconditionally guaranteed.	Enjoys an extraordinary reputation among people who know wood stoves.	The Fireplace Stove (small) The Fireplace Stove (larger)	140 lbs./27x18x34 240 lbs./31x18x39	8,000 cu. ft. 15,000 cu. ft.	24" 24"	$225 $315
Efel Kamina	Porcelained steel and cast iron Pyrex glass door; airtight type. Six colors available.	None stated	Long-burning heater. Includes cooking top and barbecue pit.	Efel	199 lbs./32¼x28x15	16,000 cu. ft.	18"	$525
Fireview	Barrel stove with viewing window.	None stated	Large surface area. Cooker, heater and fireplace.	180 360	118 lbs./20¼x18x16 279 lbs./26½x36½x22	4,000 cu. ft. 10,400 cu. ft.	18" 36"	$235 $400
Fisher	Steel, firebrick. Cast iron door. Airtight type.	25 yrs.	Highly efficient with doors closed.	Grandpapa Bear Grandma Bear	475 lbs./33x30x30 425 lbs./33x25½x28	10,500 cu. ft. 9,000 cu. ft.	24" 18"	$530 $495
Frontier	Steel double door step-stove; airtight type.	Lifetime	Efficient. Every unit made by hand.	74KR21101R 74KR21102R 74KR21103R	330 lbs./27½x26x16½ 380 lbs./29x28x18½ 440 lbs./32½x30x20½	1,000 sq. ft. 1,400 sq. ft. 1,800 sq. ft.	20" 22" 24"	$399 $439 $489
Fyrtonden	Steel and firebrick combination heater.	1 yr.	Superb design, sturdy and efficient.	A B C D	287 lbs./34x23 dia. 265 lbs./31x21 dia. 221 lbs./37x19 dia. 188 lbs./27x19 dia.	7,000-9,000 cu. ft. 6,000-8,000 cu. ft. 5,000-7,000 cu. ft. 3,000-5,000 cu. ft.	18" 16" 14"-16" 14"	$690 $660 $650 $550
Garrison	Rolled steel and firebrick convertible stove.	None stated	Octagonal shape makes this a virindestructible, most attractive, sensible stove.	Garrison One Garrison Two	390 lbs./29½x32x21 245 lbs./25½x26x19	10,000 cu. ft. 7,500 cu. ft.	24" 18"	$495 $395
Gibraltar	Steel and firebrick. Racon window.	None stated	Airtight and efficient.	Gibraltar III	350 lbs./32½x32x18½	None stated	24"	$440
Hydro-Temp	Steel plate and steel water jacket surrounding fire. Cast iron front and top.	15 yrs.	Captures heat normally lost up chimney.	Hydro-Heater	425 lbs./32x36x22	6,000 cu. ft.	28"	$660

Impression	Sheet steel, double-walled Franklin heater; airtight type. Optional blower.	5 yrs.	Can be safely installed 10 inches from any wall surface. Stove jacket will not burn out.	Impression 5	300 lbs./31x30x31	1,500 sq. ft. without blower	24"	$392
Jotul	All cast iron, firebrick lined, airtight combination heater.	2 yrs.	Fuel economy achieved by interior baffle plate.	No. 1 No. 4	183 lbs./33x19x19 286 lbs./41x22x22	7,000 cu. ft. 9,500 cu. ft.	12" 14"	$550 $720
King	Cast iron and steel.	None stated	Powerful radiant heater with doors closed; beams warmth and beauty with doors open.	98-1800 98-1830	295 lbs./29½x32½x15 350 lbs./30½x38¾x16½	None stated	22" 28"	$381 $525
Koco	Ball-shaped, wrought iron, free-standing stove.	1 yr.	Conducts heat faster than cast iron, will not warp like box stoves. Spherical shape means easy ignition and efficient burning.	Koco	175 lbs./45x26 dia.	None stated	22"	$660
Lange	All cast iron, firebrick lined, combination heater.	None stated	Airtight, powerful heater when closed.	61 MF	286 lbs./38x20½x19	5,000-7,000 cu. ft.	16"	$575
Logger	Cast iron and steel; grille attachment available.	30 days	Convenience and radiant heat in an open fireplace.	OHF350	315 lbs./35x39x27	2-3 rooms	20"	$400
Morso	Cast iron and firebrick lined combination heater.	None stated	Regulated air flow when doors sealed. Biggest heat output in its class.	1125	354 lbs./41x29x23	10,000 cu. ft.	22"	$729
Nashua	Boilerplate steel, firebrick lined. Radiant plus circulatory heat design.	Lifetime guarantee.	Alone in its field. Heats an entire house in five minutes flat.	Nashua Fireplace-1 Nashua Fireplace-2	385 lbs./29¾x27x33 525 lbs./32¾x34x33½	7,000-10,000 cu. ft. 12,000-16,000 cu. ft.	18" 24"	$595 $695
Old Timer	Welded boilerplate steel with cast iron doors. Firebrick lined, baffled.	5 yrs.	Extremely airtight, rugged and efficient. Cooking area.	Old Timer Fireplace	545 lbs./35¾x28½x30½	An average home	24"	$589
Quaker	Steel and cast iron airtight type with glass doors.	Lifetime	Designed for warmth, safety and efficiency.	Moravian Parlor Stove	510 lbs./35x31x21½	15,000 cu. ft.	18"	$599

Radke	All cast iron.	Dealer guarantee.	None stated	Imperial	approx. 250 lbs./37x38x25	None stated	20"	$160
Sears Roebuck	Porcelained steel with glass doors.	Standard Sears home trial.	Our most efficient free-standing fireplace.	42P8474N	252 lbs./42½x41¼x29	None stated	20"	$439
Scandia	All cast iron, firebrick lined; airtight when closed.	5 yrs.	A rugged, economical, efficient Scandinavian stove.	Combi 200 Combi 300	286 lbs./40x23x23 300 lbs./38x22x29	3-5 rooms 4-6 rooms	18" 20"	$298 $394
Thermo-Control	Steel jacket with firebrick bottom. Airtight type.	10 yrs.	Thermostatically controlled burner with door closed. Opens to enjoy fire. Durable and inexpensive.	Thermoplace The Franklin	240 lbs./31x24x24 190 lbs./28x30x24	800-1,200 sq. ft. None stated	20" 18"	$399 $329
Timberline	Airtight combination stove-fireplace; firebrick lined, baffled. Choice of legs or pedestal.	5 yrs.	Rugged and efficient.	TSF TLF	480 lbs./33x26x26 568 lbs./33x29½x20	1,600 sq. ft. 2,400 sq. ft.	24" 20"	$499 $550
Trolla	Cast iron and firebrick; airtight type.	1 yr.	Efficient heater with door closed.	810	300 lbs./41x25x30	8,000 cu. ft.	20"	$725
Vermont Castings	All cast iron airtight type; fully baffled and thermostatically controlled.	1 yr.	Highly efficient, beautiful cast iron heaters that convert to warm and friendly fireplaces.	Defiant Vigilant	340 lbs./34x36x22 245 lbs./32x32x25	8,000-10,000 cu. ft. 6,000- 8,000 cu. ft.	24" 18"	$545 $445
Warner	Boilerplate steel with cast iron door; baffled, airtight type.	Lifetime	Designed to be the most economical, efficient stove you can buy.	W 124 FP	475 lbs./25x30x29	16,000 cu. ft.	24"	$500
Washington	Cast iron double door heaters with optional glass doors; airtight type.	1 yr.	Economical and efficient stoves made the way they used to make them.	The Olympic The 49'er The Olympic Crest	310 lbs./31x26x25 295 lbs./31¼x24¾x25 322 lbs./32½x36x24½	None stated None stated None stated	24" 20" 24"	$385 $403 $685
Wonderwood	Steel and cast iron.	None stated	Baffle creates longer flame path, greater efficiency.	The Franklin	195 lbs./27x38x21	None stated	28"	$235
Zodiac	Cast iron, sheet metal and chrome, contemporary-styled round heater.	1 yr.	A unique design you will never cease to admire.	The Zodiac	250 lbs./38x26 dia.	None stated	12"	$500

Manufacturers and Distributors of Woodburners

For more information about any of the woodburners included in the 1979 New Hampshire Times Wood Stove Buyer's Guide, contact the manufacturers and regional distributors listed here:

AFS, Intercontinental Building Products, 620 East Main St., Orange, Mass. 01364.

All Nighter Stove Works Inc., 80 Commerce St., Glastonbury, Conn. 06033.

Alpiner, Lyons Supply Co., 1 Perimeter Rd., Manchester, N.H. 03108.

Arctic, Washington Stove Works, Box 687, Everett, Wash. 98206.

Ardenne, Lyons Supply Co., 1 Perimeter Rd., Manchester, N.H. 03108.

Ashley, P.O. Box 128, Florence, Ala. 35630.

Atlanta Stove Works Inc., P.O. Box 5254, Atlanta, Ga. 30307.

Atlantic, Portland Stove Foundry Co., 57 Kennebec St., Portland, Maine 04104.

Autocrat, New Athens, Ill. 62264.

Bellway, Perley C. Bell, Grafton, Vt. 05146.

Birmingham Stove and Range Co., P.O. Box 2647, Birmingham, Ala. 35202.

Bullard Manufacturing Co., 82 Learney Ave., Liverpool, N.Y. 13088.

Canadian Stepstove, New Hampshire Stove Co., 19 North Main St., Wolfeboro, N.H. 03894.

The **Cawley/LeMay** Stove Co., P.O. Box 561, Boyertown, Pa. 19512.

Chappee, Preston Distributing Co., 2 Whidden St., Lowell, Mass. 01852.

Charmaster Products Inc., 2307 Highway 2 West, Grand Rapids, Minn. 55744.

Cherokee, Gregg Distributing Co., Box 37, Stokesdale, N.C. 27357.

Combo Furnace Co., 1707 West 4th St., Grand Rapids, Minn. 55744.

Comforter, Abundant Life Farm, Box 175, Lochmere, N.H. 03252.

Controlled Combustion Systems, 1978 Washington St., Hanover, Mass. 02339.

Country Craftsmen, Box 3333, Santa Rosa, Calif. 95402.

Culvert Queen, L.W. Gay Stoveworks, 156 Vernon Rd., Brattleboro, Vt. 05301.

Sam **Daniels** Co. Inc., Box 868, Montpelier, Vt. 05602.

De Dietrich, The Burning Log, P.O. Box 438, Lebanon, N.H. 03766.

Dover Stove Co., Box 217, Sangerville, Maine 04479.

Dynamite Stove Co., RD 3, Montpelier, Vt. 05602.

Efel Kamina, Southport Stoves Inc., 959 Main St., Stratford, Conn. 06497.

El Fuego Co., 26 Main St., Oakville, Conn. 06779.

Elm, Vermont Iron Stove Works Inc., Warren, Vt. 05674.

Energy Harvesters Corp., Box 19, Fitzwilliam, N.H. 03447.

Fatsco Stoves, 251 Fair Ave., Benton Harbor, Mich. 49022.

Findlay Oval, Elmira Stove Works, 22 Church St., West Elmira, Ontario, Canada.

Firemagic, Del Gilbert Co., RFD 2, Highway 107, Laconia, N.H. 03246.

Fireview, Tate Equipment Inc., Horseheads, N.Y. 14845.

Fisher Stoves, 504 South Main St., Concord, N.H. 03301.

Fisher's, Rte. 1, Box 63A, Conifer, Colo. 80433.

Fjord, Lyons Supply Co., 1 Perimeter Rd., Manchester, N.H. 03108.

Free Flow Stove Works, South Strafford, Vt. 05070.

Free Heat Machine, Brookfield Fireside, Rte. 7, Brookfield, Conn. 06804.

Frontier, Montgomery Ward.

Fyrtonden, Bow and Arrow Stove Co., 14 Arrow St., Cambridge, Mass. 02138.

Garrison Stove Works, Box 412, Claremont, N.H. 03743.

Gibraltar, Sierra Stoves, 1 Appletree Square, Minneapolis, Minn. 55420.

Hearth Mate, C&D Distributors Inc., Box 766, Old Saybrook, Conn. 06475.

Heatilator Co., Mt. Pleasant, Iowa 52641.

Heatscreen Plus, Sears Roebuck.

Hede, Lyons Supply Co., 1 Perimeter Rd., Manchester, N.H. 03108.

Hinckley Shaker, Hinckley Foundry & Marine, 13 Water St., Newmarket, N.H. 03857.

Home Warmers, New Hampshire Woodstoves Inc., P.O. Box 310, Plymouth, N.H. 03264.

Hoval Boiler Arotek Corp., 1703 East Main St., Torrington, Conn. 06790.

Hunter, Integrated Thermal Systems, 379 State St., Portsmouth, N.H. 03801.

Huntsman, Sears Roebuck.

Hydroplace, Ridgeway Steel Fabricators Inc., Box 382, Ridgeway, Pa. 15853.

Hydro-Temp, RD 1, Box 257, Narvon, Pa. 17555.

Impression, KNT, Box 25, Hayesville, Ohio 44838.

Independence, L.W. Gay Stoveworks, 156 Vernon Rd., Brattleboro, Vt. 05301.

Jotul, The Burning Log, Box 438, Lebanon, N.H. 03766.

Kachelofen, Ceramic Radiant Heat, Lochmere, N.H. 03252.

Kickapoo Stove Works, Box 127, La Farge, Wis. 54639.

King, Martin Industries, P.O. Box 128, Florence, Ala. 35630.

Koco, Scandia Wood Stoves Inc., 174 Old York Rd., New Hope, Pa. 18938.

Koppe, Finest Stove Imports Inc., P.O. Box 1733, Silver Spring, Md. 20902.

Lakewood, Woodman Associates, Box 626, Wakefield, N.H. 03872.

Lange, Svendborg Co. Inc., Box 5, Hanover, N.H. 03755.

Leyden Hearth, Leyden Energy Conservation Corp., Brattleboro Rd., Leyden, Mass. 01337.

Locke Stove Co., 114 West 11th St., Kansas City, Mo. 64105.

Logger Stove Corp., 1104 Wilso Drive, Baltimore, Md. 21223.

Logwood, Marathon Heater Co., Box 265, RD 2, Marathon, N.Y. 13803.

Longwood, Masi Plumbing and Heating, 36 Otterson St., Nashua, N.H. 03060.

Lynndale, Integrated Thermal Systems, 379 State St., Portsmouth, N.H. 03801.

Martin Industries, Box 128, Florence, Ala. 35630.

Majestic, Lyons Supply Co., 1 Perimeter Rd., Manchester, N.H. 03108.

Monarch Kitchen Appliances, 316 South Perry St., Johnstown, N.Y. 12095.

Monarch, Lyons Supply Co., 1 Perimeter Rd., Manchester, N.H. 03108.

Montgomery Ward, Montgomery Ward.

Morso, Inglewood Stove Co., Rte. 4, Woodstock, Vt. 05091.

Nashua, Heathdelle Sales Associates Inc., Rte. 3, Meredith, N.H. 03253.

Newmac, Arotek Corp., 1703 East Main St., Torrington, Conn. 06790.

Norflame, Lyons Supply Co., 1 Perimeter Rd., Manchester, N.H. 03108.

Norman, Lyons Supply Co., 1 Perimeter Rd., Manchester, N.H. 03108.

Northeaster, Solar Wood Energy Corp., Fall Rd., East Lebanon, Maine 04027.

Norwester, Washington Stove Works, Box 687, Everett, Wash. 98206.

Old Timer, Midwest Stoves Inc., P.O. Box 1704, Mt. Vernon, Ill. 62864.

Olympic, Washington Stove Works, Box 687, Everett, Wash. 98206.

Oneida Royal, Edward R. Stephen Co. Inc., 78 Franklin St., Somerville, Mass. 02145.

Passat USA, Box 37, East Kingston, N.H. 03848.

Pillsbury Stove Works, 84 Hathaway St., Providence, R.I. 02907.

Pioneer Lamps and Stoves, 71 Yesler Way, Seattle, Wash. 98104.

Quaker, Woodman Associates, Box 626, Wakefield, N.H. 03872.

Radke Imports, P.O. Box 128, Emmett, Idaho 83617.

Ram Forge, Brooks, Maine 04921.

Reginald, S/A Imports, 730 Midtown Plaza, Syracuse, N.Y. 13210.

Riteway Manufacturing Co., P.O. Box 6, Harrisonburg, Va. 22801.

Scandia. Preston Fuel. Whidden St., Lowell, Mass. 01850.

Sears Roebuck, Sears Roebuck.

Sevca Stoveworks, Box 477, Saxtons River, Vt. 05154.

Shenandoah, P.O. Box 839, Harrisonburg, Va. 22801.

Sierra Stoves, 1 Appletree Square, Minneapolis, Minn. 55420.

Spaulding, Novatek Inc., 79R Terrace Hill Ave., Burlington, Mass. 01803.

Stanley, Inglewood Stove Co., Rte. 4, Woodstock, Vt. 05091.

Styria, The Merry Music Box, 20 McKown St., Boothbay Harbor, Maine 04538.

Suburban Manufacturing Co., P.O. Box 399, Dayton, Tenn. 37321.

Sunshine Stove Works Inc., Norridgewock, Maine 04957.

Tarm, Tekton, Conway, Mass. 01341.

Tasso, Tekton, Conway, Mass. 01341.

Tempwood, Mohawk Industries, 173 Howland Ave., Adams, Mass. 01220.

Thermo-Control Woodstoves, Box 640, Cobleskill, N.Y. 12043.

Thermo Pride, Thermo Products Inc., P.O. Box 217, North Judson, Ind. 46366.

Thriftchanger, Sturges Heat Recovery Co., Box 397, Stone Ridge, N.Y. 12484.

Tiba, Svendborg Co. Inc., Box 5, Hanover, N.H. 03755.

Timberline, Energysavers Inc., Parade Rd. at Rte. 3, Meredith, N.H. 03253.

Tirolia, Bow & Arrow Stove Co., 14 Arrow St., Cambridge, Mass. 02138.

Trolla, Lyons Supply Co., 1 Perimeter Rd., Manchester, N.H. 03108.

Ulefos, Scandia Wood Stoves Inc., 174 Old York Rd., New Hope, Pa. 18938.

Valley Comfort, Woodburning Specialties, P.O. Box 5, North Marshfield, Mass. 02059.

Vermont Castings Inc., Box 126, Randolph, Vt. 05060.

Vermont Downdrafter, Vermont Woodstove Co., P.O. Box 1016, Bennington, Vt. 05201.

Volcano, Anchor Industries, Rte. 12, Box 63, Fitzwilliam, N.H. 03447.

Waverly Heating Supply Co., 117 Elliot St., Beverly, Mass. 01915.

Wamsler, Logger Stove Corp., 1104 Wilso Drive, Baltimore, Md. 21223.

Warner Woodstove Co., Box 292, Warner, N.H. 03278.

Washington, Vermont Distributors, 11 Maple St., Essex Junction, Vt. 05452.

Weso, Ceramic Radiant Heat, Lochmere, N.H. 03252.

Wonderwood, U.S. Stove Co., South Pittsburg, Tenn. 37380.

Wood King, Martin Industries, Box 128, Florence, Ala. 35630.

Woodsman, Crendall-Hicks Co., Rte. 9, Southborough, Mass. 01772.

Yankee Woodstoves, P.O. Box 7, Bennington, N.H. 03442.

Yukon, Moutton Climate Control, Portsmouth, N.H. 03801.

Zodiac, Washington Stove Works, Box 687, Everett, Wash. 98206.

JACOBS
WIND
ELECTRIC CO.

Wind Products Buyer's Guide

*The wind products included here are listed in alphabetical order in three categories: wind machines, water pumping windmills, and wind system components/manufacturers. This information was prepared by the editors of the **Wind Power Digest,** published quarterly by Michael Evans, 54468 CR 31, Bristol, IN 46507. Subscriptions are available for $6 in the U.S. and Canada.*

*For wind product manufacturers who wish to be included in the next edition of **The Solar Age Resource Book,** write in care of **Solar Age Magazine,** 77 Church Hill, Harrisville, NH 03450.*

WIND MACHINES

AEROPOWER MODEL "C"

Aeropower, Inc.
2398 4th St.
Berkeley, CA 94710
Tom Conlon (415) 848–2710

The Aeropower model C is an exceptionally well-produced machine although it did meet with minor problems during its first entry into the market. Those problems now seem to be resolved. **Warranty:** One year—unconditional; parts and labor. **Maintenance:** Yearly general inspection and lubrication. **Owner's Manual:** Available for $1.00.

General Description: Three-bladed, horizontal-axis, up-wind. **Cut-in wind speed:** 6 mph. **Cut-out wind speed:** 22 mph. **Maximum RPM - Wind Speed:** 675 rpm at 80 mph. **Survival wind speed:** 100 mph. **Overspeed protection:** Centrifugally activated blade pitching. **Rated Power - Wind Speed:** 1000 watts at 20 mph. **Maximum Power - Wind Speed:** 1000 watts at 20 mph. **Rotor Materials:** Aircraft sitka spruce. **Rotor Diameter:** 10 feet. **Rotor Weight:** 30 lbs. **System Weight:** 155 lbs. **Generation component:** Three-phase alternator. **Transmission:** Helical gear drive, ratio - 3:1.

AMERENALT SYSTEMS, SERIES 1500 & 2500

Amerenalt Corporation
Box 905
Boulder, CO 80302
John P. Sayler (303) 442–0820

Three of these machines have been delivered to the city of Detroit for use in a unique downtown re-development project. Amerenalt prototypes have been in operation in the rugged winter weather of Boulder, Colo., for some two years now. A larger, 12 foot diameter Amerenalt is now in planning stages and may soon be commercially available. **Warranty:** Five years limited, parts and workmanship. **Maintenance:** Semi-annual check of fluid levels; annual check of system and lubrication. **Owner's manual:** Available for $15.00.

General Description: Multi-bladed, horizontal-axis, up-wind. **Cut-in wind speed:** 8 mph. **Cut-out wind speed:** 30–75 mph (adjustable). **Maximum RPM - Wind Speed:** 450 rpm at 42 mph. **Survival wind speed:** 100 mph. **Overspeed protection:** Mechanical; tail vane deflect (adjustable). **Rated Power - Wind Speed:** Series 1500: 1500 watts at 28 mph; Series 2500: 2500 watts at 38 mph. **Maximum Power - Wind Speed:** 1500 watts at 28 mph, Series 2500: 2500 watts at 38 mph. **Rotor Material:** Aluminum. **Rotor Diameter:** 8 feet. **Rotor Weight:** 56 lbs. **System Weight:** 280 lbs. **Generation component:** 12 volt, 0–100 amp alternator. **Transmission:** Double helical gear box, ratio - 9.51:1.

AEROWATT

Aerowatt S.A./Automatic Power, Inc.
37 Rue Chanzy
75-Paris 11, France
in U.S. contact: Automatic Power, Inc.
P.O. 18738
Houston, TX 77023
Robert Dodge (713) 228–5208

This is the largest of five Aerowatt units; others include a 28, 150, 350, and 1,100 watt units. Aerowatt is an expensive, over-engineered, over-rated, and largely unproven wind machine. The expense alone places it outside of the U.S. residential wind system market. **Warranty:** One year—parts and workmanship. **Maintenance:** Semi-annual inspection and lubrication.

General Description: Two-bladed, horizontal-axis, upwind. **Cut-in wind speed:** 7 mph. **Cut-out wind speed:** 60 mph. **Maximum RPM - Wind Speed:** 150 rpm at 16 mph. **Survival wind speed:** 125 mph. **Overspeed protection:** Centrifugally activated blade pitch control. **Rated Power - Wind Speed:** 4100 watts at 16 mph. **Maximum Power - Wind Speed:** 4100 watts at 16 mph. **Rotor Materials:** Aluminum. **Rotor Diameter:** 30.7 feet. **Rotor Weight:** not available. **System Weight:** 2000 lbs. **Generation component:** Permanent magnet alternator. **Transmission:** Gear box, ratio - 20:1.

AMERICAN WIND TURBINE, 12 FOOT DIAMETER

American Wind Turbine Co., Inc.
1016 E. Airport Rd.
Stillwater, OK 74074
Dixie Jennings (405) 377–5333

American Wind Turbine has recently installed an enlarged 30-foot diameter turbine of the same configuration as these smaller units. Field testing (actual operation) data on these units was not available at press time. **Warranty:** 90 days, parts and workmanship. **Maintenance:** Semi-annual check of drive belts; annual lubrication and system inspection.

General Description: Multi-bladed, horizontal-axis, upwind. **Cut-in wind speed:** 10 mph. **Cut-out wind speed:** 35 mph (adjustable). **Maximum RPM - Wind Speed:** not available. **Survival wind speed:** 100 mph. **Overspeed protection:** Mechanical; tail vane deflect. **Rated Power - Wind Speed:** 1000 watts at 20 mph. **Maximum Power - Wind Speed:** not available. **Rotor Materials:** Aluminum. **Rotor Diameter:** 12 feet. **Rotor Weight:** 92 lbs. **System Weight:** 320 lbs. **Generation component:** Permanent magnet, three-phase alternator. **Transmission:** Rim belt drive, ratio - 30:1.

AMERICAN WIND TURBINE, 16 FOOT DIAMETER

General Description: Multi-bladed, horizontal-axis, upwind. **Cut-in wind speed:** 10 mph. **Cut-out wind speed:** 35 mph (adjustable). **Maximum RPM - Wind Speed:** not available. **Survival wind speed:** 100 mph. **Overspeed protection:** Mechanical; tail vane deflect. **Rated Power - Wind Speed:** 2000 watts at 20 mph. **Maximum Power - Wind Speed:** not available. **Rotor Materials:** Aluminum. **Rotor Diameter:** 16 feet. **Rotor Weight:** 135 lbs. **System Weight:** 420 lbs. **Generation component:** Permanent magnet, three-phase alternator. **Transmission:** Rim belt drive, ratio - 30:1.

DAF DARRIEUS

Dominion Aluminum Fabricators
3570 Hawkestone Rd.
Mississauga, Ontario, L5C 2V8
Chuck Wood (416) 275–5300

Dominion is currently manufacturing its Darrieus on a one-by-one, custom basis. Their recent work increases the installation of a 80-foot diameter, 240,000 watt peak capacity Darrieus in Canada. **Warranty:** Varies with each installation. **Maintenance:** Semi-annual inspection of the oil bearings. **Owner's Manual:** Not available.

General Description: Two-bladed, vertical-axis, cross-flow. **Cut-in wind speed:** 12 mph. **Cut-out wind speed:** not available. **Maximum RPM - Wind Speed:** 160 rpm. **Survival wind speed:** 120 mph. **Overspeed protection:** Mechanical air-spoilers, aero-dynamics. **Rated Power - Wind Speed:** 14,000 watts at 30 mph. **Maximum Power - Wind Speed:** 14,000 watts at 30 mph. **Rotor Materials:** Aluminum. **Rotor Diameter:** 20 feet. **Rotor Weight:** 150 lbs. **System Weight:** 850 lbs. **Generation component:** not included. **Transmission:** Gear box.

DUNLITE WIND SYSTEMS, MODEL 81

Dunlite Electrical Products Co., Div. of PYE
28 Orsmond St.
Hindmarch, S. Australia
Graeme Jackson

Dunlite is a respected wind machine with over 40 years of operational experience in Australia and world-wide. The only draw-back is its relatively high cost (due to shipping costs and import duties). Dunlites are in service throughout the U.S. and are an important part of the modern micro-wave communications system in Australia. **Warranty:** One year—parts and workmanship. **Maintenance:** Oil change every five years. **Owner's Manual:** Available for $20.00.

General Description: Three-bladed, horizontal-axis, upwind. **Cut-in wind speed:** 7–8 mph. **Cut-out wind speed:** 80 mph. **Maximum RPM - Wind Speed:** 200 rpm at 35 mph. **Survival wind speed:** 80 mph. **Overspeed protection:** Centrifugally activated blade pitching. **Rated Power - Wind Speed:** 2000 watts at 25 mph. **Maximum Power - Wind Speed:** 3000 watts at 30 mph. **Rotor Materials:** Galvanized steel sheeting. **Rotor Diameter:** 13.5 feet. **Rotor Weight:** 130 lbs. **System Weight:** 500 lbs. **Generation component:** Three-phase alternator (12, 24, 48, 110 volt avail.). **Transmission:** Helical gearing, ratio - 5:1.

DUNLITE WIND SYSTEMS, MODEL 82

General Description: Three-bladed, horizontal-axis, upwind. **Cut-in wind speed:** 10 mph. **Cut-out wind speed:** 110 mph. **Maximum RPM - Wind Speed:** 200 rpm at 40 mph. **Survival wind speed:** 110 mph. **Overspeed protection:** Centrifugally activated blade pitching. **Rated Power - Wind Speed:** 2000 watts at 30 mph. **Maximum Power - Wind Speed:** 3000 watts at 35 mph. **Rotor Materials:** Galvanized steel sheeting. **Rotor Diameter:** 10 feet. **Rotor Weight:** 130 lbs. **System Weight:** 500 lbs. **Generation component:** Three-phase alternator (12, 24, 48, 110 volt available). **Transmission:** Helical gearing.

ELEKTRO GmbH

Elecktro GmbH
Winterthur, Schqeiz, St.
Gallerstrasse 27, Switzerland
Peter Kern

This is only one of several sizes of Elecktros that include a newly marketed 10,000 watt unit that was recently installed in West Virginia. The Elecktro has earned a somewhat tarnished reputation in the U.S. that may not be wholly correct. In any event, the Elecktro is a machine that does not quite deliver its full rated power in actual installations—at least not in those installations that have been monitored. (It should be noted that the 6kw Electro will deliver 5800 watts in winds in excess of 40 mph.) **Warranty:** One year—parts and workmanship. **Maintenance:** Semi-annual oil change and inspection.

General Description: Three-bladed, horizontal-axis, upwind. **Cut-in wind speed:** 7 mph. **Cut-out wind speed:** 60 mph. **Maximum RPM - Wind Speed:** 270 rpm at 44 mph. **Survival wind speed:** 120 mph. **Overspeed protection:** Mechanical; blade pitching. **Rated Power - Wind Speed:** 6000 watts at 26 mph. **Maximum Power - Wind Speed:** 6000 watts at 26 mph. **Rotor Materials:** Redwood or spruce (dependent on availability). **Rotor Diameter:** 16.4 feet. **Rotor Weight:** 110 lbs. **System Weight:** 585 lbs. **Generation component:** Permanent magnet, three-phase alternator. **Transmission:** Gear box ratio - 4:1.

KEDCO WIND SYSTEMS, MODEL 1200

Kedco, Inc.
9016 Aviation Blvd.
Inglewood, CA 90301
Wind Program Manager (213) 776–6636

The Kedco series of wind energy systems was designed by author-engineer Jack Park, who is best known for his excellent book, **Simplified Wind Systems for Experimenters.** Several important improvements have been made in the Kedco units since the first production run and the new units are performing well. **Warranty:** One year on parts and labor—warranty on the alternator is per manufacturer's specifications. **Maintenance:** Annual inspection of system and fluid level check. **Owner's manual:** Available for $7.50.

General Description: Three-bladed, horizontal-axis, down-wind. **Cut-in wind speed:** 6–8 mph. **Cut-out wind speed:** 70 mph. **Maximum RPM - Wind Speed:** 300 rpm at 22 mph. **Survival wind speed:** 110 mph. **Overspeed protection:** Centrifugal blade pitching and manual shut-off. **Rated Power - Wind Speed:** 1200 watts at 22 mph. **Maximum Power - Wind Speed:** 1200 watts at 22 mph. **Rotor Materials:** Aluminum. **Rotor Diameter:** 12 feet. **Rotor Weight:** 71 lbs. **System Weight:** 202 lbs. **Generation component:** 14.4 or 28.4 volt dc alternator. **Transmission:** Gear box, ratio - 8.76:1.

KEDCO WIND SYSTEMS, MODEL 1210

General Description: Three-bladed, horizontal-axis, down-wind. **Cut-in wind speed:** 10–12 mph. **Cut-out wind speed:** 70 mph. **Maximum RPM - Wind Speed:** 300 rpm at 26 mph. **Survival wind speed:** 110 mph. **Overspeed protection:** Mechanical; blade pitching and manual shut-off. **Rated Power - Wind speed:** 2000 watts at 26 mph. **Maximum Power - Wind Speed:** 2000 watts at 26 mph. **Rotor Materials:** Aluminum. **Rotor Diameter:** 12 feet. **Rotor Weight:** 71 lbs. **System Weight:** 252 lbs. **Generation component:** Permanent magnet generator. **Transmission:** Gear box, ratio - 9.87:1.

KEDCO WIND SYSTEMS, MODEL 1600

General Description: Three-bladed, horizontal-axis, down-wind. **Cut-in wind speed:** 7–9 mph. **Cut-out wind speed:** 60 mph. **Maximum RPM - Wind Speed:** 250 rpm at 17 mph. **Survival wind speed:** 110 mph. **Overspeed protection:** Mechanical; blade pitching and manual shut-off. **Rated Power - Wind Speed:** 1200 watts at 17 mph. **Maximum Power - Wind Speed:** 1200 watts at 17 mph. **Rotor Materials:** Aluminum. **Rotor Diameter:** 16 feet. **Rotor Weight:** 75 lbs. **System Weight:** 217 lbs. **Generation component:** 14.4 or 28.3 volt dc alternator. **Transmission:** Gear box, ratio - 8.76:1.

KEDCO WIND SYSTEMS, MODEL 1610

General Description: Three-bladed, horizontal-axis, down-wind. **Cut-in wind speed:** 9–11 mph. **Cut-out wind speed:** 60 mph. **Maximum RPM - Wind Speed:** 250 rpm at

22 mph. **Survival wind speed:** 110 mph. **Overspeed protection:** Mechanical; blade pitching and manual shut-off. **Rated Power - Wind Speed:** 2000 watts at 22 mph. **Maximum Power - Wind Speed:** 2000 watts at 22 mph. **Rotor Materials:** Aluminum. **Rotor Diameter:** 16 feet. **Rotor Weight:** 75 lbs. **System Weight:** 267 lbs. **Generation component:** Permanent magnet generator. **Transmission:** Gear box, ratio - 8.76:1.

MILLVILLE MODEL 10

Millville Windmills & Solar Equipment Co.
Box 32, 10335 Old 44 Drive
Millville, CA 96062
Devon Tassen (916) 547–4302

Millville is currently backlogged six months on orders and is now moving to larger manufacturing facilities to increase production of their units. Detailed evaluations of the machines already installed are not available at this time, although no major failures have been experienced. **Warranty:** 90 days parts and workmanship. **Maintenance:** Semi-annual inspection and lubrication.

General Description: Three-bladed, horizontal-axis, up-wind. **Cut-in wind speed:** 11 mph. **Cut-out wind speed:** 60 mph. **Maximum RPM - Wind Speed:** 72 rpm. **Survival wind speed:** 80–90 mph. **Overspeed protection:** Mechanical; blade pitching and tail vane deflect. **Rated Power - Wind Speed:** 10,000 watts at 25 mph. **Maximum Power - Wind Speed:** 10,000 watts at 25 mph. **Rotor Materials:** Aluminum. **Rotor Diameter:** 25 feet. **Rotor Weight:** 159 lbs. **System Weight:** 850 lbs. **Generation component:** Induction generator. **Transmission:** Triple helical gear box, ratio - 24:1.

NORTH WIND EAGLES, MODEL 2kw 32 VOLT

North Wind Power Co.
Box 315
Warren, VT 05674
Don Mayer (802) 496–2955

These units are completely re-manufactured Jacobs wind generators designed for maximum performance. North Wind is a company with three years experience in the re-building and installation of the Jacobs machines. In addition they have an excellent record of service to the customer after installation. The Jacobs generator speaks for itself. **Warranty:** One year—unconditional on generator and main components. **Maintenance:** Semi-annual lubrication; clean or replace brushes every three to five years; re-finish blades every five years. **Owner's Manual:** Available for $3.00.

General Description: Three-bladed, horizontal-axis, up-wind (remanufactured Jacobs windplant). **Cut-in wind speed:** 8 mph. **Cut-out wind speed:** 40 mph. **Maximum RPM - Wind Speed:** 300 rpm at 40 mph. **Survival wind speed:** 90 mph. **Overspeed protection:** Centrifugal blade pitching. **Rated Power - Wind Speed:** 2000 watts at 22 mph. **Maximum Power - Wind Speed:** 2000 watts at 22 mph. **Rotor Materials:** Sitka spruce with fiberglass leading edge. **Rotor Diameter:** 14.5 feet. **Rotor Weight:** 50 lbs. **System Weight:** 480 lbs. **Generation component:** Permanent magnet generator. **Transmission:** Direct drive.

NORTH WIND EAGLES, MODEL 2kw 110 VOLT

General Description: Three-bladed, horizontal-axis, up-wind. **Cut-in wind speed:** 8 mph. **Cut-out wind speed:** 40 mph. **Maximum RPM - Wind Speed:** 300 rpm at 40 mph. **Survival wind speed:** 90 mph. **Overspeed Protection:** Centrifugal blade pitching. **Rated Power - Wind Speed:** 2000 watts at 22 mph. **Maximum Power - Wind Speed:** 2000 watts at 22 mph. **Rotor Materials:** Sitka spruce with fiberglass leading edge. **Rotor Diameter:** 14.5 feet. **Rotor Weight:** 50 lbs. **System Weight:** 480 lbs. **Generation component:** Permanent magnet generator. **Transmission:** Direct drive.

NORTH WIND EAGLES, MODEL 3kw 32 VOLT

General Description: Three-bladed, horizontal-axis, up-wind. **Cut-in wind speed:** 8 mph. **Cut-out wind speed:** 40 mph. **Maximum RPM - Wind Speed:** 300 rpm at 40 mph. **Survival wind speed:** 90 mph. **Overspeed protection:** Centrifugal blade pitching. **Rated Power - Wind Speed:** 3000 watts at 27 mph. **Maximum Power - Wind Speed:** 3000 watts at 27 mph. **Rotor Materials:** Sitka spruce with fiberglass leading edge. **Rotor Diameter:** 14.5 feet. **Rotor Weight:** 50 lbs. **System Weight:** 480 lbs. **Generation component:** Permanent magnet generator. **Transmission:** Direct drive.

NORTH WIND EAGLES, MODEL 3kw 110 VOLT

General Description: Three-bladed, horizontal-axis, up-wind. **Cut-in wind speed:** 8 mph. **Cut-out wind speed:** 40

mph. **Maximum RPM - Wind Speed:** 300 rpm at 40 mph. **Survival wind speed:** 90 mph. **Overspeed protection:** Centrifugal blade pitching. **Rated Power - Wind Speed:** 3000 watts at 27 mph. **Maximum Power - Wind Speed:** 3000 watts at 27 mph. **Rotor Materials:** Sitka spruce with fiberglass leading edge. **Rotor Diameter:** 14.5 feet. **Rotor Weight:** 50 lbs. **System Weight:** 480 lbs. **Generation component:** Permanent magnet generator. **Transmission:** Direct drive.

PINSON CYCLOTURBINE

Pinson Energy Corporation
Box 7
Marston Mills, ME 02648
Herman Drees (617) 477–2913

The Cycloturbine is a remarkably innovative vertical-axis wind machine that overcomes the basic problem associated with vertical-axis designs, the inability to self-start without complex and expensive logic systems. Prototype Cycloturbines have performed very well in testing over the past year and the first production run of machines are now being installed in the Cape Cod area. A major plus with this machine is that it is extremely quiet in operation. **Warranty:** One year on all parts and services including retrofit for any design changes in the prototype units. **Maintenance:** Annual lubrication and general survey of hardware. **Owner's Manual:** Available—price not set.

General Description: Three-bladed, vertical-axis, cross-flow. **Cut-in wind speed:** 9 mph. **Cut-out wind speed:** Not applicable; unit continues to extract power. **Maximum RPM - Wind Speed:** 200 rpm at 30 mph. **Survival wind speed:** 110 mph. **Overspeed protection:** Centrifugal blade pitching. **Rated Power - Wind Speed:** 2000 watts at 24 mph. **Maximum Power - Wind Speed:** 4000 watts at 30 mph. **Rotor Materials:** Aluminum. **Rotor Diameter:** 12 feet. **Rotor Weight:** 100 lbs. **System Weight:** 150 lbs. **Generation component:** 120 volt, 240 volt alternator. **Transmission:** Timing belt single step, ratio - 7.5:1.

SENCENBAUGH, MODEL 500

Sencenbaugh Wind Electric
Box 11174
Palo Alto, CA 94306
Jim Sencenbaugh (415) 964–1593

Sencenbaugh wind machines have earned a good reputation as well-manufactured and durable units. We've received several reports of Sencenbaugh windplants surviving the harsh winter in Alaska last year. **Warranty:** One year—parts and workmanship. **Maintenance:** Annual inspection and lubrication. **Owner's Manual:** Available for $10.00

General Description: Three-bladed, horizontal-axis, upwind. **Cut-in wind speed:** 10 mph. **Cut-out wind speed:** 60 mph. **Maximum RPM - Wind Speed:** 1000 rpm at 30 mph. **Survival wind speed:** 80 mph. **Overspeed protection:** Tail vane deflect braking. **Rated Power - Wind Speed:** 500 watts at 24 mph. **Maximum Power - Wind Speed:** 600 watts at 29 mph. **Rotor Materials:** Sitka spruce, copper leading edge, epoxy finish. **Rotor Diameter:** 6 feet. **Rotor Weight:** 8 lbs. **System Weight:** 100 lbs. **Generation component:** 28 volt alternator. **Transmission:** Direct drive.

SENCENBAUGH, MODEL 1000

General Description: Three-bladed, horizontal-axis, upwind. **Cut-in wind speed:** 6 mph. **Cut-out wind speed:** 60 mph. **Maximum RPM - Wind Speed:** 350 rpm at 27 mph. **Survival wind speed:** 80 mph. **Overspeed protection:** Tail vane deflect braking. **Rated Power - Wind Speed:** 1000 watts at 23 mph. **Maximum Power - Wind Speed:** 1200 watts at 28 mph. **Rotor Materials:** Sitka spruce, copper leading edge, epoxy finish. **Rotor Diameter:** 12 feet. **Rotor Weight:** 15 lbs. **System Weight:** 120 lbs. **Generation component:** 14 or 28 volt alternator. **Transmission:** Helical gear box, ratio - 3:1.

WINCHARGER WINDPLANTS

Winco, Division of Dyna Technology
Box 3263, East 7th & Division St.
Sioux City, IA 51102
Tom Poe (612) 853–8400

This 200 watt windplant is the only remaining remnant of a large line of Wincharger machines that was produced until 1957. With its low output (no more than 10–15 kw/hrs. per month) it is suited best for extremely low-demand applications such as a one light hunting cabin. **Warranty:** One year—parts and labor. **Maintenance:** Annual inspection and lubrication. **Owner's Manual:** Available for 50 cents.

General Description: Two-bladed, horizontal-axis, upwind. **Cut-in wind speed:** 7 mph. **Cut-out wind speed:** 70 mph. **Maximum RPM - Wind Speed:** 900 rpm at 70 mph. **Survival wind speed:** 70 mph. **Overspeed protection:** Mechanical air brake and manual brake. **Rated Power - Wind Speed:** 200 watts at 23 mph. **Maximum Power - Wind Speed:** 200 watts at 23 mph. **Rotor Materials:** Varnished redwood, copper leading edge. **Rotor Diame-**

ter: 6 feet. **Rotor Weight:** 20 lbs. **System Weight:** 134 lbs. **Generation component:** DC generator. **Transmission:** Direct drive.

WIND WIZARD

Aerolectric
13517 Winter Lane
Cresaptown, MD 21502
Kevin Moran (609) 547–3488

The Wind Wizard C-9 has been in operation for two years and evaluation and testing are ongoing. **Warranty:** Limited one year warranty. **Maintenance:** General inspection required semi-annually. **Owner's Manual:** Available upon request for $3.00.

General Description: Three-bladed, horizontal-axis, upwind. **Cut-in wind speed:** 27–35 mph. **Maximum RPM - Wind Speed:** 720 rpm at 40 mph. **Survival wind speed:** 60 mph. **Overspeed protection:** Mechanical - vane deflect, with a manual crank for tail fold out. **Rated Power - Wind Speed:** 600 watts at 26 mph (measured at the hub). **Maximum Power - Wind Speed:** 2000 watts at 30 mph (measured at the hub). **Rotor Materials:** Urethane coated wood. **Rotor Diameter:** 9 feet. **Rotor Weight:** 22 lbs. **System Weight:** 51 lbs. **Generation component:** 12 volt, 61 amp alternator. **Transmission:** V-belt, ratio - 10:1.

WINDSTREAM 25/PROTOTYPE

Grumman Energy Systems
4175 Veterans Memorial Hwy.
Ronkonkoma, NY 11779
Kenneth Speiser (516) 575–6205

The Windstream 25 wind machine is an off-shoot of Grumman's work with the Sailwing rotor design. Windstream prototypes have been in service for over a year now, and one is undergoing operational testing at the Rocky Flats Wind System Testing Center at this time. **Warranty:** Limited. **Maintenance:** Not Available. **Owner's Manual:** Not available.

General Description: Three-bladed, horizontal-axis, down-wind. **Cut-in wind speed:** 8 mph. **Cut-out wind speed:** 50 mph. **Maximum RPM - Wind Speed:** 125 rpm at 26 mph. **Survival wind speed:** 130 mph. **Overspeed protection:** Blade tip spoilers and dc servo motor. **Rated Power - Wind Speed:** 15,000 watts at 26 mph. **Maximum Power - Wind Speed:** 20,000 watts at 29 mph. **Rotor Materials:** Aluminum. **Rotor Diameter:** 25 feet. **Rotor Weight:** 750 lbs. **System Weight:** 2,000 lbs. **General component:** Self-excited alternator. **Transmission:** Oil bath gear box, ratio - 14.5:1.

ZEPHYR WIND DYNAMO

Zephyr Wind Dynamo Co.
Box 241, 21 Stamwood St.
Brunswick, ME 04011
Bill Gillette (207) 725–6534

The 15kw Zephyr is still essentially a prototype machine with several of the units now installed and gathering operational data. One important feature of this design is the peak output at relatively low alternator rpms.

General Description: Three-bladed, horizontal-axis, down-wind. **Cut-in wind speed:** 8 mph. **Cut-out wind speed:** 45 mph. **Survival wind speed:** 120 mph. **Overspeed protection:** Aero spoilers and automatic yawing. **Rated Power - Wind Speed:** 15,000 watts at 30 mph. **Maximum Power - Wind Speed:** 15,000 watts at 30 mph. **Rotor Materials:** Plastic composite. **Rotor Diameter:** 20 feet. **Rotor Weight:** 100 lbs. **System Weight:** 675 lbs. **General component:** Permanent magnet alternator. **Transmission:** Direct drive.

WATER PUMPING WINDMILLS

AERMOTOR MODEL 702

Aermotor Division, Braden Industries
Box 1364
Conway, AZ 72032
Stan Anderson (501) 329–9811

This is the largest of five Aermotor models. Aermotors have been in service for some 44 years. All prospective buyers of water-pumping windmills should take great care to determine which machine is best suited to any particular application. Basic criteria are: pumping capacity at a given windspeed, maximum well depth possible, and durability in unattended operation. **Warranty:** One year—parts and workmanship. **Maintenance:** Annual lubrication. **Owner's Manual:** Available in the near future.

General Description: Multi-bladed, horizontal-axis, water pumper. **Cut-in wind speed:** 9 mph. **Cut-out wind speed:** 28 mph. **Maximum RPM - Wind Speed:** 53 rpm at 20 mph. **Survival wind speed:** 45 mph. **Overspeed protection:** Trail vane deflect. **Rated Power - Wind Speed:**

Horsepower unavailable. **Maximum Power - Wind Speed:** Horsepower unavailable. **Rotor Materials:** Galvanized steel. **Rotor Diameter:** 16 feet. **Rotor Weight:** 830 lbs. **System Weight:** 2,450 lbs. **Generation component:** none. **Transmission:** Single reduction, double gear, ratio - 3.29:1.

AMERICAN WIND TURBINE

American Wind Turbine Co., Inc.
1016 E. Airport Road
Stillwater, OK 74074
Dixie Jennings (405) 377–5333

The three-phase alternator used in the AWT water-pumping machines is designed specifically to interface with a three-phase electric submersible deep-well pump. A 12 foot diameter machine is available as well. **Warranty:** 90 days parts and workmanship. **Maintenance:** Semi-annual check of the drive belts; annual lubrication and system inspection.

General Description: Multi-bladed, up-wind, horizontal-axis, water-pumper. **Cut-in wind speed:** 10 mph. **Cut-out wind speed:** 35 mph (adjustable). **Maximum RPM - Wind Speed:** Not available. **Survival wind speed:** 100 mph. **Overspeed protection:** Mechanical; tail vane deflect. **Rated Power - Wind Speed:** 2000 watts at 20 mph. **Maximum Power - Wind Speed:** Not available. **Rotor Materials:** Aluminum. **Rotor Diameter:** 16 feet. **Rotor Weight:** 135 lbs. **System Weight:** 420 lbs. **Generation component:** Permanent magnet, three phase alternator. **Transmission:** Rim belt drive, ratio - 30:1.

BAKER WINDMILLS

Heller Aller Company
Perry & Oakwood St.
Napoleon, OH 43545
James Bradner (419) 592–1856

This is the largest of four models of Baker mills which have been in service for 50 years. Prospective buyers of water-pumping mills should take great care to determine which machine is best suited for the anticipated application. Basic criteria are: pumping capacity at a given windspeed, maximum well depth possible, and durability in unattended operation. **Warranty:** One year parts and workmanship. **Maintenance:** Annual inspection and oil change.

General Description: Multi-bladed, horizontal-axis, up-wind. **Cut-in wind speed:** 7 mph. **Cut-out wind speed:** 25 mph. **Maximum RPM - Wind Speed:** Unavailable. **Survival wind speed:** 70 mph. **Overspeed protection:** Spring loaded tail vane deflect. **Rated Power - Wind Speed:** Unavailable. **Maximum Power - Wind Speed:** Unavailable. **Rotor Materials:** Galvanized steel. **Rotor Diameter:** 12 feet. **Rotor Weight:** Unavailable. **System Weight:** 800 lbs. **Generation component:** None. **Transmission:** Gear box, ratio 3:1.

DEMPSTER MODEL 16

Dempster Industries, Inc.
Box 848
Beatrice, NE 68310
Roy Smith (402) 223–4026

Dempster has been manufacturing water-pumping windmills for 80 years with various model and design changes. Prospective buyers of water-pumping windmills should take great care to determine which machine is best suited for the anticipated application. Basic criteria are: pumping capacity at a given windspeed; maximum well depth possible, and durability in unattended operation. **Warranty:** Five years—parts and workmanship. **Maintenance:** Annual inspection and oil change. **Owner's Manual:** Available free.

General Description: Multi-bladed, horizontal-axis, up-wind water pumper. **Cut-in wind speed:** 5 mph. **Cut-out wind speed:** 20 pump strokes per minute. **Maximum RPM - Wind Speed:** Unavailable. **Overspeed protection:** Tail vane deflect. **Rated Power - Wind Speed:** .88 horsepower at 18–20 mph. **Maximum Power - Wind Speed:** Unavailable. **Rotor Materials:** Galvanized steel. **Rotor Diameter:** 16 feet. **Rotor Weight:** Unavailable. **System Weight:** Unavailable. **Generation Component:** None. **Transmission:** Single reduction gear.

PONDMASTER

Wadler Manufacturing Co., Inc.
Rt 2, Box 76
Galena, KS 66739
Jerry Wade (316) 783–1355

Although this is not a water-pumping mill, it does serve a useful function for lake and pond aeration and as a means to prevent lake and pond freeze over.

General Description: Two-bladed, vertical-axis, Savonious rotor. **Cut-in wind speed:** 2 mph. **Cut-out wind speed:** Not applicable. **Maximum RPM - Wind Speed:** Unavailable. **Survival wind speed:** 90 mph. **Overspeed protection:** None. **Rated Power - Wind Speed:** Not applicable. **Maximum Power - Wind Speed:** Not applicable. **Rotor Materials:** Aluminum. **Rotor Diameter:** Unavailable. **Rotor Weight:** Unavailable. **System Weight:** 16 lbs. **Generation component:** None. **Transmission:** Direct drive, ratio - 1:1.

SPARCO MODEL D

Sparco
Box 420
Norwich, VT 05055
Edmund Coffin (802) 649–1145

This simple machine is best suited for modest pumping requirements such as a residential water supply and is definitely not designed for heavy agricultural use. The Sparco has been in service for little over a year at this point. **Warranty:** One year—parts and workmanship. **Maintenance:** Semi-annual inspection and lubrication. **Owner's Manual:** Available for $2.00.

General Description: Two-bladed, horizontal-axis, up-wind water pumper. **Cut-in wind speed:** 5 mph. **Cut-out wind speed:** Unavailable. **Maximum RPM - Wind Speed:** 150 rpm at 25 mph. **Survival wind speed:** 80 mph. **Overspeed protection:** Mechanical blade pitching. **Rated Power - Wind Speed:** Unavailable. **Maximum Power - Wind Speed:** Unavailable. **Rotor Materials:** Aluminum. **Rotor Diameter:** 4.17 feet. **Rotor Weight:** 8 lbs. **System Weight:** 50 lbs. **Generation component:** None. **Transmission:** Full eccentric drive, ratio - 1:1.

ANEMOMETERS

Bendix Environmental & Process Instruments Div.

1400 Taylor Avenue
Baltimore, MD 21204
(301) 825–5200

Climatronics Corp.

1324 Motor Parkway
Hauppauge, NY 11787
(516) 234–2772

Climet, Inc.

1620 West Colton Avenue
Redlands, CA 92373
(714) 793–2788

Danforth-Eastern

500 Riverside Industrial Parkway
Portland, ME 04103
(207) 797–2791

Davis Instrument Manufacturing Co., Inc.

513 East 36th Street
Baltimore, MD 21218
(301) 243–4301

Dwyer Instruments, Inc.

P.O. Box 373
Michigan City, IN 46360
(219) 872–9141

Hightstown, NJ
(609) 448–9200

Marietta, GA
(404) 427–9406

Anaheim, CA
(714) 991–6720

Entertech Corp.

P.O. Box 420
Norwich, VT 05055
(802) 649–1145

Kahl Scientific Instrument Corp.

P.O. Box 1166
El Cajon (San Diego), CA 92022
(714) 444–2158, 444–5944

Kenyon Marine Co.

P.O. Box 308
Guilford, CT 06437
(203) 453–4374

2730B South Main Street
Santa Ana, CA 92702
(714) 546–1101

Maximum, Inc.

42 South Avenue
Natick, MA 01760
(617) 785–0113

Meteorology Research, Inc.

P.O. Box 637
Altadena, CA 91001
(213) 791–1901

Natural Power, Inc.
New Boston, NH 03070
(603) 487–5512

Sencenbaugh Wind Electric Co.
P.O. Box 11174
Palo Alto, CA 94306
(415) 964–1593

Sign X Laboratories, Inc.
Essex,CT 06426
(203) 767–1700

R. A. Simerl Instrument Div.
238 West Street
Annapolis, MD 21401
(301) 849–8667

M. C. Stewart Co.
Ashburnham, MA 01430
(617) 827–5840

Texas Electronics, Inc.
P.O. Box 7225
Inwood Station
Dallas, TX 75209
(214) 631–2400

Thermo-Systems, Inc.
500 Cardigan Road
P.O. Box 3394
St. Paul, MN 55165
(612) 483–0900

Westberg Manufacturing Co.
3400 Westach Way
Sonoma, CA 95476
(707) 938–2121

BATTERIES

C & D Batteries
3043 Walton Road
Plymouth Meeting, PA 19462
(215) 828–9000

Century Storage Battery Company, LTC.
Birmingham Street
Alexandria, Australia

ESB, Inc.
Wisco Div.
2510 North Blvd.
Raleigh, NC 27604
(919) 834–8465

Globe Battery
Div. of Globe-Union, Inc.
5757 North Green Bay Avenue
Milwaukee, WI 53201
(414) 228–2581

Gould, Inc.
Industrial Battery Div.
2050 Cabot Blvd. West
Langhorn, PA 19047
(215) 752–0555

Keystone Battery Corp.
35 Holton Street
Winchester, MA 01890
(617) 729–8333

Mule Battery Co.
325 Vallet Street
Providence, RI 02908
(401) 421–3773

Surrette Storage Battery Co., Inc.
P.O. Box 3027
Salem, MA 01970
(617) 745–4444

Trojan Batteries, Inc.
1125 Mariposa Street
San Francisco, CA 94107
(415) 864–1565

GENERATORS/ ALTERNATORS

Georator Corp.
P.O. Box 70, 9016 Prince William Street
Manassas, VA 22110
(703) 368–2101

Zephyr Wind Dynamo Co.
21 Stanwood Street
Brunswick,ME 04011
(207) 715–6534

INSTRUMENTS

Natural Power, Inc.
New Boston, NH 03070
(603) 487–5512

Prairie Sun and Wind Co.
4408 62nd Street
Lubbock, TX 79414
(806) 795–1412

Propeller Engineering Duplicating Co.
403 Avenida Teresa
San Clemente, CA 92672
(714) 498–3739

Universal Manufacturing Co.
900 Cedar Ridge Road
Duncanville, TX 75116
(214) 298–0531

West Wind
P.O. Box 1465
Farmington, NM 87401

INVERTERS

ATR Electronics, Inc.
300 East 4th Street
St. Paul, MN 55101
(612) 222–3791

Carter Motor Company
2711 West George Street
Chicago, IL 60618
(312) 588–7700

Dynamote Corp.
1130 N.W. 85th Street
Seattle, WA 98117
(206) 784–1900

Elgar Corp.
8225 Mercury Court
San Diego, CA 92111
(714) 565–1155

Soleq Corp.
5969 Elston Avenue
Chicago, IL 60646
(312) 792–3811

Topaz Electronics
3855 Ruffin Road
San Diego, CA 92123
(714) 279–0831

Windworks
P.O. Box 329
Route 3
Mukwanago, WI 53149
(414) 363–4408

TOWERS

American Tower Co.
Shelby, OH 44875
(419) 347–1185

Astro Research Corp.
P.O. Box 4128
Santa Barbara, CA 93103
(805) 684–6640

Natural Power, Inc.
New Boston, NH 03070
(603) 487–5512

Rohn Tower
Div. of Unarco Industries, Inc.
P.O. Box 2000
Peoria, IL 61601
(309) 697–4400

Solar Products Buyer's Guide

The following section includes over 20 different solar product categories totaling 370 products with each manufacturer's name, address, phone number, and contact person. For easy cross-referencing of products and manufacturers, we have included an alphabetical listing of solar manufacturers beginning on page 000. Each product is described here, but more technical information is available in the ***Solar Products Specification Guide,*** *published by* ***Solar Age Magazine.*** *For further information on the* ***Guide,*** *write care of* ***Solar Age*** *magazine, 77 Church Hill, Harrisville, NH 03450. Please also write if you wish to be included in the* ***Spec Guide*** *or in the next edition of* ***The Solar Age Resource Book.***

Solar Packages

GREENHOUSES

ENERGY FACTORY SOLAR GREENHOUSE

The Energy Factory
5622 East Westover
Suite 105
Fresno, CA 93727
Thomas W. Kristy (209) 292–6622

These kit greenhouses are constructed on a modular framework using high-quality clear redwood and glass or plastic glazing. Preassembled components are ready for bolt-together construction, tools for assembly are included. Water barrels used for passive heat storage and temperature maintenance are not included.

Features: basic kit includes redwood construction, transparent redwood-framed roof, silicone rubber-sealed glass modules, pre-hung door (combination clear redwood and twin plastic glazing), pre-hung vents, weatherstripping, vent screens, foundation (2 × 4 in. redwood), and assembly brackets, nuts, bolts, sealers, caulks and adhesives. Freestanding models are available in 2-ft. modules, 6 × 8 ft. to 8 × 12 ft. Attached models are available in 2-ft. modules, 6 x 8 ft. to 8 x 12 ft. **Installation requirements/considerations:** unit must face within 30 degrees of due South for best results. **Guarantee/warranty:** 90 days against factory defects. **Maintenance requirements:** redwood must be water sealed. **Manufacturer's technical services:** advice upon request. **Availability:** 3 wk between order and delivery. **Suggested retail price:** from $612 (freestanding 6 × 8 ft).

SOLAR HEAT COLLECTING GREENHOUSE

Vegetable Factory, Inc.
100 Court St.
Copiague, NY 11726
Fred Schwartz (516) 842–9300

The Solar Heat Collecting Greenhouse has double-wall fiberglass-reinforced polyester or GE Lexan solar panels that reduce energy needs up to 60 percent. Lean-to and free-standing models up to 24 ft wide (any length). All have aluminum frameworks and no foundations are required. Lean-to models may be used for solar heat gathering applications. Do-it-yourself kits. Backup systems not provided, but the manufacturer has a complete line of accessories for greenhouse operation.

Features: rigid panels are bonded permanently to aluminum I-beam frame. **Options:** stock panels or custom panels, available in acrylic/fiberglass or GE Lexan. **Installation requirements/considerations:** lean-to models may be used to gather solar heat. **Guarantee/warranty:** 5 yr. **Manufacturer's technical services:** special technical services and assistance available. **Suggested retail price:** from $699; write for free color catalog.

SOLAR HEATED GREENHOUSE

Energy Shelters/Egge Research (joint venture)

Energy Shelters Inc.
2162 Hauptman Rd.
Saugerties, NY 12477
Morton Schiff (914) 246–3135

Egge Research
RFD 6, Box 394B
Kingston, NY 12401
Morton Schiff (914) 336–5937
Tamil Bauch

This greenhouse is a well-insulated shell with only the south wall glazed with a Beadwall. The north wall is a curved reflector composed of molded fiberglass, reinforced-cement-urethane-foam-core sandwich panels, and east and west walls are made of sheetrock interior finish, sprayed urethane-foam insulation, and textured plywood sheathing. Wood studs are used. Passive model has no heat pump, but north wall is covered inside with a mirrored film which reflects the radiation incident on the wall into a water tank located at its base, and blowers are used that draw air from a header running along the ridge of the house and bubble the air through the water via a second header that runs along the bottom of the water storage tank. The active model uses a heat pump to remove surplus heat and humidity from the air and put heat into the water storage tank; its north wall is painted white to reflect diffuse radiation.

Features: active model will recycle all transpired moisture back to plants and will also heat an additional space whose requirements are 250 Btu's/hr/°F. **Options:** active or passive model; size of building; kit/installed; color of GFRC panels. Available in any length in 4-ft increments and any width up to 20 ft. **Installation requirements/considerations:** consult manufacturer. **Guarantee/warranty:** 1-yr equipment; 3-yr structural (installed only). **Maintenance requirements:** wood exterior to be oil-stained every 7-yr; occasional addition of anti-static solution to Beadwall; and additional beads as slight shrinkage occurs. **Manufacturer's technical services:** design service and installation. **Regional applicability:** above 36 degrees North latitude, and below 36 degrees South latitude, depending on altitude. **Availability:** 4 to 8 wk between order and delivery. **Suggested retail price:** installed 15 × 24-ft, passive model, $9500; as a kit, $6000.

SOLAR ROOM

Solar Room Co.
Box 1377
Taos, NM 87571
Leah Alexander (505) 758–9344

A twin skin air-inflated tension structure designed for retrofit attachment to the south wall of a building to collect solar Btu and circulate heat through an existing window or other aperture. Manufactured in kit-form for owner-installation.

Features: kit includes two Dutch-door ends and 20-in. window fan to circulate heat. **Options:** 8 or 9-ft height, 12 to 39-ft length, in 3-ft increments; high-low thermometer. **Guarantee/warranty:** 100 percent cash refund for 6 mo, 2 yr free replacement of defective materials. **Availability:** 2 wk. **Suggested retail price:** $849 to $1299.

SOLERA/SOLAR GARDEN

Solar Technology Corp.
2160 Clay St.
Denver, CO 80211
Richard Speed (303) 455–3309

A walk-in passive solar collector of modular structure, adaptable as an enclosure for a solar greenhouse, hot tub, pool or home addition.

Options: flat plate collector panels are also available (see Flat Plate Collectors). **Installation requirements/considerations:** available in kit form with detailed installation instructions. **Guarantee/warranty:** 1 yr on all parts except glass breakage. **Manufacturer's technical services:** installation, engineering design and backup. **Regional applicability:** Colorado and adjacent states. **Availability:** 4 to 8 wk. **Suggested retail price:** varies from $15 to $30/sq ft of floor area according to size and options.

HOT WATER SYSTEMS

COMPLETE HOT WATER SYSTEM, DHW64-82FP

Solar Living, Inc.
P.O. Box 12
Netcong, NJ 07857
Richard Bonte (201) 691–8483

This draindown system includes two 4 × 8 ft solar panels, an 82-gal storage tank, and all necessary controls, valves, pump and thermostats. System is direct.

Features: fully automatic. **Options:** other sizes available. **Installation requirements/considerations:** knowledge of household plumbing. **Guarantee/warranty:** 2 yr, 5 yr on tank. **Manufacturer's technical services:** technical assistance provided. **Availability:** 30 days. **Suggested retail price:** $1100.

CONSERDYNE SOLAR WATER SYSTEM

Conserdyne Corp.
4437 San Fernando Rd.
Glendale, CA 91204
Howard Kraye (213) 246–8408

This system consists of a double-walled, glass-lined storage tank by American Appliance, a Southwest Ener-Tech Tempered Glass stainless steel collector designed by Conserdyne, pump, sensors, hardware, and all necessary connections to complete a standard installation.

Features: flush mount collectors on new construction. **Options:** 66, 82, or 120-gal storage tank; size and number of collectors. **Guarantee/warranty:** 5 yr. **Manufacturer's technical services:** installation and maintenance. **Regional applicability:** southern California. **Availability:** 3 to 4 wk between order and delivery. **Suggested retail price:** from $1900.

DIXON

Dixon Energy Systems, Inc.
47 East St.
Hadley, MA 01035
Jane Nevin (413) 584–8831

A complete system for domestic hot water with a Drain-back feature that works without moving parts.

Guarantee/warranty: 5-yr warranty. **Availability:** 4 wk. **Suggested retail price:** $2500 to $2900.

DOMESTIC HOT WATER PACKAGE

Champion Home Builders Co.
Solar Products Div.
118 Walnut St.
Waynesboro, PA 17268
Al Cool (717) 762–3113

This air system has heat-exchange coil for water heating; complete with collector, heat exchanger, solar water pump, differential thermostat controls, face and bypass damper. Automatically preheats domestic water when space heating is not required.

Installation requirements/considerations: available space for water storage tank; available space for heat exchanger between inlet and outlet furnace ducts. **Maintenance requirements:** consult manufacturer. **Availability:** stock. **Suggested retail price:** consult manufacturer.

DOMESTIC WATER HEATING SYSTEM 2DEH

Solar Unlimited, Inc.
4310 Governors Drive, W.
Huntsville, AL 35805
Larry Frederick (205) 837–7340

This model uses the existing hot water as storage in retrofit installations. One 34.4 sq ft collector, double-wall heat exchanger assembly, sensors and heat transfer fluid complete the package to supply hot water for a family of two.

Features: easy filling procedure. **Options:** flat roof mounting or tilted mounting kits available. **Installation requirements/considerations:** install heat exchanger near existing hot water heater; sweat interconnecting copper tubing. **Guarantee/warranty:** 5-yr limited warranty on collector, 1 yr on other components. **Availability:** from stock. **Suggested retail price:** $1195.

DUMONT HW-1 SOLAR DOMESTIC HOT WATER SYSTEM

Dumont Industries
Main St.
Monmouth, ME 04259
A. Douglas Scott (207) 933–4811

A closed-loop liquid system that uses aluminum, flat plate, integral tube absorber housed in double-glazed, insulated, galvanized steel cases. Collectors feed 20 to 40 sq ft finned copper tube heat exchangers in stone-lined tank.

Options: 2 to 4 panels; 80, 100, 120-gal tanks with one or more electric back-up elements. **Guarantee/warranty:** 5 yr limited guarantee on collectors; other components as warranted by manufacturers. **Maintenance requirements:** maintenance of transfer fluid level as indicated by system operations, annual check of transfer fluid pH, preservation of exposed surfaces. **Manufacturer's technical services:** custom and commercial design/fabrication, installation assistance, periodic check-ups on systems. **Availability:** 2 wk or less; only in Maine at present for servicing/monitoring. **Suggested retail price:** $1200 (2 panels, 80-gal tank).

ECOTHERM SOLAR SYSTEMS 24, 48, 72

Horizon Enterprises, Inc.
P.O. Box V 1011 NW 6th St.
Homestead, FL 33030
Ed Glenn (305) 245–5145

A solar domestic hot water system using durable collectors, a water-cooled and lubricated stainless steel pump, a variable controller with immersible probe and integral freeze protection.

Options: number of collectors and size of tank. **Guarantee/warranty:** 5 yr on collector, 1 yr on all other components. **Regional applicability:** areas that seldom freeze. **Availability:** from stock. **Suggested Retail Price:** $700 to $1350.

HELIOSYSTEMS DOMESTIC HOT WATER, 82 AND 120

Heliosystems Corp.
3407 Ross Ave.
Dallas, TX 75204
Gary deLarios (214) 824–5971

These kit-form hot water systems include two 3 × 8 ft

solar collectors, a stainless steel pump, automatic temperature controller, collector mounting brackets, valving and fittings, and an 82 or 120-gal storage tank with heating element and thermostat.

Features: recirculation of drain down freeze protection for warm or cold climates. **Options:** size; kit form. **Installation requirements/considerations:** southern exposure for collector; protected area for tank and pumps. **Guarantee/warranty:** 5-yr limited warranty on panel, tank and controller; 18-mo warranty on pump. **Maintenance requirements:** acid wash collector every 5 yr; clean pump filter twice a year; drain tank sediment every 3 mo. **Manufacturer's technical services:** sizing, design and installation assistance. **Availability:** 2 to 3 wk between order and delivery. **Suggested Retail Price:** Model 82, $1375; Model 120, $1475.

HOT WATER AND SPACE HEATING SYSTEM

Vulcan Solar Industries, Inc.
200 Conant St.
Pawtucket, RI 02893
J. Michael Levesque (401) 725–6061

This closed-loop system consists of three Sunline collectors, an 80-gal storage tank with a 20-sq-ft heat exchanger and circulator pump, differential thermostat control, and various components. The recommended heat transfer fluids are propylene glycol and water. Depending on energy efficiency and need, various tanks are used.

Features: custom design. **Options:** drain-down system and various transfer mediums. **Installation requirements/considerations:** varies with each system. **Manufacturer's technical services:** available for the life of the system. **Regional applicability:** Continental United States. **Availability:** 2 to 3 wk between order and delivery. **Suggested retail price:** depends on order, consult manufacturer.

HOT WATER SYSTEM KIT

Solar Usage Now, Inc.
Box 306
Bascom, OH 44809
Joseph Deahl (419) 937–2226

A solar hot water preheating system in kit form, including collector parts, heat exchanger/storage tank, circulating pump, differential controller, collector fluid, absorber paint, silicone glazing seal and miscellaneous installation components, such as, expansion tank, check valve, pressure relief valve, auto air eliminator and 50 ft of low-loss sensor wire.

Features: installer needs to furnish only collector box, piping and fittings, backing and pipe insulation. **Options:** preassembled collector available. **Guarantee/warranty:** 5 yr on absorber plate and tank. **Availability:** from stock. **Suggested retail price:** $650.

HOT WATER SYSTEMS

Solar and Geophysical Engineering
P.O. Box 576
Sparta, NJ 07871
Gary Bubb (201) 729–7287

Features: Redwood-framed, all-copper collectors. **Guarantee/warranty:** 5 yr. **Regional applicability:** Eastern US. **Suggested retail price:** $1750 to $2150.

JOULE BOX™

InterTechnology/Solar Corp.
100 Main St.
Warrenton, VA 22186
N. L. Beard (703) 347–9500

A system designed to provide 60–75 percent of the annual hot water requirements of a given family size in a given region of the country. The Joule Box™ uses distilled or deionized water as a heat transfer fluid in a closed-loop-coil heat exchange system. Freeze protection is provided by drain down when pump cuts off.

Features: modular system can be sized for region and demand. **Options:** roof or ground-support mounting of collectors. **Guarantee/ warranty:** 1-yr, full-system warranty; 5-yr limited warranty on collectors, tank and heat exchanger. **Maintenance requirements:** evaporation losses in the drain-down tank must be made up occasionally. **Manufacturer's technical services:** solar system design and specification services. **Availability:** 4 to 6 wks. **Suggested retail price:** $789 to $3160.

PIONEER SOLAR DOMESTIC HOT WATER SYSTEM

Pioneer Energy Products
Rt. 1 Box 189
Forest, VA 24551
Timothy M. Hayes (804) 239–9020

A potable water, open-type hot water system with flow

and drain-down freeze protection controlled by one panel. The system includes all-copper waterways, bronze head pump, controller and valves, pre-plumbed and mounted.

Options: closed, antifreeze loop with heat exchanger; additional collectors and variable sizes of storage tanks are available. **Installation requirements/considerations:** requires access to open drain and to 110-vac grounded outlet. **Guarantee/warranty:** 1-yr limited on control box; 5-yr limited warranty on storage tank and collector. **Manufacturer's technical services:** each job is sized including computed payback. **Availability:** 2 wk. **Suggested retail price:** $1295.

PRIMA PAC 2001

Prima Industries, Inc.
P.O. Box 141
Deer Park, NY 11729
A. L. Gruol (516) 242–6347

A complete, domestic, hot water system including two flat plate, liquid, closed-type, solar collectors, an 82-gal storage tank and heat exchanger, a fully proportional, solid-state controller, a circulator and heat transfer fluid.

Features: collectors have removable/refillable desiccant cartridges. **Installation requirements/considerations:** do not use dissimilar metals. **Maintenance requirements:** change heat transfer fluid every 2 years. **Guarantee/warranty:** 5 yr for collectors, 1 yr for other components. **Availability:** 4 to 8 wk. **Suggested retail price:** $889.

REFRIGERANT-CHARGED WATER HEATING PACKAGE 6200

Refrigeration Research, Inc., Solar Research Div.
525 N. Fifth St.
Brighton, MI 48116
Jerry Kay (313) 227–1151

Hot refrigerant vapor from the top of the collectors moves by its own vapor pressure into the heat exchanger, gives up its heat to the continuously finned copper water coil, condenses and returns to the bottom of the collectors.

Features: no freezing problems; UL listed for five most common refrigerants. **Installation requirements/considerations:** heat exchanger package must be mounted above the tops of the collectors. **Guarantee/warranty:** 5-yr, prorata, limited warranty on Solar-Research-manufactured, components; 1 yr on controls, pumps, valves, etc. **Manufacturer's technical services:** limited assistance available by mail or telephone. **Availability:** 4 to 6 wk. **Suggested retail price:** consult manufacturer.

RAYPAK DHWS-2-T80

Raypak, Inc.
31111 Agoura Rd.
Westlake Village, CA 91361
H. Byers (213) 889–1500

A domestic, hot-water package with two all-metal collectors and differential thermostatic controls. Fully automatic.

Features: high limit and freeze protection, 80-gal storage tank is standard. **Options:** 52 or 100-gal tank. **Guarantee/warranty:** 1-yr, materials and workmanship. **Maintenance requirements:** pump motor must be oiled once or twice a year. **Availability:** from stock. **Suggested retail price:** $1490.

SOLAFERN HOT WATER SYSTEM S-100

Solafern, Ltd.
536 MacArthur Blvd.
Bourne, MA 02532
Philip Levine (617) 563–7181

This domestic hot water system consists of two Solafern Model-30 collectors. Blower heat exchanger prepackaged with insulation. Components (circulator, tank and ducting) available separately.

Installation requirements/considerations: heat exchanger should be as close to collector as possible. **Guarantee/warranty:** 5 yr on collectors; 1 yr on system. **Maintenance requirements:** tight joints. **Regional applicability:** northern and midwest United States. **Availability:** 6 wk. **Suggested retail price:** $1000 base.

SOLAR DOMESTIC HOT WATER SYSTEM

Southeastern Solar Systems, Inc.
2812 New Spring Rd.
Suite 150
Atlanta, GA 30339
Joe Cooper (404) 434–4447

With the DHW-1 System, cold incoming water on its way to the existing water heater is routed through a large copper heat exchanger in the storage tank. This step preheats the cold water before it reaches the water heater, which now must carry only a portion of its former heating load.

Features: solar collectors, storage tank, solid-state controls. **Options:** recirculation loop from the water heater to the storage tank. **Installation requirements/considerations:** southerly exposure for the collectors, proper angle. **Guarantee/warranty:** collectors (exclusive of glass), controller, tank, 5 yr; pump, 1½ yr; switches and other controls, 1 yr. **Manufacturer's technical services:** engineering assistance available from manufacturer. **Regional applicability:** State of Illinois. **Availability:** systems are warehoused in central Illinois at Solarflame Systems, Inc., P.O. Box 99, LeRoy, IL, 61752 (309) 962–2861. **Suggested retail price:** consult manufacturer.

SOLAR HOT WATER SYSTEM, N-SERIES AND R-SERIES

Energy Systems, Inc.
4570 Alvarado Canyon Rd.
San Diego, CA 92120
Terrence R. Caster (714) 280–6660

N-series models for new installations and R-series models for retrofit installations include tank, pump, controls, valves, and double-glazed collectors. All models operate directly on potable water without heat exchanger and have automatic controls. Units are prepackaged.

Options: automatic drain, freeze protection. **Installation requirements/considerations:** consult manufacturer. **Guarantee/warranty:** 5-yr limited warranty. **Maintenance requirements:** cleaning of collector surface. **Availability:** 1 to 2 wk between order and delivery. **Suggested retail price:** $1378 to $3510; depends on system.

SOLARKIT™

Solarkit of Florida, Inc.
1102 139th Ave.
Tampa, FL 33612
Wm. Denver Jones (813) 971–3934

A kit which consists of one or more collectors, a circulating pump, a solid-state controller with both freeze- and upper-temperature limit-controls, all necessary valves and a special, leakproof roof feed-through.

Features: kits feature redwood collector cases. **Options:** single or double glazing; storage tanks, mounting frames, heat exchangers, pipe insulation available as accessories. **Installation requirements/considerations:** pre-drilled legs, optional mounting frames; straightforward plumbing, low-voltage-dc power requirements. **Guarantee/warranty:** 5-yr limited warranty on collector(s), 1-yr OEM warranty on pump and controller. **Maintenance requirements:** minimal; oil pump every few months, refinish collector case every 5 yr. **Availability:** 1 to 4 wk. **Suggested retail price:** $358 for kit with one collector, $189 for each additional collector.

SOLAROLL HOT WATER SYSTEM

Bio-Energy Systems, Inc.
Box 489 Mountaindale Rd.
Spring Glen, NY 12483
Michael F. Zinn (914) 434–7858

This system featured EPDM elastomer tubing and extrusions with a semi-closed-loop drain-down system.

Features: low-cost, on-site installation. **Options:** size, layout, glazing, insulation, pump, tank, heat transfer system. **Guarantee/warranty:** 5 yr limited warranty against defects in materials and workmanship, or cracking or leaking of EPDM material due to weathering or heat aging; others carry OEM warranty from manufacturer. **Maintenance requirements:** minimal. **Manufacturer's technical services:** design engineering, comprehensive manual and support literature. **Availability:** 2 wk. **Suggested retail price:** consult manufacturer.

SOLAR SAVER WATER HEATER

W.L. Jackson Manufacturing Co., Inc.
Box 11168
Chattanooga, TN 37401
P.G. Para (615) 867–4700

The Jackson Solar Saver solar water heater is a complete system including 80-gal, insulated, jacketed storage tank, wrap-around heat exchanger, two collector panels, differential thermostat with panel and tank sensors, 50 ft of wire, circulating pump, and auxiliary heating element.

Features: heat exchanger is double-wall aluminum; drain-back design; temperature and pressure-relief valve. **Options:** element wattage from 1 to 6 kw, additional panel, glass collector glazing, or more powerful pump. **Installation requirements/considerations:** use CPVC pipe or aluminum tubing. **Guarantee/warranty:** 5-yr limited warranty on storage tank, panels, heat exchanger, surge tank; 1-yr limited warranty on electrical components. **Maintenance requirements:** change heat exchange fluid every 2 yr; keep panels clean; drain storage tank regularly. **Availability:** in stock in many areas; special orders take 3 wk. **Suggested retail price:** $1,400.

SOLARSTREAM

American Appliance Manufacturing Corp.
2341 Michigan Ave.
Santa Monica, CA 90404
Paul Hegg (213) 870–8541

A complete packaged system, pre-wired and pre-plumbed, using Reynolds aluminum collectors. Available in three sizes, 66, 82 and 120 gal using 4 × 8 ft or 4 × 12 ft collectors.

Features: the heat exchanger in this closed system is UL listed and IAPMO approved. **Options:** available with or without electrical back-up unit. **Guarantee/warranty:** 5 yr on collector(s) and tank. **Availability:** 3 wk. **Suggested retail price:** consult manufacturer.

SOLATRON HOT WATER HEATING

General Energy Devices, Inc.
P.O. Box 5679
Clearwater, FL 33516
Clyde Bouse (800) 237–0137

This system is designed to provide 75 to 90 percent residential hot water heating north of the freeze line. Exchanger system complete with flat plate liquid collectors, differential control, Grundfos 1/12-hp pumps, backup electrical elements, expansion tanks and exchanger.

Features: collectors feature ambient air Pre-heaters and aluminum fluid travel. **Options:** integrated wrap-around, all-copper counterflow heat exchanger. **Guarantee/warranty:** 5-yr warranty on collector; 10-yr warranty on exchanger; 1-yr warranty on electrical. **Maintenance requirements:** circulating pump to be oiled annually. **Manufacturer's technical services:** F-chart computer program analysis of system analysis; system design. **Availability:** 1 to 2 wk between order and delivery after receipt of order. **Suggested retail price:** according to installation; consult manufacturer.

SOLATRON HOT WATER HEATING

This system is designed to provide 75 to 90 percent of customer's usage for residential hot water heating in areas that have fewer than 10 days of freezing conditions per year. Flat plate liquid collector is fiberglass encased with copper Solar Bond absorber; collector features ambient air preheater. Direct water system complete with collector, differential controller, 1/220-hp pump, storage tank, air-bleed valves, PT valves, and recirculation freeze sensor.

Features: ambient air preheater; all-copper water trough. **Guarantee/warranty:** 10-yr warranty on collector; 5-yr warranty on storage tank; 1-yr warranty on electrical components. **Maintenance requirements:** oil circulating pump annually. **Manufacturer's technical services:** F-chart computer program analysis of system requirements; system design. **Regional applicability:** for use in areas that have fewer than 10 days of freezing conditions per year. **Availability:** 1 to 2 wk between order and delivery after receipt of order. **Suggested retail price:** according to installation; consult manufacturer.

SOLECTOR[R] PAK 1000 SERIES

Sunworks Div., Enthone, Inc.
PP.O. Box 1004
New Haven, CT 06508
Floyd C. Perry (203) 934–6301

A closed-loop, forced circulation, domestic hot water system that circulates heat transfer fluid from collector to storage tank.

Options: 65, 80 or 120-gal storage tank. **Guarantee/warranty:** 5-yr on collector and tanks, 1-yr on all other items. **Maintenance requirements:** heat transfer/freeze inhibitor should be replaced every 4 or 5 yr. **Availability:** 10 days. **Suggested retail price:** consult manufacturer.

SOLECTOR[R] PAK 2100

An open-loop, domestic, water heating system for use in summer residences and in areas with fewer than 30 freezing days per year.

Features: designed to circulate storage tank water through the collectors to keep them from freezing. **Installation requirements/considerations:** not recommended for areas where prolonged freezing temperatures occur. **Guarantee/warranty:** 5-yr on collectors, 1-yr on all other components. **Regional applicability:** areas with fewer than 30 freezing days or for summer-only applications. **Availability:** 10 days. **Suggested retail price:** consult manufacturer.

SOLECTOR[R] PAK 2200

An open-loop, domestic water heating system for use in regions where outside temperatures frequently fall below freezing. The collectors are protected from freezing by automatically shutting off the water supply and simultaneously draining them when the water approaches freezing.

Features: can be retrofitted to any existing electric water heating tank or in series with gas- or oil-fired heaters. **Guarantee/warranty:** 5-yr on collectors, 1-yr on all other components. **Regional applicability:** designed for areas with more than 30 freezing days each year. **Availability:** 10 days. **Suggested retail price:** consult manufacturer.

SOLERGY HOT WATER PREHEATER (SG-1000, SG-2000, SG-3000)

Solergy Co.
7612 Boone Ave. N.
Minneapolis, MN 55428
Vince Grimaldi (612) 535–0305

This hot water preheater consists of three Solergy SG-100 flat plate liquid collectors (active area of 68.1 sq ft), a 110-gal tank-in-tank heat exchanger and storage system, 5 gal of Suntemp fluid, an expansion tank, and a differential thermostat with a 160° F cutoff and three sensors. This preheater is designed to provide approximately 60 percent annual hot water needs of a family of four in a Minnesota-type climate.

Features: collectors can be double glazed in field; kit available. **Options:** aluminum or copper collectors; Ford, stone-lined tank or Solergy 110-gal tank-in-tank. **Installation requirements/considerations:** installation kit available; installation instructions supplied. **Guarantee/warranty:** 5-yr limited warranty. **Maintenance requirements:** coat glazing with Kalwall weatherable surface every 5 yr. **Manufacturer's technical services:** design and engineering services available. **Availability:** 4 to 6 wk between order and delivery (FOB). **Suggested retail price:** consult manufacturer.

SOLITE SYSTEM, MODELS 308, 408, 508, 512, 612 AND 712

Solar Alternative, Inc.
30 Clark St.
Brattleboro, VT 05301
Jim Kirby, Alain Ratheau (802) 254–6668

Solite Systems are designed for the HUD Hot Water Initiative and are approved by the Polytechnic Institute of New York (PINY). Solite collectors are low cost, efficient, durable, and easy to install. All system components are high quality.

Features: a light-weight, durable collector with concave glazing to shed snow easily; ideal for roof-top installation. **Options:** system may include three to seven collectors, 80- or 120-gal storage with or without the electric element connected. **Guarantee/warranty:** 5-yr limited warranty on collectors and tanks. **Maintenance requirements:** heat transfer fluid must be checked once a year. **Availability:** 3 wk. **Suggested retail price:** $1200 to $2000, depending on size of system.

SUMMERSUN

Solar American Corp.
106 Sherwood Drive
Williamsburg, VA 23185
R.J. Pegg (804) 874–0836

A collector/storage system that requires no controls, pumps or electrical provisions. Each unit has 30-gal capacity.

Features: may be mounted on building (roof) or ground. System has an adjustable leg to aid in obtaining optimum solar orientation. **Options:** unit may be connected in parallel to increase storage capacity. **Installation requirements/considerations:** system must be drained during freezing periods. **Guarantee/warranty:** 5-yr on tank; 2-yr parts and workmanship. **Maintenance requirements:** infrequent cleaning of glazing. **Regional applicability:** areas and seasons not affected by freezing. **Availability:** approximately 3 wk. **Suggested retail price:** $425.

SUNFIRED™ ENERGY SYSTEMS, CS-66-2, CS-82-2, CS-82-3, CS-120-4

Solar Energy Products, Inc.
1208 N.W. 8th Ave.
Gainesville, FL 32601
Jack C. Ryals (904) 377–6527

SEP's closed domestic hot water systems offer several important advantages: Freeze protection for the colder climates, panel fluid loop is free from corrosion, and there is no possibility of mineral buildup in the collectors if harsh water conditions exist. These Sunfired Energy Systems were among the first to be approved by Polytechnic Institute of New York for use in HUD's Solar Hot Water Initiative Grant Program.

Features: one of only a few systems which, with as few as two collectors, will achieve the percent efficiency required for HUD's Grant Program in the northern states. **Options:** proportional or fixed-flow controllers with various sensor packages; from two to four collectors in various mounting configurations. **Installation requirements/considerations:** warranty requires licensed installer and has an authorized dealership network of sales and service solar specialists. **Guarantee/warranty:** 1-yr limited warranty: approved for HUD Solar Hot Water. **Maintenance requirements:** initiative; 30-day and 1-yr warranty validation inspection required during which routine cleaning and system performance is evaluated. **Manufacturer's technical services:** technical assistance, regional panel performance analysis and life cycle cost analysis, federal grant assistance. **Regional applicability:** optional features allow cost effective regional applications. **Availability:** 60 days minimum order time. Contact Solar Energy Products, Inc. for production schedule. **Suggested retail price:** $1,162 to $4,883.

SUN TECH SOLAR HOT WATER SYSTEM STS-100W, STS-100HW

Sun Tech Solar Industries Corp.
P.O. Box 203
Chester, NY 10918
Laurence T. Wansor (914) 469–4212

A hot water system suitable for ground level, roof or wall installations, with air collectors, heat exchangers, storage tank, pump and controllers.

Features: non-freeze system. **Options:** support stand for remote installations, double-glazed, selective surface. **Guarantee/warranty:** 5-yr limited warranty. **Maintenance requirements:** lubrication of pump and air-handler motors and adjustment of belts. **Manufacturer's technical services:** systems and technical design service; computer program available for design, performance and economic analysis. **Availability:** 2 wk. **Suggested retail price:** consult manufacturer.

SUNEARTH DHW PACKAGES

Sunearth Solar Products Corp.
Box 5155A
Montgomeryville, PA 18936
H. Katz (215) 699–7892

Complete domestic hot water systems for residential or light commercial use. Drain-down or antifreeze operation available. Systems designated by number of collectors and size of tank.

Options: drain-down, drain-back, antifreeze. **Installation requirements/considerations:** should be done by persons knowledgeable in hydronics. **Guarantee/warranty:** 10-yr on collectors, 5-yr on tank, 1-yr on other components (copies available on request). **Maintenance requirements:** change anti-freeze or inspection for drain-down systems; pump lubrication. **Manufacturer's technical services:** seminars; design manual available for $3. **Regional applicability:** system can be sized for all regions by number of collectors and tank size. All systems are automatically freeze proof. **Availability:** from stock. **Suggested retail price:** $1100 to $2000.

SUNSYSTEM I, II, III

Solar American Corp.
106 Sherwood Drive
Williamsburg, VA 23185
R.J. Pegg (804) 874–0836

The Sunsystem series of solar water heaters are open-loop systems designed for use in a variety of climates. Fail-safe freeze protection is provided by an automatic drain-down cycle.

Features: direct transfer of solar energy to potable water with no losses due to heat exchangers or heat transfer fluids. **Options:** 52-gal (I), 66-gal (II) or 80-gal (III) tank. **Installation requirements/considerations:** none. Manual comes with each kit for proper installation. **Guarantee/warranty:** 5-yr on collector; 1-yr on electrical components. **Maintenance requirements:** infrequent cleaning of window. **Availability:** approximately 3 wk. **Suggested retail price:** $1650 (I).

VANTECH HOT WATER SYSTEMS

Vantech Corp.
P.O. Box 26790
San Diego, CA 92126
John van Geldern (714) 271–7933

Non-tracking, concentrating, double-glazed collector with own design sensor/controller and Grundfos stainless-steel circulating pump. Designed to be retrofitted to existing hot water heaters without need for heat exchangers. Comes with mounting hardware, installation manual and schematics. Sells as a kit including collectors, pump, sensor controller.

Features: concentrating reflector troughs are used instead of flat plates. **Options:** ¾-in. tubing for swimming pool heaters. **Installation requirements/considerations:** meets all plumbing code requirements and is recommended to be installed with tube and tank insulation and tempering valves. **Guarantee/warranty:** 5-yr. **Manufacturer's technical services:** plumber/dealer outlets supply technical service and warranty service in Southern California, Nevada and Arizona. **Regional applicability:** Southern California, Nevada and Arizona. **Availability:** from stock when available, no longer than 7 days when backlog exists. **Suggested retail price:** $617.

SPACE HEATING SYSTEMS

BRYANT SOLAR ENERGY SYSTEM

BDP Co.
7310 West Morris St.
Indianapølis, IN 46231
Robert J. Johnson (317) 243–0851

The BDP Solar Energy System is designed to provide optimum indoor space heating and domestic water heating at a minimum installed cost.
Features: complete system design including auxiliary equipment, pumps, blowers, electronic air filters and humidifiers. **Options:** domestic water heating, rock or water storage and swimming pool heating. **Guarantee/warranty:** limited 1-yr warranty on all components and additional limited 5-yr warranty on collectors. **Availability:** 8 wk. **Suggested retail price:** established by local dealer.

CHAMPION SOLAR FURNACE

Champion Home Builders Co.
Solar Division
5573 E. North St.
Dryden, MI 48428
Henry H. Leck (313) 796–2211

System consists of Champion VertaFin air collector, Honeywell solid state controller, air handler, rock bed storage, and reflector shield. Unit is prepackaged and free-standing. Has completely automatic controls.

Installation requirements/considerations: southern exposure. **Guarantee/warranty:** 5-yr warranty on structure; 1-yr warranty on component parts. **Maintenance requirements:** consult manufacturer. **Manuracturer's technical services:** computerized performance project for city where weather data are available. **Availability:** stock. **Suggested retail price:** consult manufacturer.

MINI SYSTEM

Southeastern Solar Systems, Inc.
2812 New Spring Rd.
Suite 150
Atlanta, GA 30339
Joe Cooper (404) 434–4447

A complete, supplementary heating system consisting of solar collectors, a storage tank, pumps, solid-state controls, and a heat extraction grate for the fireplace. The system extracts the heat from the sun and your fireplace then stores it for short-term use on nights and cloudy days.

Features: solar collectors, water circulating fireplace grate, storage tank, solid-state controls. **Options:** additional capacities; double glazing. **Installation requirements/considerations:** southerly exposure for collectors, proper angle. **Guarantee/warranty:** collectors (exclusive of glass), grate, controller, tank, 5-yr; pumps, 1½-yr; switches and other controls, 1-yr. **Manufacturer's technical services:** engineering assistance available from manufacturer. **Regional applicability:** state of Illinois. **Availability:** systems are warehoused in central Illinois at Solarflame Systems, Inc., P.O. Box 99, LeRoy, IL, 61752 (309) 962–2861. **Suggested retail price:** consult manufacturer.

SOLAFERN 600

Solafern, Ltd.
536 MacArthur Blvd.
Bourne, MA 02532
Philip Levine (617) 563–7181

The complete space and domestic hot water heating Solafern-600 system includes collectors, control panel, air handler with heat exchanger, water-storage tanks and controls. Mounting hardware is included.

Options: number of collectors; omit direct space heating mode for lower cost model. **Installation requirements/considerations:** roof structure, duct auxiliary interfaces. **Guarantee/warranty:** 5-yr, collector; 1-yr, system. **Maintenance requirements:** tight joints. **Regional applicability:** northern and midwest United States. **Suggested retail price:** $5560 base.

SOLAR AIRE SOLAR FURNACE, SA96 - SA192

SunSaver Corp.
Box 276
North Liberty, IA 52317
D. Dunlavy (319) 626–2343

This self-contained air collector with pebble-bed storage includes blower, dampers and all controls. Collector is vertical-vane black-painted aluminum with low-iron glass double glazing. Motor is ½ hp. 115-vac, 60-cycle permanent split capacitor; controller is solid-state.

Features: 10-in duct outlets; positive acting screw-operated dampers. **Installation requirements/considerations:** southern exposure; unit to sit within 100 ft of house. **Maintenance requirements:** annual belt check; clean glazing. **Availability:** 7 to 10 days between order and delivery. **Suggested retail price:** $3725 to $5775, FOB Iowa.

SOLAR SPACE HEATING

Conserdyne Corporation
4437 San Fernando Rd.
Glendale, CA 91204
Howard Kraye (213) 246–8408

Solar space heating systems custom designed to meet the requirements of individual installations, depending upon heat load calculations and other variables. All connections necessary for standard installation are supplied. Sun Temp heat transfer fluids used; heat exchangers are by American Appliance (designed by Conserdyne). Collectors are Southwest EnerTech (designed by Conserdyne). System is liquid-to-air space heating.

Features: heat pump interface capability. **Options:** size of tank; heat pump. **Installation requirements/considerations:** sized with regard to passive solar heating potential and heat load calculations. **Guarantee/warranty:** 5-yr on parts. **Manufacturer's technical services:** design and installation. **Regional applicability:** southern California. **Availability:** 3 to 4 wk between order and delivery. **Suggested retail price:** depends on installation.

SOLAROLL HEATING SYSTEMS

Bio-Energy Systems, Inc.
Box 489 Mountaindale Rd.
Spring Glen, NY 12483
Michael F. Zinn (914) 434–7858

This system features EPDM elastomer tubing and extrusions permitting cost-effective, on-site, do-it-yourself design and installation.

Features: low cost per sq ft. **Options:** liquid or air, with or without heat-pump, choice of hot-air, convection, or radiant heat distribution, glazing, size, dimensions. **Guarantee/warranty:** 5-yr limited warranty against defects in materials and workmanship and against cracking or leaking of EPDM due to weathering or heat aging. **Manufacturer's technical services:** design engineering, comprehensive manual and support literature. **Availability:** 2-wk. **Suggested retail price:** consult manufacturer.

SOLARON SOLAR ENERGY SYSTEMS

Solaron Corp.
300 Galleria Tower
720 South Colorado Blvd.
Denver, CO 80222

Air-type solar collector panels integrated into a system using all matched components from the same manufacturer to include air handlers, controls, heat exchanger and all installation accessories. Manufacturer also provides full engineering manual for sizing, layout and installation of entire system.

Features: Space heating direct air-to-air, air-to-water heat exchanger provides domestic hot water. **Options:** Solaron system is used for space heating, domestic water heating, industrial process heating, agricultural drying process fluid heating and outside air heating. **Installation requirements/considerations:** factory training provided for Solaron dealers who are licensed mechanical contractors. **Guarantee/warranty:** 10-yr limited performance guarantee on collectors; 1-yr parts and workmanship on other components. **Maintenance requirements:** Filter change, motor lubrication, fan belt adjustment. **Manufacturer's technical services:** full design and engineering staff; extensive factory training distributors have local hardware availability. **Suggested retail price:** contact local dealer.

SPACE HEATING SYSTEM

General Energy Devices, Inc.
1751 Ensley Ave.
Clearwater, FL 33516
Clyde Bouse (800) 237–0137

These systems are designed to provide 50 to 70 percent annual heating load to residential or commercial buildings. Exchanger system consists of Solatron flat-plate liquid collector, with aluminum Solar Bond absorber and fiberglass casing; circulation pumps, expansion tank, balance valves, and DT-110 differential controller. Water-filled storage tank will store up to 5 million Btu's for use at night and cloudy days.

Guarantee/warranty: 5-yr warranty on collectors; 1-yr warranty on electrical components. **Maintenance requirements:** pumps to be oiled annually. **Manufacturer's technical services:** F-Chart computer program analysis of system requirements; system design. **Availability:** 1 to 2 wk between order and delivery. **Suggested retail price:** according to installation; consult manufacturer.

SUN STONE SYSTEMS

Sun Unlimited Research Corp.
P.O. Box 941
Sheboygan, WI 53081
Glenn F. Groth (414) 452–8194

Systems include collectors, blowers, controls, and dampers. Also included are rock storage or modular canned distilled water storage and ductwork, electric, and plumbing hookups. Also available are heat exchangers and storage tanks for hot water.

Options: collectors can be roof or ground mounted; domestic hot water preheater. **Installation requirements/considerations:** must be dealer installed to validate warranty. **Guarantee/warranty:** 1-yr on parts and labor for entire system; 1-yr on labor and 10-yr on parts for collectors. **Maintenance requirements:** change air filters; inspect dampers and blowers annually; wash collector glass when necessary. **Manufacturer's technical services:** architectural design; solar engineering and design. **Availability:** 3 to 4 wk. **Suggested retail price:** contact dealers.

SUN TECH SOLAR STS-100M, STS-100C

Sun Tech Solar Industries Corp.
P.O. Box 203
Chester, NY 10918
Laurence T. Wansor (914) 469–4212

A unitized, modular system suitable for space and hot water heating system in 20,000-Btu/hr-or-less design conditions. Ground level installation requires 25 to 30 man hours.

Features: integral ducting; after assembly, only final utility connections required. **Options:** double glazed; selective surface. **Guarantee/warranty:** 5-yr limited warranty. **Maintenance requirements:** lubrication of motors and adjustment of fan belts as recommended. **Manufacturer's technical services:** systems and technical design assistance; computer program available for design performance and economic analysis. **Availability:** 2-wk. **Suggested retail price:** consult manufacturer.

SUNPUMP SOLAR THERMAL ENERGY SYSTEM

Entropy Limited
5735 Arapahoe Ave.
Boulder, CO 80303
Hank Valentine (303) 443–5103

Sunpump collector modules vaporize water within heat-pipe absorber; steam is self-transported to heat exchanger/storage tank where it recondenses and transfers latent heat of vaporization to storage medium (water).

Features: no moving parts; no auxiliary electrical equipment needed for operation; distilled water by-product; collector modules are vented through a filter. **Options:** distilled water may be recirculated to collector modules. **Installation requirements/considerations:** collectors may be mounted on horizontal, vertical, or sloping surfaces (roof or ground); connection to storage by small-diameter pipe. **Guarantee/warranty:** 1-yr limited warranty for materials and workmanship. **Maintenance requirements:** occasional external cleaning of collector windows. **Manufacturer's technical services:** total system design; technical/installation training to distributor/dealers. **Regional applicability:** not practical in parts of Pacific Northwest and Alaska. **Availability:** 30 to 90 days, depending upon quantities. **Suggested retail price:** consult manufacturer; varies with installation type.

SUN*TRAC COMPLETE HEATING SYSTEM (S96, S128, S160)

Future Systems, Inc.
12500 W. Cedar Drive
Lakewood, CO 80228
Bill Thompson (303) 989–0431

The Sun*Trac solar furnace is a completely factory-assembled, unitized structure containing an air collector, air handler, rock storage chamber, and control system. Normally ground mounted, the structure is retrofitted easily (but not limited to) existing forced air heating systems without major alterations. For domestic hot water, accessory external heat exchanger is Dunham Bush, Model SPW47 or SPW60 with hydronic coil. Reflector is a fin air collector: fins are .019-in aluminum.

Features: ground mounted, unitized (air collectors, air handler and rock storage assembled into one unit), reflector augmented collection. **Options:** add on domestic hot water preheat system as accessory. **Installation requirements/considerations:** ground space with southern exposure as close to dwelling as possible; interfaces best with existing forced air systems; adaptability to retrofit. **Guarantee/warranty:** 5-yr warranty on materials and workmanship on structural components (except glass and reflector surfaces); 1-yr on mechanical components. **Maintenance requirements:** normal maintenance of one ½-hp electric motor; annual check of seals on collectors. **Availability:** 3 to 4 wk between order and delivery. **Suggested retail price:** S96, $3990; S128, $4490; and S160, $4990.

SWIMMING POOL HEATERS

AM-SOL COMPLETE POOL HEATING SYSTEM

American Solar Power Inc.
715 Swann Ave.
Tampa, FL 33606

A swimming pool heating system that uses medium-temperature collectors for high efficiency and long life, reducing the cost over the long term. Collectors are patented with a unique design.

Features: copper collectors. **Options:** heating or cooling modes; white or brown enclosures. **Guarantee/warranty:** 5-yr on collectors, except glass. **Maintenance requirements:** wash glass covers occasionally. **Manufacturer's technical services:** limited to advice on installation practices and siting advisability. **Regional applicability:** nonfreezing regions. **Availability:** immediate. **Suggested retail price:** $2700 and up.

AQUASOLAR POOL HEATER

Aquasolar, Inc.
1232 Zacchini Ave.
Sarasota, FL 33578
Gerald J. Zella (813) 366–7080

A low-temperature, high-volume solar pool heater. The collector is an open-air tube system, that uses 1.5 in. internally finned black Cycolac tubes that are laid out in a serpentine design.

Features: easy installation for new or existing pools. **Options:** can be fully automatic. **Installation requirements/considerations:** adaptable to most situations. **Guarantee/warranty:** tubes and fittings are guaranteed for 10-yr. **Maintenance requirements:** drain during prolonged freezing. **Manufacturer's techncial services:** installation manual. **Availability:** 10 working days outside Florida and California. **Suggested retail price:** $1400 (not installed).

FAFCO SOLAR SWIMMING POOL HEATERS

Fafco, Inc.
235 Constitution Drive
Menlo Park, CA 94025
Alex Battey (415) 321–3650

Flat plate solar collector for swimming pool and other low-temperature applications.

Features: a closed system; no separate pipes needed to return water to the pool; panels can withstand all the pressures a pool pump can exert. **Options:** a fully automated control system that can maintain water at the desired temperature level. **Installation requirements/considerations:** factory-trained crew can usually install the entire system in less than a day. **Guarantee/warranty:** 10-yr pro-rated warranty on materials and labor. **Maintenance requirements:** panels should be drained in the winter areas where the temperature drops below 25° F. **Manufacturer's technical services:** factory-trained dealer-installers. **Availability:** approximately 3 wk depending on location. **Suggested retail price:** consult manufacturer.

SEPI SWIMMING POOL HEATER

Solar Energy Products, Inc.
Mountain Pass
Hopewell Junction, NY 12533
B. R. Kryzaniwsky (914) 226–8596

A do-it-yourself swimming pool heating system designed around a modular collector using all-copper absorber plates.

Features: automatic operation with solid-state controller and solenoid valves. **Options:** glazed or unglazed collectors. **Guarantee/warranty:** 5-yr on collector assembly. **Availability:** 3 to 4 wk. **Suggested retail price:** $245 to $295.

S.I. SWIMMING POOL HEAT SYSTEM

Solar Industires, Inc.
Monmouth Airport Industrial Park
Farmingdale, NJ 07727
Norman Reitman (201) 938–7000

A complete system, including collector(s) and controls.

Features: not subject to corrosion or scale accumulation. **Options:** electronic differential controls with heating and cooling logic; solar collector mounting racks. **Installation requirements/considerations:** connects to pool filtration system. **Guarantee/warranty:** 10-yr limited warranty on collector, 1-yr on controls. **Manufacturer's technical service:** application engineering; technical publications. **Availability:** from stock. **Suggested retail price:** consult manufacturer.

INDOOR POOL HEATING SYSTEM, SUNMAT-70

Calmac Manufacturing Corp.
150 S. Van Brunt St.
Englewood, NJ 07631
John Armstrong (201) 569–0420

The SUNMAT-70 System is specifically designed to heat indoor pools. The system features a non-metallic collector, which is connected directly into the pool line as in conventional summer pool heating systems. But the collector is glazed, so it operates efficiently at winter-time temperatures and is also freeze-tolerant.

Features: lightweight, freeze-tolerant collector with non-metallic absorber. **Options:** none. **Installation requirements/considerations:** collector mounts directly on roof or other supporting structure. **Guarantee/warranty:** 5-yr limited warranty on collector. **Regional applicability:** distribution only in Northeast. **Availability:** 3 to 4 wk. **Suggested retail price:** depends on size.

SOLAROLL

Bio-Energy Systems, Inc.
Box 489
Mountaindale Rd.
Spring Glen, NY 12483
Michael F. Zinn (914) 434–7858

Flexible, roll-out EPDM elastomer solar collector designed to use a minimum of plumbing connections to create continuous collectors up to 150-ft long for permanent or roll-up-and-store systems. SolaRoll is the only non-metallic collector material that will tolerate 400° F. It can be glazed for year-round use and conforms to unusual shapes and bends around obstructions.

Features: UV resistant. **Options:** copper or plastic headers. **Installation requirements:** relatively even and sound surface required if adhesive mounting desired. **Guarantee/warranty:** 5-yr limited warranty against defects in materials and workmanship or cracking or leaking of EPDM material due to weathering or aging; other components carry OEM warranty of manufacturer. **Main-**

tenance requirements: minimal; manual and repair splice kit included. **Manufacturer's technical services:** design engineering, comprehensive and support literature. **Availability:** 2 wk. **Suggested retail price:** under $3/sq ft for all components needed.

SOLAR POOL/SPA HEATING

Conserdyne Corp.
4437 San Fernando Rd.
Glendale, CA 91204
Howard Kraye (213) 246–8408

Custom design and installation of swimming pool heating systems include collectors, controls, sensors, hardware and pump. The number of collectors to be used depends on the heat requirements of the pool and the number of months the pool is to be heated. All collectors are made of aluminum and copper tubing. Systems are designed for 70 percent of the swim season.

Options: number of collectors to heat pool to desired temperature. **Installation requirements/considerations:** desired pool temperature and number of months pool is to be heated. **Guarantee/warranty:** 5-yrs. **Manufacturer's technical services:** design and installation. **Regional applicability:** Southern California. **Availability:** 3 to 4 wk between order and delivery. **Suggested retail price:** average system, $3000.

SOL-HEATER 10-101 AND 10-102

Solar Research Systems
3001 Redhill Ave. I-105
Costa Mesa, CA 92626
Dr. Joseph Farber (714) 545–4941

Unglazed, low-temperature panels made by extruding a special long-life polypropylene with a serrated upper surface and containing nearly 200 channels per 4 ft width; tested to 50 psi; accessories and automatic controls; for use in swimming pool or with solar-assisted heat pumps.

Features: basic material tested in Arizona sun for over 9-yr with no deterioration. Designed for high-efficiency heat transfer. **Options:** panels are 4 × 10 ft or 4 × 8 ft. **Installation requirements/considerations:** can be installed horizontally or on angle with only consideration being drainage. **Guarantee/warranty:** 5-yr limited warranty. **Manufacturer's technical services:** engineering staff will help in system design and installation and operating problems. **Availability:** 30 days from receipt of order. **Suggested retail price:** $140 and $120.

SOLAR SWIMMING POOL HEATER

Solar Living, Inc.
P.O. Box 12
Netcong, NJ 07857
Richard Bonte (201) 691–8483

This system consists of all-copper solar collectors, differential thermostat, and electric bypass valves; system operation is fully automatic.

Features: panels can be custom sized. **Options:** kit or fully assembled models. **Guarantee/warranty:** 2-yr. **Manufacturer's technical services:** technical assistance. **Availability:** 30 days between order and delivery. **Suggested retail price:** approximately $10 sq ft; consult manufacturer.

SOLATRON POOL HEATING SYSTEM

General Energy Devices, Inc.
P.O. Box 5679
Clearwater, FL 33516
Clyde Bouse (800) 237–0137

This swimming pool heating exchanger system, designed to raise pool temperature 10 to 15° F, includes a flat plate liquid collector with aluminum Solar Bond absorber and fiberglass case, aquastats, circulating pump on collector loop, expansion tank and balance valves. May be installed in conjunction with or independent of existing filter system.

Guarantee/warranty: 5-yr warranty on collector; 1-yr warranty on electrical. **Maintenance requirements:** circulating pump should be oiled annually. **Manufacturer's technical services:** system sizing. **Availability:** 1 to 2 wk between order and delivery. **Suggested retail price:** according to installation; consult manufacturer.

Solar Components

ABSORBER COATINGS

ENERSORB[R] ABSORBER COATING

DeSoto, Inc.
1700 S. Mt. Prospect Rd.
Des Plaines, IL 60018
Kenneth Lawson (312) 391–9000

Enersorb[R] is a durable, flat-black, non-selective absorber coating easily applied with conventional spray equipment, air dry or bake.

Features: absorptivity, 96 to 97 percent. **Options:** Super Koropon[R] Primer recommended for achieving best adhesion and corrosion protection. **Installation requirements/considerations:** applied with conventional paint spray equipment, air or baked dry. **Manufacturer's technical services:** technical data and application details available; technical service lab available to commercial users. **Availability:** stock. **Suggested retail price:** $21 to $33/gal, depending on volume purchased.

MICROSORB

Refrigeration Research, Inc., Solar Research Div.
525 N. Fifth St.
Brighton, MI 48116
Jerry Kay (313) 227–1151

A special collector coating having high heat absorptivity and relatively low emissivity: $\alpha/\epsilon = 1.56$.

Guarantee/warranty: guaranteed against defects in material and workmanship. **Manufacturer's technical services:** limited assistance available by mail or telephone. **Availability:** 4 to 6 wk. **Suggested retail price:** consult manufacturer.

PERMALOY SOLAR COAT

Permaloy Corporation
P.O. Box 1559
Ogden, UT 84402
Harry G. James (801) 731–4303

This inorganic, crystalline solar coating is applied electrochemically to aluminum and will last 50 yrs in solar system. It has an absorptivity of 91 to 97 percent, and its emissivity range is 64 to 69 percent.

Features: corrosion resistant. **Installation requirements/considerations:** applied only to aluminum sheets and shapes. **Guarantee/warranty:** 25-yr. **Availability:** 4 to 6 wk between order and delivery. **Suggested retail price:** consult manufacturer.

SELEX ABSORBER COATING

Lambda Selective Coatings
580 Alexander Road
Princeton, NJ 08540
Roger Mulock (609) 921–3330

Selex is a wavelength selective coating for either aluminum- or nickel-plated copper absorber plates.

Options: coatings applied at Lambda facilities or application lines designed and installed for high-volume panel manufacturers. **Guarantee/warranty:** 25-yr life design, 10-yr warranty. **Manufacturer's technical services:** consulting services on selective coatings. **Availability:** 1 month between order and delivery. **Suggested retail price:** consult manufacturer.

SUN-IN ABSORBER PAINT

Solar Usage Now, Inc.
350 E. Tiffin St.
Bascom, OH 44809
Joseph Deahl (419) 937–2226

A dark gray, graphite, oil-base absorber paint. Good adhesion to metal and other surfaces without special primer. Absorptivity 90 percent, emissivity 68 percent.

Guarantee/warranty: 5-yr. **Availability:** from stock. **Suggested retail price:** $17.50/gallon.

ABSORBER PLATES

ABSORBER PLATE (LIQUID)

Shelley Radiant Ceiling Co., Inc.
456 W. Frontage Rd.
Northfield, IL 60093
William Shelley (312) 446–2800

This liquid absorber plate of straight or serpentine copper

tube metallurgically bonded with solder onto an .040-in aluminum plate can use any fluid compatible with copper. The absorber surface is coated with two coats of baked enamel paint on back and customer's choice of coating on front. Normal flow rate for this closed system is 1/min per eight panels.

Features: metallurgical bond between aluminum sheet and copper tube. Custom sizes in quantity. **Design flow rate and pressure drop:** 1 gpm per eight panel (2 × 4 ft) circuit; pressure drop approximately 12 ft of water column. **Options:** copper or steel sheets in lieu of aluminum. **Installation requirements/considerations:** all interconnections made with copper couplings (½-inch ID) or ½-inch type-L as female. **Guarantee/warranty:** 1 to 5 yr, depending on application. **Maintenance requirements:** water treatment for copper tube. **Manufacturer's technical services:** design and application engineering. **Availability:** 6-wk delivery after tooling. **Suggested retail price:** approximately $6/ft.

CHEVRON ABSORBER PLATE

Mid-West Technology
P.O. Box 26238
Dayton, OH 45426
Vern L. Huffines (513) 837–8551

This is an abosrber plate made of light-gauge aluminum for use in flat plate, air-cooled collectors.

Features: the .005-in thick aluminum is pressed into a 1.5-in deep sawtooth pattern that increases the absorbing area and assists heat transfer by causing a more turbulent air flow. Standard size is 3 × 7 ft. Available in natural finish only; experiments are underway with various coatings. **Availability:** 30 days. **Suggested retail price:** 1 to 10, $75 each; 11 to 25, $40; 26 to 50, $30; 51 to 100, $20; 101 to 500, $15; 501 to 1000, $12; over 1000, $10.

COPPERQUEEN

Solarkit of Florida, Inc.
1102 139th Ave.
Tampa, FL 33612
Wm. Denver Jones (813) 971–3934

An absorber plate of continuous copper tubing soldered in a serpentine pattern to a copper sheet, which is grooved to wrap partially around the tubing. A high-temperature, flat-black enamel coating is applied after copper is thoroughly cleaned and chemically treated.

Options: available unpainted or with any coating of your choice. **Guarantee/warranty:** 1-yr limited warranty. **Availability:** 2 to 4 wk. **Suggested retail price:** $50.

MEGA SUNPLATE

Mega Engineering
1717 Elton Rd.
Silver Spring, MD 20903
Richard E. Dame (301) 445–1110

This absorber plate of .06-in galvanized steel features copper tubing cemented into stamped troughs for increased plate-to-fluid thermal conductivity and high bond strength at elevated temperatures with 3M Nextel Black Coating.

Options: copper plate with selective coating. **Guarantee/warranty:** 1-yr with optional extension. **Availability:** 6 to 8 wk between order and delivery. **Suggested retail price:** consult manufacturer.

MR. SUN

Wojcik Industries, Inc.
301 N. Brandon Rd.
Suite 7
Fallbrook, CA 92028
Warren Wojcik (714) 728–0553

Unglazed solar panels designed to heat swimming pools or to preheat up to 140° F large quantities for commercial, agriculture, or industrial uses.

Features: electron accelerator used in the manufacture of the material. **Installation requirements/considerations:** insulated wire forms hold panels. 2 lb/sq ft wet load factor. **Guarantee/warranty:** 5-yr limited guarantee; 1-yr parts and labor, 5-yr parts. **Maintenance requirements:** drain headers during freezing conditions. **Manufacturer's technical services:** system application and design. **Regional applicability:** sun belt. **Availability:** from stock. **Suggested retail price:** $345.

SOLAR BOND[R] ABSORBERS

Olin Brass
East Alton, IL 62024
J.I. Barton (618) 258–2443

For air and liquid solar collectors with intermediate heat exchangers, these tube-in-sheet absorber plates are available in either aluminum or copper (stock and custom design). Height and width dimensions are variable.

Features: integral flow passageways and header system. **Design flow rate and pressure drop:** system design and application dependent; consult manufacturer. **Options:** stock panels or custom design. **Guarantee/warranty:** against defects in workmanship and materials. **Manufacturer's technical services:** custom design, technical literature and data back-up. **Suggested retail price:** consult manufacturer.

SOLAR COLLECTOR PLATE

Refrigeration Research, Inc., Solar Research Div.
525 N. Fifth Street
Brighton, MI 48116
Jerry Kay (313) 227–1151

Collector plates suitable for use with many commonly used refrigerants; designed to combine efficiency with low cost. Custom designs in the construction of collector boxes, insulation and glazing can be accomplished using these plates.

Features: a strong, positive bond is accomplished when the plates are hydrogen-copper brazed. UL approved for the five most common refrigerants. **Installation requirements/considerations:** installation should be made only by a qualified refrigeration service engineer following all local codes. **Guarantee/warranty:** 5-yr, pro rata, limited warranty. **Maintenance requirements:** when using R-11, use only unpressurized grade; the pressurized grade is sold for cleaning systems and is not pure. **Manufacturer's technical services:** limited assistance available by mail or telephone. **Availability:** 4 to 6 wk. **Suggested retail price:** consult manufacturer.

SOLAR COLLECTOR PLATE

Refrigeration Research, Inc.,Solar Research Division
525 N. Fifth Street
Brighton, MI 48116
Jerry Kay (313) 227–1151

Flat collector plates designed to combine efficiency with low cost. Custom designs in the construction of collector boxes, insulation, and glazing can be accomplished when using these plates. Reduced costs may result when several plates are used in one collector case.

Features: a strong, positive bond is accomplished when the plates are hydrogen-copper brazed. UL approved. **Installation requirements/considerations:** suitable collector fluid must be used to protect against freezing and corrosion. **Design flow rate and pressure drop:** 1 gal/sq ft/hr; less than 1 ft water. **Guarantee/warranty:** 5-yr, pro rata, limited warranty. **Maintenance requirements:** must be protected against freezing and corrosion. **Manufacturer's technical services:** limited assistance available by mail or telephone. **Availability:** 4 to 6 weeks. **Suggested retail price:** consult manufacturer.

SOLAROLL

Bio-Energy Systems, Inc.
Box 489
Mountaindale Rd.
Spring Glen, NY 12483
Michael F. Zinn (914) 434–7858

Flexible extrusions of EPDM elastomer are used both as

the absorber plate and the panel enclosure. The T-2 tubing extrusion is the absorber plate, said to be as efficient as copper panels; the GR-1 extrusion holds the glazing, protects the sidewalls and acts as waterproof flashing.

Features: on-site capability allows the freeze-proof material to be pre-assembled to 150-ft lengths, rolled up, then installed in frame enclosure built from local material. **Options:** size; dimensions; choice of glazing, insulation and plumbing connections. **Installation requirements/considerations:** a relatively even and sound surface for adhesive mounting. **Guarantee/warranty:** 5-yr limited warranty against defects in material (and workmanship, if installed by authorized installer) and cracking or leaking of EPDM due to weathering or heat aging. **Maintenance requirements:** minimal; outlined in manual. **Availability:** 2-wks. **Suggested retail price:** under $8/sq ft for all needed components.

SOLAROLL T-2

Bio-Energy Systems, Inc.
Box 489 Mountaindale Rd.
Spring Glen, NY 12483
Micahel F. Zinn (914) 434–7858

Flexible, roll-out EPDM elastomer absorber plates allow fabrication of continuous collectors up to 150-ft long by any width in 4.4-in increments. Collectors may be fastened to substrate with thermosetting mastic, are freeze-proof, and will withstand 375° F continuously.

Features: material may be used as heat exchanger, coiled inside a water tank, or embedded in walls or floors as a means of radiant heating. **Design flow rate and pressure drop:** .2 gpm; .04 psi drop/sq ft. **Options:** may be glazed or unglazed, copper or plastic manifolds, counterflow or one-way flow. **Installation requirements/considerations:** relatively even and sound substrate required for adhesive mounting; may be used in removable installations. **Guarantee/warranty:** 5-yr limited warranty against defects in materials and workmanship. **Manufacturer's technical services:** design engineering, comprehensive manual and support literature. **Availability:** from stock. **Suggested retail price:** $2.27/sq ft (/32.7 in. running).

SOLARSTRIP

Berry Solar Products
P.O. Box 327
Edison, NJ 08817
Calvin C. Beatty (201) 549–3800

Coil-plated, selective black chrome on copper strip and foil. Thicknesses from 0.0014 to 0.016 in. Maximum width 26 in. Nickel underlayment provided. Optical properties: .93 min., .10 max. emissivity. 2,000-lb coils can be handled.

Features: continuous plating versus batch plating; unrestricted length; absorber plate material and selective coating combined together. **Options:** can be formed to make either liquid or air-type absorbers; can be adhesively bonded to wall systems as a passive solar absorber; slitting to small widths. **Installation requirements/considerations:** absorber plate material for collector modules and on-site construction application. **Guarantee/warranty:** consult manufacturer. **Optical properties:** .93 min. absorptivity and .10 max. emissivity. **Manufacturer's technical services:** optical property analysis. **Availability:** 2-wk. **Suggested retail price:** $.70/sq ft.

SOLAVAL

S.P.L. Industries Ltd.
400 West Main St.
Babylon, NY 11702

Stainless steel absorber plate for use by itself for swimming pool heat or to be incorporated in a housing or hot water application. Two waffle sheets of stainless steel; 22 × 69 in. and 44 × 69 in. available dimensions.

Options: both low-pressure and high-pressure envelopes available in various sizes. **Guarantee/warranty:** 5-yr when used as specified. **Availability:** 3 to 6 wk. **Suggested retail price:** pproximately $4/sq ft.

SUNBURST APC-48, APC-410

Sunburst Solar Energy, Inc.
123 Independence Drive
Menlo Park, CA 94025
Brian E. Lanhston (415) 327–8022

Unglazed metal absorber plate. Can be used directly as a pool heater or installed in a box and glazed. All copper construction.

Features: aircraft-type, black epoxy coating for long weatherability. **Design flow rate and pressure drop:** 1.75 psi at 5 gpm. **Options:** Stainless-steel-jacketed neoprene clamps for panel connection in pool heating systems. **Guarantee/warranty:** 5-yr limited; will conform to California Tax Incentive Program. **Availability:** 1 to 3 wk. **Suggested retail price:** APC-410, $250.

SUNLOCK

Approtech
770 Chestnut St.
San Jose, CA 95110
Kent Alfred Dogey (408) 297–6527

The Sunlock absorber plate has hermaphroditically interlocking aluminum extrusions about type-L copper waterway and black chrome or semi-selective paint surface. Glazed or unglazed, these panels are used for flat plate and concentrating collectors.

Features: silver soldered type-L copper tubing, one-piece manifold. **Design flow rate and pressure drop:** consult manufacturer. **Options:** semi-selective paint; panels available in any length up to 10 ft; custom sizes. **Installation requirements/considerations:** use with copper or brass supply; may be integrated into roof skin or wall framing. **Guarantee/warranty:** 5-yr limited warranty. **Maintenance requirements:** periodic assessment of transfer fluid quality. **Manufacturer's technical services:** sizing/installation procedures, microclimate analysis, and payback assessment. **Availability:** 1 to 3 wk between order and delivery. **Suggested retail price:** consult manufacturer.

THRUFLO™

Park Energy Company
Star Route, Box 9
Jackson, WY 83001
Frank D. Werner (307) 733–4950

A thin (.005-in), flat, black aluminum sheet, with very narrow slits, which heats the passing air to within a few degrees of the absorber sheet.

Features: heat transfer nearly as good as liquid-type collectors. **Design flow rate and pressure drop:** 3 cfm/sq ft delta p of .05 in H_20. **Options:** 22 by 24-in interlocking sections or 24-in wide rolls. **Installation requirements/ considerations:** use well filtered air; provide for thermal expansion. **Guarantee/warranty:** 5-yr. **Availability:** from stock. **Suggested retail price:** $3 to $4/sq ft.

TRANTER ECONOCOIL ABSORBER PLATES

Tranter, Inc.
735 East Hazel St.
Lansing, MI 48909
Robert Rowland (517) 372–8410, ext. 243

These wet plate design absorber plates with 90 percent internally wetted surfaces come in standard sizes of 34¼ × 82¼ in. and 34¼ × 94¼ in. in 18- or 20-ga carbon steel for use in closed systems; designed for hot water heating and air conditioning.

Features: wet-plate design with distribution headers. **Design flow rate and pressure drop:** at 6 gpm, pressure drop is about 2 psi. **Options:** other sizes; stainless steel available. **Installation requirements/considerations:** use inhibited fluids. **Guarantee/warranty:** 1-yr against defects in materials and workmanship. **Maintenance requirements:** use of suitable inhibitor (nitrite preferable) with periodic monitoring. **Availability:** stock to 8 wk. **Suggested retail price:** consult manufacturer.

TRUE BOND ABSORBER PLATE

Solar Technologies of Florida
Reynolds Industrial Park
P.O. Box 40485
Jacksonville, FL 32203
Jack W. Hoover (904) 269–3264

The True Bond Absorber Plate consists of 3 × 10 ft of .021-in. copper sheet grooved to accept one half of pipe diameter (½ in. L-type copper tube). All fluid joints are silver soldered. 100 percent capillary flow soldered to plate with 50/50 solder.

Features: water fittings silver soldered. **Design flow rate and pressure drop:** depends on application; consult manufacturer. **Options:** manifold upon request. **Guarantee/warranty:** 5-yr limited warranty. **Manufacturer's technical services:** upon request. **Availability:** consult manufacturer. **Suggested retail price:** consult manufacturer.

AIR HANDLERS

SOLAR AIR BLOWER

BDP Co.
7310 W. Morris St.
Indianapolis, IN46231
Robert J. Johnson (317) 243–0851

The Solar Air Blower has a multi-speed direct-drive motor, washable air filter, and air-filter rack. Has hi-static air-delivery capability.

Options: multi-speed taps for selecting air flow. **Installation requirements/considerations:** consult manufacturer.

Guarantee/warranty: 1-yr limited warranty. **Availability:** 8-wk between order and delivery. **Suggested retail price:** consult manufacturer.

SOLAR AIR DIRECTOR

BDP Co.
7310 W. Morris St.
Indianapolis, IN 46231
Robert J. Johnson (317) 243–0851

The Solar Air Director allows switching of the air direction during various modes of operation and reduces field installation costs by combining three sets of dampers into one package. Unit includes dampers, 2-position spring return motor, lined 1-in. insulation, removable cover plate.

Features: matches BDP auxiliary equipment. **Guarantee/warranty:** 1-yr limited warranty. **Maintenance requirements:** consult manufacturer. **Availability:** 8 wk between order and delivery. **Suggested retail price:** consult manufacturer.

UNIVERSAL SWITCHING UNIT USU-A, USU-B

Contemporary Systems, Inc.
68 Charlonne St.
Jaffrey, NH 03452
John C. Christopher (603) 532–7972

Compact, high-efficiency air moving systems which contain all necessary valving to generate the basic solar operating modes: collectors to storage, collectors to living area or storage to living area. Easy to design with and install, they use standard HVAC techniques.

Features: uses GE Energy Saver motor. **Option:** 24-vac or 115-vac air valves; 115-vac or 230-vac drive motor. **Guarantee/warranty:** 1-yr materials and workmanship. **Maintenance requirements:** annual check, belt alignment, belt wear, valve linkage adjustment. **Manufacturer's technical services:** technical design service. **Availability:** 30 days. **Suggested retail price:** A, $2275; B, $850.

COLLECTORS—FLAT PLATE

AMSOLHEAT SOLAR PANEL

American Solar Heat Corp.
7 National Place
Danbury, CT 06810
Joseph Heyman (203) 792–0077

Aluminum case, all-copper absorber plate and tubing, cover sheet of impact-resistant fiberglass. Spray polyurethane insulation of R-10 or better.

Dimensions and net absorber area: Gross, 4 × 8 ft (1.2 × 2.4 m). **Features:** passed HUD ASHRAE 93–77 test; qualifies for HUD solar initiative hot water grant. **Guarantee/warranty:** meets HUD warranty. **Availability:** 28 working days. **Suggested retail price:** $350 for 4 × 8 ft panel.

CHAMBERLAIN SOLAR COLLECTOR PANEL, MODEL 711101

Chamberlain Manufacturing Corp.
845 Larch Ave.
Elmhurst, IL 60126
John E. Balzer (312) 279–3600, ext. 355

Liquid flat plate collectors featuring copper or steel absorbers with selective coatings. Glazing is single, ⅛-in, low-iron, tempered glass. Tested per ASHRAE 93–77 and certified by independent test.

Dimensions and net absorber area: gross, 84¼ × 36¼ × 4⅜ in; net, 21 sq ft. **Features:** can be used as an integral roof system. **Options:** size, glazing, copper or steel absorber plates, black chrome or paint absorber coating. **Installation requirements/considerations:** closed-loop, inhibited system only, series piping not to exceed three units. **Guarantee/warranty:** 5-yr limited warranty. **Maintenance requirements:** system must be properly inhibited. **Manufacturer's technical services:** design assistance, computer analysis, technical support during installation and system start-up. **Availability:** 30 to 60 days. **Suggested retail price:** $8 to $12/sq ft.

CHAMBERLAIN SOLAR COLLECTOR PANEL, MODEL 711301

Liquid flat plate collectors featuring copper or steel absorbers with selective coatings. Glazing is double, ⅛-in, low-iron, tempered glass. Tested per ASHRAE 93–77 and certified by independent test.

Dimensions and net absorber area: gross, 84¼ × 36¼ × 5¹/₁₆ in; net, 21 sq ft. **Features:** can be used as an integral roof system. **Options:** size, glazing, copper or steel absorber plates, black chrome or paint absorber coating. **Installation requirements/considerations:** closed-loop, inhibited system only, series piping not to exceed three units. **Guarantee/warranty:** 5-yr limited warranty. **Maintenance requirements:** system must be properly inhibited. **Manufacturer's technical services:** design assistance, computer analysis, technical support during installation and system start-up. **Availability:** 30 to 60 days. **Suggested retail price:** $8 to $12/sq ft.

CHAMBERLAIN SOLAR COLLECTOR PANEL, MODEL 712101

Liquid flat plate collectors featuring copper or steel absorbers with non-selective coatings. Glazing is single, ⅛-in, low-iron, tempered glass. Tested per ASHRAE 93–77 and certified by independent test.

Dimensions and net absorber area: gross, 84¼ × 36¼ × 4⅜ in, net, 21 sq ft. **Features:** can be used as an integral roof system. **Options:** size, glazing, copper or steel absorber plates, black chrome or paint absorber coating. **Installation requirements/considerations:** closed-loop, inhibited system only, series piping not to exceed three units. **Guarantee/warranty:** 5-yr limited warranty. **Maintenance requirements:** system must be properly inhibited. **Manufacturer's technical services:** design assistance, computer analysis, technical support during installation and system start-up. **Availability:** 30 to 60 days. **Suggested retail price:** $8 to $12/sq ft.

CHAMBERLAIN SOLAR COLLECTOR PANEL MODEL 712301

Liquid flat plate collectors featuring copper or steel absorbers with non-selective coatings. Glazing is double, ⅛-in, low-iron, tempered glass. Tested per ASHRAE 93–77 and certified by independent test.

Dimensions and net absorber area: gross, 84¼ × 36¼ × 5¹/₁₆ in; net, 21 sq ft. **Features:** can be used as an integral roof system. **Options:** size, glazing, copper or steel absorber plates, black chrome or paint absorber coating. **Installation requirements/considerations:** closed-loop, inhibited system only, series piping not to exceed three units. **Guarantee/warranty:** 5-yr limited warranty. **Maintenance requirements: system must be properly inhibited. **Manufacturer's technical services:** design assistance, computer analysis, technical support during installation and system start-up. **Availability:** 30 to 60 days. **Suggested retail price:** $8 to $12/sq ft.

CHAMPION VERTAFIN COLLECTOR, MODEL 1032

Champion Home Builders Co.
Solar Div.
5573 E. North St.
Dryden, MI 48428
Henry Leck (313) 796–2211

Manufactured in 4 × 8-ft models, these flat plate collectors have single or double glazing of low-iron glass, aluminum-fin absorber plate with black paint coating, urethane insulation, and wood-grain aluminum case.

Dimensions and net absorber area: gross, 4 × 8, 8 × 12, 8 × 20 ft; net, 29 to 32 sq ft per collector. **Options:** side or rear ports; reflector shield. **Installation requirements/considerations:** southern exposure. **Guarantee/warranty:** 5-yr warranty on structure; 1-yr warranty on components parts. **Maintenance requirements:** consult manufacturer. **Manufacturer's technical services:** computerized performance projection for any city where weather data is available. **Availability:** stock. **Suggested retail price:** consult manufacturer.

CHAMPION VERTAFIN COLLECTOR, MODEL 2032

Manufactured in 4 × 8-ft modules, these flat plate collectors have single or double glazing of low-iron glass, aluminum fin abosrber plate with black paint coating, urethane insulation, and wood grain aluminum case.

Dimensions and net absorber area: gross, 4 × 8, 8 × 12, 8 × 16, 8 × 20 ft; net aperture area, 29 sq ft/32 sq ft collector. **Options:** side or rear ports; reflector shield. **Installation requirements/considerations:** southern exposure. **Guarantee/warranty:** 5-yr warranty on structure; 1-yr warranty on component parts. **Maintenance requirements:** consult manufacturer. **Manufacturer's technical services:** computerized performance projection for any city where weather data is available. **Availability:** stock. **Suggested retail price:** consult manufacturer.

COLT ALUMINUM PLATE COLLECTOR

Colt, Inc. of Southern California
71–590 San Jacinto Drive
Rancho Mirage, CA 92270
Charles Barsamiam (714) 346–8033

This liquid collector uses an aluminum absorber plate and is insulated with 2 in. of SI-100 spun glass. The case material is structural aluminum with a dark-bronze anodized finish. Temperature limitation of the collector is 600° F. Collector can be mounted directly on framing rafters or installed on a truss support on flat roofs. Glazing material is ⅛-in, low-iron, low-glare glass.

Dimensions and net absorber area: gross, 35 × 96 × 3¼ in; net, 22.43 sq ft. **Features:** flashing adaptor snaps on to collector to facilitate flashing for watertight roof membrane. **Guarantee/warranty:** 5-yr limited warranty. **Manufacturer's technical services:** system engineering, computer generated solar sizing, estimated cost and system payback. **Availability:** 2 to 4 wk between order and delivery. **Suggested retail price:** $281.50.

COLT COPPER PLATE COLLECTOR

This liquid collector uses a copper absorber plate and is insulated with 2 in. of SI-100 spun glass. The case material is structural aluminum with a dark-bronze anodized finish. Temperature limitation of the collector is 600° F. Collector can be mounted directly on framing rafters or installed on a truss support on flat roofs. Glazing material is ⅛-in, low-iron, low-glare glass.

Dimensions and net absorber area: gross, 35 × 96 × 3¼ in; net, 22.43 sq ft. **Features:** flashing adaptor snaps on to collector to facilitate flashing for water-tight roof membrane. **Guarantee/warranty:** 5-yr limited warranty. **Manufacturer's technical services:** system engineering, computer generated solar sizing, estimated cost and system payback. **Availability:** 2 to 4 wk between order and delivery. **Suggested retail price:** $281.

DUAL HYBRID COLLECTOR

Solar Power West
709 Spruce St.
Aspen, CO 81611
Raymond N. Auger (303) 925–4698

Designed for swimming pool heat applications and low-temperature space heating, this collector uses thin plastic-enclosed aluminum channels between each 9.6-in wide absorber module. It contains the water flow within the plastic envelope and uses aluminum for the absorber surface. Collector backing might be plywood or plywood-aluminum laminate in any color. The collectors are self-draining and require no antifreeze or drain-down valves. Each plate is 4 × 8 ft.

Features: modular construction, seamless butt matting with adjacent collectors. **Options:** high-temperature versions for use with glazing and mounting arrangements, also kit available with Mylar liner only. **Installation requirements/considerations:** can be through-nailed to supporting wooden beams or bolted to metallic supports at any point. **Guarantee/warranty:** 10-yr against leaks and corrosion. **Maintenance requirements:** may require repainting in 5yrs. **Manufacturer's technical services:** economic analyses and installation booklets. **Availability:** 60 days between order and delivery; for small orders, from stock. **Suggested retail price: $3/sq ft.**

ENERGY SYSTEMS COLLECTOR, MODEL 1111D

Energy Systems Inc.
4570 Alvarado Canyon Rd.
San Diego, CA 92120
James C. Senn (714) 280–6660

Flat plate liquid solar collector with non-selective surface and single glazing. Absorber is copper tube in aluminum plate.

Dimensions and net absorber area: gross, 36 × 78 × 4⅛ in; net, 16.61 sq ft. **Features:** operates directly with utility water. **Options:** connections left-right or right-left; optional side manifolding headers; internal temperature sensor. **Installation requirements/considerations:** must be protected against freezing by automatic circulation. **Guarantee/warranty:** 5-yr limited. **Maintenance requirements:** occasional glass cleaning. **Availability:** up to 100 from stock, immediate delivery; larger quantities, 6 wk maximum. **Suggested retail price:** $379.

ENERGY SYSTEMS COLLECTOR, MODEL 1111S

Flat plate liquid solar collector with non-selective surface and single glazing. Absorber is copper tube in aluminum plate.

Dimensions and net absorber area: gross, 36 × 78 × 4⅛ in; net, 16.61 sq ft. **Features:** operates directly with utility water. **Options:** connections left-right or right-left; optional side manifolding headers; internal temperature sensor. **Installation requirements/considerations:** must be protected against freezing by automatic circulation. **Guarantee/warranty:** 5-yr limited. **Maintenance requirements:** occasional glass cleaning. **Availability:** up to 100 from stock, immediate delivery; larger quantities 6 wk maximum. **Suggested retail price:** $350.

ENERGY SYSTEMS COLLECTOR, MODEL 1111U

Flat plate liquid solar collector with non-selective surface and single glazing. Absorber is copper tube in aluminum plate.

Dimensions and net absorber area: gross, 48 × 120 in. **Features:** operates directly with utility water. **Options:** connections left-right or right-left; optional side manifolding headers; internal temperature sensor. **Installation requirements/considerations:** must be protected against freezing by automatic circulation. **Guarantee/warranty:** 5-yr limited. **Maintenance requirements:** occasional glass cleaning. **Availability:** up to 100 from stock, immediate delivery; larger quantities, 6 wk maximum. **Suggested retail price:** $323.

ECONOSOL, SOLARCO-45

Alpha Solarco
1014 Vine St., Suite 2230
Cincinnati, OH 45202
M. Uroshevich (513) 621–1243

This flat plate collector has a copper-to-copper, adhesive-bonded absorber plate and low-iron oxide, tempered-glass glazing. Insulation is 2-in glass wool batting; case material is 0.125-in aluminum without contact points to the copper parts. The plate is tempered against hail damage. The overall size of the collector is 78 × 36 in. The EconoSol can be used for space and process heat, domestic hot water, pools, and as a preheat for the Sun Trek ATH.

Dimensions and net absorber area: gross, 78 × 36 in; net, 76 × 34 in. **Features:** high solar transparency glass, black chrome receiver. **Options:** double glazing, special insulations, non-selective finish. **Installation requirements/considerations:** solder connections in field. **Guarantee/warranty:** 90 days on parts, 5-yr limited warranty. **Manufacturer's technical services:** full instructions; computer system evaluations. **Availability:** 2 to 3 wk between order and delivery. **Suggested retail price:** $195 to $285.

ECOTHERM SOLAR COLLECTOR HE-2400A

Horizon Enterprises, Inc.
P.O. Box V
1011 NW 6th St.
Homestead, FL 33030
Ed Glenn (305) 245–5154

An aluminum-framed, copper absorber plate collector, engineered and tested to withstand hurricane-force wind loads.

Dimensions and net absorber area: gross, 36.4 × 96.5 × 3.9 in (24.4 sq ft); net, 22.9 sq ft. **Features:** all-copper construction in interior. **Options:** painted or mill-finish frame. **Guarantee/warranty:** 5-yr on collector. **Availability:** from stock or within 2 wk. **Suggested retail price:** $285.

ENER-TECH 376 AND 377 SERIES COLLECTORS

Southwest Ener-Tech, Inc.
3030 S. Valley View Blvd.
Las Vegas, NV 89102
Gary Halderson (702) 876–5444

A flat plate liquid collector with a stainless steel absorber, black chrome selective coating, ASG low-iron glass outer glazing, Teflon inner glazing, and extruded aluminum housing.

Dimensions and net absorber area: model 376: gross, 20.2 sq ft; net, 17.2 sq ft. Model 377: gross, 31.3 sq ft; net, 26.5 sq ft. **Features:** 85 percent wetted surface; on the job replacement of any component. **Options:** flat black coating (instead of black chrome); Sunadex glass (instead of ASG low iron). **Installation requirements/considerations:** plan for freeze protection and thermal expansion. **Guarantee/warranty:** 5-yr warranty. **Availability:** 4 wk. **Suggested retail price:** approximately $15/sq ft.

FLAT PLATE AIR COLLECTORS, MODEL 170A/SCF

BDP Co.
7310 West Morris St.
Indianapolis, IN 46231
Robert J. Johnson (317) 243–0851

These flat plate air collectors have aluminum absorber plates with baked-on flat black paint, low-iron tempered glass glazing, fiberglass-board insulation, and aluminum frames; can be installed between roof rafters or surface mounted.

Dimensions and net absorber area: gross, 2 × 8, 2 × 12, 2 x 18 ft.; net, 14, 21, 32 sq. ft. respectively. **Installation requirements/considerations:** 170A/SCF can be mounted between roof rafters; consult manufacturer. **Guarantee/warranty:** 5-yr limited warranty. **Maintenance requirements:** consult manufacturer. **Suggested retail price:** consult manufacturer.

FLAT PLATE COLLECTOR

Solarsystems Industries, Inc.
5511 128th Street
Surrey, B.C. V3W 4B5
Erich W. Hoffmann (604) 596–2665

Flat plate collectors featuring large panel size to reduce installation costs. Available 4 ft wide up to 26 ft long, with or without insulation and with or without main header. The main header connects all panels in an array and is integral with the collector; no external connecting piping required.

Dimensions and net absorber area: gross, width 4 ft, length in 2-ft increments 8 to 26 ft; net, approximately 93% of gross. **Installation requirements/considerations:** can be mounted directly to roof sheathing; collector and flashing supplied for raintight surface, replacing the roofing materials. **Guarantee/warranty:** materials and workmanship, 1-yr. **Maintenance requirements:** occasional washing in dry climates. **Manufacturer's technical services:** system design available. **Availability:** 4 to 8 wk. **Suggested retail price:** $13.90/sq ft.

FLAT PLATE SOLAR COLLECTOR-AIR

Sunworks Div., Enthone, Inc.
P.O. Box 1004
New Haven, CT 06508
Floyd C. Perry (203) 934–6301

An air-cooled, factory-assembled, solar energy collector for systems that require high thermal efficiency, long-term performance and minimum installed cost per Btu delivered.

Dimensions and net absorber area: gross, 35½ × 84 × 4 in; net, 18.5 sq ft. **Options:** single or double glazing; three finishes; mounting clips. **Guarantee/warranty:** 5-yr material and workmanship, except glazing. **Availability:** 2-wk. **Suggested retail price:** $430.

HELIOSYSTEMS LIQUID FLAT PLATE COLLECTOR

Heliosystems Corp.
3407 Ross Ave.
Dallas, TX 75204
Gary deLarios (214) 824–5911

This liquid flat plate collector has an Olin Solar-Bond copper absorber plate, one-piece fiberglass and polystyrene-foam insulation (back and sides, respectively). Sturdy, universal mounting brackets are supplied.

Dimensions and net absorber area: gross, 36 × 98 × 6 in; net, 22.67 sq ft. **Features:** extra-heavy insulation. **Options:** selective or non-selective coating; custom mounting brackets; single glaze. **Installation requirements/considerations:** standard roof will carry weight. **Guarantee/warranty:** 5-yr limited warranty. **Maintenance requirements:** must acid-flush potable water systems every 5 yr. **Manufacturer's technical services:** heat loss evaluations and system sizing. **Availability:** 2 to 3 wk between order and delivery. **Suggested retail price:** copper, $400.

INSOLARATOR SA120–125

Specialty Manufacturing Inc.
DBA Insolarator
7926 Convoy Court
San Diego, CA 92111
Frank B. Ames (714) 292–1857

A medium-temperature collector with a copper and aluminum absorber plate, mounted in a monolithic case of high-temperature polycarbonate.

Dimensions and net absorber area: gross, 18.25 sq ft ; net, 17.25 sq ft^2. **Features:** no shear fasteners, no nuts, bolts or rivets. **Options:** single or double glazing. **Guarantee/warranty:** 5-yr limited warranty. **Maintenance requirements:** occasional cleaning of glazing. **Manufacturer's technical services:** full design and engineering services. **Availability:** from stock. **Suggested retail price:** $268.

INSOLARATOR SA130–125U

Specialty Manufacturing, Inc.
DBA Insolarator
7926 Convoy Court
San Diego, CA 92111
Frank B. Ames (714) 292–1857

A low-temperature solar collector for swimming pool heating, this closed system unit incorporates all-copper waterways in an aluminum heat sink.

Dimensions and net absorber area: 30 sq ft , nominal. **Guarantee/warranty:** 5-yr limited warranty. **Availability:** from stock. **Suggested retail price:** $169.95.

HYPERION SOLECTIVE VEE

Hyperion, Inc.
7209 Valtec Ct.
Boulder, CO 80301
John Eatwell, Larry Brand (303) 449–9544

20-ga. galvanized housing; combined fiberglass-urethane insulation; low-iron tempered glass cover;EPDM and neoprene gaskets; silicone sealants; high-efficiency Solective Vee corrugated 24-ga. mild steel with iron-oxide surface.

Dimensions and net absorber area: 77⅞ × 35⅞ × 5$^1/_{16}$ in; net, 17.2 sq ft. **Features:** Solective Vee absorber, spectrally selective iron-oxide surface on corrugated steel. **Options:** available with extruded aluminum or galvanized steel glass holder/flashing, flat mount, off-angle or adjustable mount, union connection fittings, double pane. **Guarantee/warranty:** 5-yr storage tank and collectors, 1-yr other components. **Maintenance requirements:** occasionally rinse dust off collector glazing. **Manufacturer's technical services:** design work for installations, custom collector design and manufacture. **Regional applicability:** greater Southwest and Central US. **Availability:** small quantities shipped immediately or within 5 days; large orders vary, depending on requirements. **Suggested retail price:** $258.71.

ISIS SOLAR PANEL

Impac Corp.
312 Blondeau St.
P.O. Box 365
Keokuk, IA 52632
Duff Decker, Paul Hosemann (319) 524–3304

This solar panel can be installed as a roof, wall, or window mount. Its net absorber area is 19.4 sq ft; absorber plate material is 0.003-in. aluminum with black epoxy paint. Glazing material is acrylic-modified fiberglass reinforced polyester; insulation provided is 2-in. foamed-in-place urethane foam. No ducting or air mover is supplied with the unit.

Dimensions and net absorber area: gross, 38 × 86 × 8 in; net, 19.4 sq ft. **Features:** weighs approximately 50 lb.

Installation requirements/considerations: components for roof mounting must be supplied by contractor. **Guarantee/warranty:** 1-yr limited warranty. **Availability:** 2 to 4 wk between order and delivery. **Suggested retail price:** $465.

MCKIM SOLAR COLLECTORS F-100

McKim Solar Energy Systems, Inc.
1142 East 64th St.
Tulsa, OK 74132
Eric Paschall (918) 749–8896

These collectors are designed for roof or box mounting.

Dimensions and net absorber area: gross, 17.53 sq ft; net, 15 sq ft. **Features:** components and designs may be purchased separately. **Options:** single or double glazing, with or without selective coating. **Guarantee/warranty:** 5-yr, if installed by McKim. **Availability:** 30 days (components). **Suggested retail price:** $7.50/sq ft (components).

MEDIUM TEMPERATURE COLLECTOR

Semco Corp.
1054 N.E. 43rd St.
Ft. Lauderdale, FL 33334
Jeff Prutsman (305) 565–2516

A medium-temperature flat plate liquid collector available with flat black enamel coating or black chrome selective coating. Tempered low-iron glass cover plate. Copper tubes soldered to grooved-fin absorber.

Dimensions and net absorber area: gross, 49 × 121 × 3.5 in; net, 37.8 sq ft. **Features:** approved for use in areas subject to hurrican wind loads. **Options:** Teflon inner glazing. **Installation requirements/considerations:** contact factory. **Guarantee/warranty:** 5-yr limited warranty covering defects in materials and workmanship. **Manufacturer's technical services:** full system sizing and engineering capability. **Availability:** variable, contact factory. **Suggested retail price:** available on request from manufacturer.

MEGA[R] SUNWATT

Mega Engineering
1717 Elton Rd.
Silver Spring, MD 20903
Richard E. Dame (301) 622–4030

This liquid flat plate collector of standard 0.06-in. steel with type-L copper tubes bonded into stamped troughs has an extruded aluminum case and polyurethane foam and fiberglass insulation. Glazing is tempered ASG Water White Glass.

Dimensions and net absorber area: gross 96 × 48 in; net, 29.7 sq ft. **Features:** low-iron glass glazing. **Options:** copper plate with selective coating. **Installation requirements/considerations:** HUD Minimum Properties Studies, NBSIR-77-1272 SMACNA installation standards. **Guarantee/warranty:** 1-yr warranty; optional extension coverage. **Maintenance requirements:** test pH of fluid and correct. **Availability:** 6 to 8 wk between order and delivery. **Suggested retail price:** consult manufacturer.

LOW-TEMPERATURE COLLECTOR

Solar Industries, Inc.
Monmouth Airport Industrial Park
Farmingdale, NJ 07727
Normal Reitman (201) 938–7000

Open plate, low-temperature collector panels for pool heating and solar assisted heat pumps.

Dimensions and net absorber area: gross, 48 × 96 in; net, 31.4 sq ft. **Features:** not subject to corrosion or scale accumulation. **Options:** electronic differential controls, mounting components and racks of different sizes. **Installation requirements/considerations:** panels require supporting surface. **Guarantee/warranty:** 10-yr limited warranty. **Availability:** from stock. **Suggested retail price:** consult manufacturer.

MARK V COLLECTOR

InterTechnology/Solar Corp.
100 Main St.
Warrenton, VA 22186
N.L. Beard (703) 347–9500

A single-glazed liquid collector with all-copper water passages and internal manifolds. Flat black coating and low-iron glass are standard.

Dimensions and net absorber area: gross, 35⅝ × 77⅜ × 5¼ in (19.0 sq ft); net, 17.3 sq ft. **Options:** selective black chromium absorber coating ($\alpha = 0.94$, $\epsilon = 0.1$), water-white glass ($\tau = .91$). **Installation requirements/considerations:** collectors should be sloped to drain and installed with the short dimension horizontal. **Guarantee/warranty:** 5-yr limited warranty. **Maintenance requirements:** normal precipitation is sufficient to keep glazing clean; clean with water if glass becomes excessively dirty. **Manufacturer's

technical services: solar system design and specification services. **Availability:** 4 to 6 wk. **Suggested retail price:** $216.

MIROMIT SOLAR COLLECTOR, MODEL 200

Miromit American Heliothermal Corp.
2625 S. Santa Fe Drive
Denver, CO 80223
Bill Phillips (303) 778–0650

A steel, tube-in-plate design liquid collector with Tabor selective black absorber surface.

Dimensions and net absorber area: gross, 23.3 sq ft; net, 20.5 sq ft. **Options:** end, side or back-piped collectors; 1, 2, or 3 in. of insulation. **Guarantee/warranty: 5-yr limited warranty. Maintenance requirements/considerations:** closed systems recommended; yearly monitoring of pH of heat transfer fluid. **Availability:** from stock. **Suggested retail price:** $385.

NATIONAL SOLAR COLLECTOR PANEL/NSS 3476

National Solar Supply
2331 Adams Drive, N.W.
Atlanta, GA 30318
Sid Stansell (404) 352–3478

Flat plate collector with all-copper absorber plate with copper tubes; double glazed with glass; extruded aluminum frame.

Dimensions and net absorber area: gross, 76.25 × 34.25 in; net, 17 sq ft. **Guarantee/warranty:** 1-yr on material and workmanship (limited to purchase price). **Manufacturer's technical services:** assistance with installation. **Regional applicability:** United States. **Availability:** traveling time only, immediate delivery. **Suggested retail price:** $210.

NORTHRUP FLAT PLATE COLLECTOR, MODEL FP1G

Northrup
302 Nichols Dr.
Hutchins, TX 75141
(214) 225–4291

This is a modular collector suitable for any type of low- or medium-temperature solar system. Collectors may be installed individually or in multiple banks in a variety of configurations. Construction allows for installation in drain-down systems as well as antifreeze solution systems.

Dimensions and net absorber areas: gross, 103$^{9}/_{16}$ in × 37¼ in × 5½ in; net, 98.38 × 32.13 in. **Options:** special collector support frames. **Guarantee/warranty:** yes. **Availability:** stock. **Suggested retail price:** consult manufacturer.

NORTHRUP FLAT PLATE COLLECTOR, MODEL FP2G

Northrup
302 Nichols Dr.
Hutchins, TX 75141
(214) 225–4291

This is a modular collector suitable for any type of low- or medium-temperature solar system. Collectors may be installed individually or in multiple banks in a variety of configurations. Construction allows for installation in drain-down systems as well as antifreeze solution systems.

Dimensions and net absorber area: gross, 103$^{9}/_{16}$ in × 37¼ in × 5½ in; net, 98.38 × 32.13 in. **Options:** special collector support frames. **Guarantee/warranty:** yes. **Availability:** stock. **Suggested retail price:** consult manufacturer.

REDI-MOUNT SOLAR COLLECTOR, Model 3494-1

Columbia Chase Corp.
Solar Energy Div.
55 High St.
Holbrook, MA 02243
Walter H. Barrett (617) 767–0513

The Redi-Mount Solar Collector has a seamless monolithic fiberglass frame and back assembly, an all-copper absorber plate, fiberglass-reinforced polyester glazing, and urethane-foam insulation. The net absorber area is 22.2 sq ft. The collector can be used for domestic hot water heating, space heating, and pool heating.

Dimensions and net absorber area: gross, 44 × 107 × 4½ in; net, 22.2 sq ft. **Features:** full-perimeter mounting flange; lightweight at 62 lb. **Options:** accessory double-glazing; black-chrome coating; special manifold options. **Installation requirements/consideration:** roof- or rack-installed according to manufacturer's instructions. **Guar-**

antee/warranty: 5-yr limited warranty. **Manufacturer's technical services:** installation manual and full engineering provided with system purchase. **Availability:** 2 wk. between order and delivery. **Suggested retail price:** consult manufacturer.

REDI-MOUNT SOLAR COLLECTOR, MODEL 3494-2

Columbia Chase Corp.
Solar Energy Div.
55 High St.
Holbrook, MA 02243
Walter H. Barrett (617) 767–0513

The Redi-Mount Solar Collector has a seamless monolithic fiberglass frame and back assembly, on all-copper absorber plate, fiberglass-reinforced polyester glazing, Teflon inner glazing, and urethane-foam insulation; net absorber area, 22.2 sq ft. Collector can be used for domestic hot water heating, space heating, and pool heating.

Dimensions and net absorber area: gross, 44 × 107 × 4½ in; net, 22.2 sq ft. **Features:** full-perimeter mounting flange; light-weight at 62 lb. **Options:** accessory double-glazing; special manifold options. **Installation requirements/considerations:** roof- or rack-installed according to manufacturer's instructions. **Guarantee/warranty:** 5-yr limited warranty. **Manufacturer's technical services:** installation manual and full engineering provided with system purchase. **Availability:** 2-wk between order and delivery. **Suggested retail price:** consult manufacturer.

ROM-AIRE COLLECTOR

Solar Energy Products Co.
121 Miller Road
Avon Lake, OH 44012
Frank Rom (216) 933–5000

This collector has a corrugated, aluminum absorber plate and black absorptive coating, foam polyisocyanurate insulation, tempered glass cover plate, and an extruded-aluminum frame. Collectors can be installed 2 ft or 4 ft on center between rafters or mounted on roof.

Dimensions and net absorber area: gross, 2 × 12 ft, 4 × 8 ft, 2 × 18 ft. **Options:** collector available in 2 × 12 ft, 4 × 8 ft, or 2 × 18 ft sizes. **Installation requirements/considerations:** 310° F maximum stagnation. **Guarantee/warranty:** 5-yr limited warranty. **Manufacturer's technical services:** complete systems analysis. **Availability:** 6 to 8 wk between order and delivery. **Suggested retail price:** $9.90/sq ft.

SOLAR COLLECTOR 2001

Prima Industries, Inc.
P.O. Box 141
Deer Park, NY 11729
A.L. Gruol (516) 242–6347

A relatively low-cost collector with Olin Roll-Bond absorber plate in copper or aluminum.

Dimensions and net absorber area: gross, 34½ × 97 in (23.24 sq ft); net, 22.66 sq ft. **Features:** removable, refillable desiccant cartridge; reinforced corner capping. **Installation requirements/considerations:** do not use dissimilar metals; do not drop or twist collector. **Guarantee/warranty:** 5-yr against defects in materials or workmanship. **Maintenance requirements:** change heat transfer fluid every 2 yr. **Availability:** 4 to 8 wk. **Suggested retail price:** aluminum, $245; copper, $319.

SOLAR COLLECTOR 5849

Solar Research Div.
Refrigeration Research, Inc.
525 N. Fifth Street
Brighton, MI 48116
Jerry Kay (313) 227–1151

These flat plate collectors combine the low cost of steel with good performance and rugged construction. The steel tubes are hydrogen copper brazed to the steel plate and copper is fused to the steel surface at over 2000° F.

Dimensions and net absorber area: gross, 21⅜ × 73⅝ × 4⅛ in; net, 10 sq ft. **Features:** the only collector with a metallurgical bond of pure copper between the steel tubes and steel plate, resulting in better heat transfer. **Guarantee/warranty:** 5-yr, pro rata, limited warranty on components manufactured by Solar Research, 1-yr on glazing, etc. **Manufacturer's technical services:** limited assistance available by mail or telephone. **Availability:** 4 to 6 wk. **Suggested retail rpice:** $112.55.

SOLAR COLLECTOR PANEL, SCCU-XX

Solar Living, Inc.
P.O. Box 12
Netcong, NJ 07857
Richard Bonte (201) 691–8483

Panel in kit or fully assembled models consists of all copper absorber plate, extruded-aluminum frame, 1 in. urethane insulation, and Kalwall Sunlite glazing.

Dimensions and net absorber areas: gross up to 4 × 20 ft. **Features:** assembled panels ready for mounting with integral mounting clips; kits contain all necessary precut components for assembly. **Options:** models available in sizes up to 4 × 20 ft. **Guarantee/warranty:** 2-yr. **Manufacturer's technical services:** technical assistance. **Availability:** 1-wk between order and delivery. **Suggested retail price:** $5.50/sq ft, kit; $8.50/sq ft, assembled.

SERIES V INTEGRATED WALL/ROOF SOLAR COLLECTOR

Contemporary Systems, Inc.
68 Charlonne St.
Jaffrey, NH 03452
John C. Christopher (603) 532–7972

The CSI Series V is designed as a structural unit, which can be incorporated with ease into a wall or roof. The collectors form a watertight assembly, which is easy to install and maintain. Provides a significant cost saving over conventional systems.

Dimensions and net absorber area: gross, 2 × 12 ft to 20 ft; net, 87% of gross area. **Features:** functions as an active solar collector as well as a structural building component. **Options:** available in any length from 12 to 20 ft. **Installation requirements/considerations:** framing must be designed on 14-in centers. **Guarantee/warranty:** 1-yr on parts and workmanship. **Maintenance requirements:** yearly cleaning; spray finish every 4 to 5 yr. **Manufacturer's technical services:** computer analysis of proposed designs; design service. **Regional applicability:** temperate latitudes. **Availability:** 60 days. **Suggested retail price:** $7.83/sq ft of active area.

SOLAFERN, MODEL 30

Solafern, Ltd.
536 MacArthur Blvd.
Bourne, MA 02532
Philip Levine (617) 563–7181

The Model-30 collector has two-pass ducting, flange and gasketed connections, single, low-iron glass glazing, a selective copper and black chrome absorber plate, and an aluminum enclosure.

Dimensions and net absorber area: gross, 33.9 sq ft; net, 30 sq ft. **Features:** installed in continuous rows with inlet and return on same end. **Options:** sizes and number of collectors; collector sensor assembly; installation hardware. **Guarantee/warranty:** 5-yr warranty. **Maintenance requirements:** maintain tight joints. **Regional applicability:** northern and midwestern United States. **Availability:** 4 to 6 wk. **Suggested retail price:** $450.

SOLAR COLLECTOR, MODEL 8-D

Solar Systems by Sun-Dance, Inc.
13939 N.W. 60th Ave.
Miami Lakes, FL 33014
Thomas L. Abell (305) 947–4456

This flat plate collector can be used for domestic hot water and swimming pools. Glazed with Sun-Lite Premium II fiberglass-reinforced polyester sheeting, the absorber plate is 36-ga. copper tooling foil with ¾-in. copper tubing. Insulation is urethane 210 CPR; case material is gray dipped galvanized steel; sealant is Top Seal Plastic Asbestos (water-proof).

Dimensions and net absorber area: gross, 39.33 sq ft. **Features:** ¾-in. copper tubing completely soldered to flat copper absorber plate. **Options:** thermosyphon or mechanical systems for hot water and swimming pool heating. **Installation requirements/considerations:** varies depending on area according to longitude and latitude. **Guarantee/warranty:** limited 5-yr. **Maintenance requirements:** occasional washing. **Suggested retail price:** $450.

SOLAR COLLECTOR-REFRIGERANT, MODEL 5821

Solar Research Div.
Refrigeration Research, Inc.
525 N. Fifth Street
Brighton, MI 48116
Jerry Kay (303) 227–1151

Flat plate collectors for use with the most commonly used refrigerants combine the low cost of steel with good performance and rugged construction. The steel tubes are hydrogen copper brazed to the steel plate and copper fused to the steel surface.

Dimensions and net absorber area: gross, 21⅜ × 73⅝ × 4½ in; net, 10 sq ft. **Features:** freezing, the corrosion problems are eliminated; UL approved for use with five most common refrigerants. **Installation requirements:** should be installed by a qualified refrigeration-service engineer following all local codes. **Guarantee/warranty:** 5-yr pro-rata, limited warranty on Solar Research manufactured components, 1-yr on glazing material, etc. **Maintenance requirements:** when R-11 is used, use only unpressurized grade to avoid toxicity. **Manufacturer's**

technical services: limited assistance by mail or telephone. **Availability:** 4 to 6 wk. **Suggested retail price:** $112.55.

SOLAR-KAL AIRHEATER/ COLLECTOR

Kalwall Corp., Solar Components Div.
P.O. Box 237
Manchester, NH 03105
Scott F. Keller (603) 668–8186

Designed for space heating, this air collector has a Sunlite fiberglass-reinforced polymer cover sheet and a 0.032-in black coated aluminum absorber plate. Spacers are aluminum and the collector is cased in an aluminum I-beam frame which is mill finished. Collector operates in either thermosyphon or power-assisted modes; air flows from cold-air intake into panels through in-line duct ports and through the panel on both sides of the absorber plate, thus removing and distributing the heat to the building spaces. The Solar-Kal can be easily fastened outside existing walls or roofs or installed as the wall itself.

Dimensions and net absorber area: gross, 47⅜ × 95½ × 2¾ in; net, 30 sq ft. **Features:** operates in thermosyphon or powered modes. **Options:** double-glazed cover; two double options available. **Installation requirements/considerations:** hardware available; contact manuracturer. **Guarantee/warranty:** contact manufacturer. **Maintenance requirements:** cleaning and field applied weatherable surface coating at 10-yr intervals suggested. **Manufacturer's technical services:** customer requirements discussed. **Availability:** 4 to 6 wk between order and delivery. **Suggested retail price:** $192.

SOLAR PAK II

Chicago Solar Corp.
1772 California St.
Rolling Meadows, IL 60008
T. Crombie (312) 358–1918

Designed to supplement daytime heating, the Solar Pak II is a packaged solar system that consists of an inflatable collector, flexible connecting ducts, duct insulation, hardware, and an air handling unit. The air handling unit can be installed in a standard window unit and consists of the mounting plate, squirrel-cage fan, hot-air inlet and thermometers that indicate air temperature entering the collector and in the room. No component assembly is required. Absorber plate is black polyethylene or vinyl; glazing is clear polyethylene, vinyl, or other. UV inhibitors are added to glazing materials.

Dimensions and net absorber area: 100-, 200-, or 300-sq ft active area. **Features:** easy to install; lightweight; inflatable with single blower. **Options:** black polyethylene or vinyl, or other glazing. **Installation requirements/considerations:** orient south and attach to frame on 9-in. centers. **Guarantee/warranty:** 1-yr, limited. **Maintenance requirements:** suggest removal and storage in summer. **Manufacturer's technical services:** dealer installation instructions and technical backup. **Regional applicability:** continental United States, except Southwest. **Availability:** stock. **Suggested retail price:** $459.50.

SOLARIS

Thomason Solar Home, Inc.
609 Cedar Ave.
Fort Washington, MD 20022
Jack Thomason, Jr. (301) 839–1738

The Thomason collector has a fiberglass base of insulation in an insulating (wood) frame, corrugated aluminum absorber with proprietary coating system, distributor pipe at top and hot water collector trough at bottom and glass glazing.

Options: safety glass, double glazing, custom-made sizes, extra insulation. **Guarantee/warranty:** warranted by the Licensee who manufactures the collector (usually 5-yr warranty). **Maintenance requirements:** none, for general usage. **Availability:** 30 to 60 days from most Licensees. **Suggested retail price:** $5 to $8/sq ft.

SOLARISTOCRAT

Approtech
770 Chestnut St.
San Jose, CA 95110
Kent Alfred Dogey (408) 297–6527

This collector has an aluminum/copper interlocking absorber, outgassed silicon seals, binderless fiberglass insulation, Water White or low-iron glass glazing, and a black chrome selective surface. Case is anodized aluminum.

Dimensions and net absorber area: gross, 48 × 78 in; net, 23.44 sq ft. **Features:** anchor bracket permits panel to swing through 60-degree arc for seasonal adjustment. **Options:** semi-selective paint; glass (low-iron or Water White); clear, bronze or black anodized housing; collector shape. **Installation requirements/considerations:** install on 48-in centers. **Guarantee/warranty:** 5-yr limited warranty. **Maintenance requirements:** monitor transfer fluids. **Manufacturer's technical services:** sizing and installation instructions, microclimate analysis, and payback assessment. **Availability:** 1 to 3 wk between order and delivery. **Suggested retail price:** $288.

SOLAR-PAK DG-18P

Raypak, Inc.
31111 Agoura Rd.
Westlake Village, CA 91361
A. Boniface (213) 889–1500

Collectors with copper waterways, aluminum heat absorbing surface; balanced grid waterway for minimum pressure drop and efficient heat transfer. Fully compatible with water or aqueous glycol solutions.

Dimensions and net absorber area: gross, 20.7 sq ft; net, 17.3 sq ft. **Features:** floated absorber, heavy-duty insulation, replaceable desiccant. **Guarantee/warranty:** 1-yr materials and workmanship, 5-yr against galvanic corrosion. **Availability:** from stock. **Suggested retail price:** $400.

SOLAR-PAK SG 18P

Raypak, Inc.
31111 Agoura Rd.
Westlake Village, CA 91361
A. Boniface (213) 889–1500

Collectors with copper waterways, aluminum heat absorbing surface; balanced grid waterway for minimum pressure drop and efficient heat transfer. Fully compatible with water or aqueous glycol solutions.

Dimensions and net absorber area: gross, 37½ × 79½ × 3¾ in; net, 17.3 sq ft. **Features:** floated absorber, heavy-duty insulation, replaceable desiccant. **Guarantee/warranty:** 1-yr materials and workmanship, 5-yr against galvanic corrosion. **Availability:** from stock. **Suggested retail price:** $340.

SOLAR PANEL CUS 30

Gulf Thermal Corp.
629 17th Ave. W.
Bradenton, FL 33505
Dudley Slocum (813) 748–3433

A flat plate collector with copper tubes brazed to copper headers; extruded aluminum framewall. Overall size, 48½ × 98½ in.

Dimensions and net absorber area: gross, 48½ × 98½ in. (33.17 sq ft); net, 45½ × 95½ in (30.17 sq ft). **Features:** all components accessible through cover with standard hand tools. **Options:** custom mounting systems for fixed or adjustable installations; copper or aluminum fins on absorber tubes. **Guarantee/warranty:** 5-yr limited warranty. **Maintenance requirements:** occasional exterior cleaning. **Availability:** 2 to 5 wk. **Suggested retail price:** $336.

SOLARAY, MODEL 02

SolaRay, Inc.
324 S. Kidd St.
Whitewater, WI 53190
Robert K. Skrivseth (414) 473–2525

All-aluminum construction for corrosion resistance, coupled with an absorber plate scientifically designed for efficient heat transfer to the air flow.

Dimensions and net absorber area: gross, 48 × 143 in; net, 42.8 sq ft. **Options:** 4 × 8 ft, 4 × 10 ft, 4 × 12 ft or custom-designed sizes available; tempered glass available at extra cost. **Guarantee/warranty:** 5-yr limited warranty. **Maintenance requirements:** wash with mild detergent once a year and rinse; keep system air filters clean. **Manufacturer's technical services:** complete professional systems engineering. **Regional applicability:** especially designed for cold northern climates. **Availability:** 60 to 90 days. **Suggested retail price:** $420.

SOLARPACK COLLECTOR

Solar Devices, Inc.
G.P.O. Box 3727
San Juan, Puerto Rico 00936
Frank Casa (809) 783–1775

The flat aluminum collecting plate of the Solarpack is forced-fit to copper fluid pipes, and the collector surface is treated with granular black coating. The collector cover is a flat sheet of low-iron-content glass; case material is extruded aluminum.

Dimensions and net absorber area: gross, 34 × 76 × 3½ in; net, 17½ sq ft. **Guarantee/warranty:** 5-yr. **Manufacturer's technical services:** designing of application systems for customers. **Regional applicability:** tropical and subtropical climates. **Availability:** consult manufacturer. **Suggested retail price:** $245.

SOLARQUEEN SOLAR COLLECTOR

Solarkit of Florida, Inc.
1102 139th Ave.
Tampa, FL 33612
Wm. Denver Jones (813) 971–3934

A streamlined collector, 3 × 5 ft over-all, 50 lb, features a solid-copper absorber plate, aluminum backplate, closed-cell insulation, integral aluminum leg mounts and a pre-sealed redwood case.

Dimensions and net absorber area: gross, 36 × 60 × 3½ in; net, 14 sq ft. **Features:** no auxiliary mounting brackets, frames or roof reinforcing required for most installations. **Options:** single or double glazing, length of mounting legs, location of tubing entry and exit from case. **Installation requirements/considerations:** pre-drilled 2-in. leg at each corner of the case allows the collector to be bolted directly to a roof or any sturdy frame. **Guarantee/warranty:** 5-yr limited warranty against defects in materials or workmanship. **Maintenance requirements:** minimal; recoating with polyurethane every 5-yr recommended. **Availability:** 1 to 4 wk. **Suggested retail price:** $189 to $239.

SOLARVAK

Solarsystems Inc.
507 West Elm St.
Tyler, TX 75702
Robert F. Faulkner (214) 592–5343

Evacuated flat plate collector. Copper absorber plate with black chrome selective coating and copper tubing attached in a serpentine pattern on back. Absorber plate suspended between 0.250-in.-thick acrylic housings; self draining; weight, 160 lb.

Dimensions and net absorber area: gross, 4 × 8 ft × 5 in. **Features:** by suspending absorber plate in a vacuum, convective currents are suppressed to obtain higher efficiencies and higher temperatures. **Guarantee/warranty:** 1-yr. **Maintenance requirements:** check vacuum first 9 mo. until experienced pattern is established. **Manufacturer's technical services:** design recommendations; collector technical data. **Availability:** 15 to 20 days. **Suggested retail price:** $525 FOB Tyler, TX.

SOLECTOR LIQUID COLLECTOR

Sunworks Division
Enthone, Inc.
P.O. Box 1004
New Haven, CT 06508
Floyd C. Perry (203) 934–6301

A liquid cooled, factory assembled solar energy collector for systems that require high thermal efficiency, long-term performance and minimum installed cost per Btu delivered.

Dimensions and net absorber area: gross, 35½ × 84 × 4 in; net, 18.68 sq ft. **Options:** two sizes, drain-down or internal manifold; single or double glazing; three finishes; mounting clips. **Guarantee/warranty:** 5-yr material and workmanship except glazing. **Availability:** 2 wk. **Suggested retail price:** $405.

SOLERA COLLECTOR

Western Energy, Inc.
454 Forest Ave.
Palo Alto, CA 94302
Norman Rees (415) 327–3371

This packaged unit consists of a copper absorber plate, serpentine copper water passage, extruded-aluminum frame, 3/16 in. tempered glass as single glazing, and urethane insulation. Nailing strip is built-in around the frame.

Dimensions and net absorber area: gross, 24 sq ft; net, 21 sq ft. **Features:** no internal solder connections. **Options:** horizontal or vertical panels; internal manifolding. **Installation requirements/considerations:** four PSF loading, predrilled nailing flange around collector. **Guarantee/Warranty:** 6-yr warranty. **Maintenance requirements:** clean glass occasionally. **Manufacturer's technical services:** complete engineering and system design service. **Regional applicability:** Northern California. **Availability:** 4 to 6 wk between order and delivery. **Suggested retail price:** $288.

SOLERGY FLAT PLATE COLLECTOR, MODEL SG-100, SG-200

Solergy Co.
7216 Boone Ave. N.
Minneapolis, MN 55428
Vince Grimaldi (612) 535–0305

These flat plate collectors consist of a single layer of 0.040-in. Kalwall Premium II, thermally isolated Olin aluminum or copper absorber plates, 0.060-in. aluminum extrusions with a silicone adhesive. Net absorber area is 22.7 sq ft.

Dimensions and net absorber area: gross, 35¼ × 97¾ × 3 in; net, 22.7 sq ft. **Features:** installation kit available from manufacturer, lightweight and easy to retrofit. **Options:** aluminum or copper plate, black chrome. **Installation requirements/considerations:** instructions provided; Suntemp fluid only to be used with aluminum plate collectors. **Guarantee/warranty:** 5-yr limited warranty.

Maintenance requirements: coat glazing with Kalwall weatherable surface every 5-yr. **Manufacturer's technical services:** installation instructions; design and engineering services. **Availability:** 4 to 6 wk between order and delivery (FOB). **Suggested retail price:** consult manufacturer.

SOLARAY, MODEL 04

SolaRay, Inc.
324 S. Kidd St.
Whitewater, WI 53190
Robert K. Skrivseth (414) 473–2525

Corrosion-resistant aluminum enclosure with efficient corrosion-resistant copper absorber plates and high-impact acrylic/glass covers.

Options: Olin Brass Solar-Bond or Terra-Light, Inc. copper absorber panels may be specified. **Guarantee/warranty:** 5-yr limited warranty. **Maintenance requirements:** wash with mild detergent once a a year and rinse; monitor heat transfer fluid if non-permanent type is used for acidity. **Manufacturer's technical services:** complete professional system engineering. **Regional applicability:** especially designed for cold northern climates. **Availability:** 60 to 90 days. **Suggested retail price:** $485.

SUN SAVE LIQUID COLLECTOR PANEL

Northern Solar Power Co.
311 South Elm St.
Moorhead, MN 56560
Bruce Hilde (218) 233–2515

Panels are double-glazed with Teflon and Kalwall Sunlite II; aluminum or copper Roll-Bond absorber is painted with 3M Nextel. 1½-in. urethane foam back and edge insulation. 1-in. fiberglass batt behind absorber. The envelope is aluminum coil stock with an enameled finish. Tedlar tape is used for an air-tight seal.

Dimensions and net absorber area: gross, 37 × 99 × 5 in; net, 22.5 sq ft. **Features:** effective area: 22.5 sq ft; 70 lb. **Options:** color of exterior envelope. **Installation requirements/considerations:** for retrofit installation with mounting brackets. Mounting brackets not supplied (galvanized angle suggested). **Guarantee/warranty:** 5-yr. **Manufacturer's technical services:** engineering and installation. **Regional applicability:** northern United States. **Availability:** 1-mo between order and delivery. **Suggested retail price:** $288.

SOLITE COLLECTOR

Solar Alternative, Inc.
30 Clark St.
Brattleboro, VT 05301
Jim Kirby (802) 254–6668

This lightweight collector has a copper absorber plate and tubing, Kalwall Sunlite Premium II glazing, fiberglass-board insulation and aluminum-sheet casing. Construction allows the collector materials to expand and contract independently as the collector heats and cools.

Dimensions and net absorber area: gross, 24 × 5 × 104 in; net, 15.4 sq ft. **Features:** lightweight design (2 lb/sq ft). **Installation requirements/considerations:** consult manufacturer. **Guarantee/warranty:** 5-yr limited warranty. **Maintenance requirements:** heat transfer fluid must be checked for pH annually. **Availability:** 3 wk between order and delivery. **Suggested retail price:** $135.

SUN SPONGE

A-1 Prototype, Inc.
1288 Fayette St.
El Cajon, CA 92020
Jerry Hull (714) 449–6726

4 × 8 ft flat plate collector, double glazed with polycarbonate covers; aluminum exterior and flat-black coated absorber plate; 1-in. headers, copper manifold.

Dimensions and net absorber area: gross, 4 × 8 ft; net, 30 sq ft. **Features:** lightweight. **Options:** sensor mounted in panel. **Installation requirements/considerations:** reverse-return plumbing design; flow per panel, 2½ gpm for medium-temperature applications, 4½ gpm for low-temperature applications. **Guarantee/warranty:** 5-yr. **Maintenance requirements:** resurface absorber plate at approximately 5-yr intervals. **Manufacturer's technical services:** design assistance. **Availability:** 2 wk. **Suggested retail price:** $260.

SUN-AID, MODEL 132, 211, 322

Revere Solar and Architectural Products, Inc.
P.O. Box 151
Rome, NY 13440
(315) 328–2401

The Revere Sun-Aid collector is factory assembled in a weathertight enclosure. It features a Tube-in-Strip copper absorber plate. The tubes are an integral part of the plate.

Dimensions and net absorber area: gross, 77 × 35 × 4½ in; net, 17.2 sq ft. **Options:** different glazing and absorber coating. **Guarantee/warranty:** yes. **Availability:** stock. **Suggested retail price:** consult manufacturer.

SUN-AIRE SOLAR AIR COLLECTORS

Gem Manufacturing Co.
Star Route No. 18
Bascom, OH 44809
Joseph Deahl (419) 937–2225

A solar air collector with an extruded-aluminum frame, corrugated aluminum absorber plate with a black, absorptive coating, polyisocyanurate insulation and single, tempered-glass cover.

Dimensions and net absorber area: gross, 25⅛ × 146⅞ × 3⁵⁄₁₆ in; net, 19.9 sq ft. **Features:** drop-in roof mount. **Options:** duct locations, frame colors. **Installation requirements/considerations:** 24-in. rafter spacing. **Guarantee/warranty:** 10-yr. **Availability:** 4-wk. **Suggested retail price:** $280.

SUNBURST BG-48

Sunburst Solar Energy, Inc.
123 Independence Dr.
Menlo Park, CA 94025
Brian E. Langston (415) 327–8022

A single-glazed flat plate collector with copper tube-aluminum fin absorber in with a baked enamel finish, an aluminum box, insulated with 1-in foam. Single glazed with Glasteel.

Dimensions and net absorber area: gross, 48 × 96 in; net, 45.6 × 94 in. **Features:** lightweight. **Guarantee/warranty:** 5-yr limited; will conform to California Tax Incentive program. **Availability:** 1 to 3 wk. **Suggested retail price:** consult manufacturer.

SUNBURST BG-410

Sunburst Solar Energy, Inc.
123 Independence Dr.
Menlo Park, CA 94025
Brian E. Langston (415) 327–8022

A single-glazed flat plate collector with copper tube-aluminum fin absorber in with a baked enamel finish, an aluminum box, insulated with 1-in foam. Single glazed with Glasteel.

Dimensions and net absorber area: gross, 48 × 120 in; net, 45.6 × 118 in. **Features:** lightweight. **Guarantee/warranty:** 5-yr limited; will conform to California Tax Incentive program. **Availability:** 1 to 3 wk. **Suggested retail price:** $400.

SUNCATCHER H-1

Solar Unlimited, Inc.
4310 Governors Drive, W.
Huntsville, AL 35805
Larry Frederick (205) 837–7340

A double-glazed collector with serpentine design, all-copper waterways.

Dimensions and net absorber areas: gross, 35⅜ × 77⅜ in (19 sq ft); net, 16.7 sq ft. **Features:** patented vertically finned absorber plate; Sunadex anti-reflective glazing. **Options:** mounting kits available. **Guarantee/warranty:** 5-yr against defects in materials or workmanship. **Availability:** from stock. **Suggested retail price:** $380.

SUNCATCHER H-2

A single-glazed collector with serpentine design, all-copper waterways.

Dimensions and net absorber area: gross, 35⅜ × 77⅜ in (19.0 sq ft); net, 16.7 sq ft. **Features:** patented vertically finned absorber plate; Sunadex anti-reflective glazing. **Options:** mounting kits available. **Guarantee/warranty:** 5-yr against defects in materials or workmanship. **Availability:** from stock. **Suggested retail price:** $280.

SUNCEIVER II

Halstead and Mitchell
P.O. Box 1110
Scottsboro, AL 35768
Troy Barkley (205) 259–1212

A flat plate liquid collector: 35⅛ × 77⅜ × 4 in. deep; net glass area 17.2 sq ft; double glazed with two ⅛-in. Sunadex tempered glass covers; 1½-in. Foamglas insulation in back; enclosed by 0.051-in. aluminum frame. Absorber is aluminum-finned copper tube coated with Nextel Velvet Black, 0.002-in. coating.

Dimensions and net absorber area: gross, 37 × 77 × 4 in; net, 17.2 sq ft. **Guarantee/warranty:** 5-yr. **Maintenance requirements:** none. **Availability:** stock to 6 wk. **Suggested retail price:** $380.

SUNDEC™

Direct Energy Corp.
16221 Construction Circle West
Irvine, CA 92714
Karl E. Sterne (714) 552–6211

This low-temperature collector is made of concrete; its

panels are 32 × 64 in. and can be laid down as part of the deck around the pool. The panels are also light enough to be placed on the roof and can be used as part of a wall structure. Glazing is optional and not normally supplied.

Dimensions and net absorber area: gross, 32 × 64 in; net, 30 × 62 in. **Options:** glazing accessory available for heating spas. **Installation requirements/considerations:** should be installed by competent pool deck subcontractor. **Guarantee/warranty:** 10-yr guarantee. **Regional applicability:** Southern California. **Suggested retail price:** $80/ panel.

SUNEARTH SOLAR COLLECTOR/ MODEL 3597

Sunearth Solar Products Corp.
Box 515SA
Montgomeryville, PA 18936
H. Katz (215) 699–7892

Flat plate, double-glazed, copper water-way collector with internal manifolding. Primary use: residential water heating. Incorporates mounting systems and plumbing systems designed for easy installation. Distributed through plumbing and heating wholesalers or authorized solar distributors.

Dimensions and net absorber area: gross, 32.5 × 97.5 × 4 in; net, 19.78 sq ft. **Features:** lightweight, 75-lb filled per 22 sq ft; all interconnections supplied to complete headering; no soldering. **Options:** limited to attachment hardware for roof pitches and sensor placement for system control. **Installation requirements/considerations:** should be done by persons knowledgeable in hydronics. **Guarantee/warranty:** 10-yr, limited to inspection of connections and periodic cleaning in dusty areas. **Manufacturer's technical services:** design manual available for $3. **Regional applicability:** designed for northeast United States, but suitable in all climate areas. **Availability:** quantities up to 50 available from stock. **Suggested retail price:** $297.

SUNERATOR

Solar Energy Systems, Inc.
One Olney Ave.
Cherry Hill, NJ 08003
Nathan E. Brussels (609) 386–7700

A line of flat plate solar collectors designed to be used with both closed and drain-down systems.

Options: single or double glazing; flat black or black chrome coating. **Guarantee/warranty:** 2-yr; 5-yr against corrosion. **Availability:** 2-wk. **Suggested retail price:** consult manufacturer.

SUNLINE COLLECTOR

Vulcan Solar Industries, Inc.
200 Conant St.
Pawtucket, RI 02680
J. Michael Levesque (401) 725–6061

This medium-temperature solar collector can be used for domestic hot water, space and swimming pool heating systems. It has an all-copper absorber plate with grid-pattern copper tubing spaced 2 in. on center and is housed in a one-piece fiberglass frame that uses Filon glazing. The full-perimeter mounting flange is an integral part of the unit, and the collector is offered with internal manifolding if desired.

Dimensions and net absorber area: gross, 32.1 sq ft; net, 22.2 sq ft. **Features:** lightweight (60 lb); one-piece fiberglass enclosure, fully insulated, full-perimeter mounting flange. **Options:** double glazing and selective coating available; also choices of absorbers and glazings as well as internal manifolding. **Insulation requirements/ considerations:** complete installation manuals are available; recommend installation by a certified installer. **Guarantee/warranty:** HUD requirements met. **Maintenance requirements:** schedule is offered with system. **Manufacturer's technical services:** available as needed. **Regional applicability:** all regions, but depends on transfer medium. **Availability:** 2 to 3 wk between order and delivery. **Suggested retail price:** $260.

SUNMAT

Calmac Manufacturing Corp.
150 S. Van Brunt St.
Englewood, NJ 07631
John Armstrong (201) 569–0420

The Sunmat is a single-glazed, non-selective, liquid solar collector. It features the use of a non-metallic absorber system. It is lightweight, comes in lengths to 25 ft and will allow water to freeze in it without damage.

Dimensions and net absorber area: gross, from 32 to 100 sq ft; net, from 30 to 95 sq ft. **Features:** non-metallic, EPDM absorber is key to the design. **Options:** condensation dryer system. **Installation requirements/ considerations:** mounts directly on roof or other supporting structure. **Guarantee/warranty:** 5-yr limited warranty. **Regional applicability:** distribution only in Northeast. **Availability:** 3 to 4 wk. **Suggested retail price:** $13/sq ft.

SUNNY DAYSTAR COLLECTOR, MODEL 21-B

Daystar Corp.
90 Cambridge St.
Burlington, MA 01803
C. Greely (617) 272–8460

This collector is designed and engineered for applications where output temperature requirements will be less than 60° F above average ambient condition and is used for solar assisted water-to-air heat pumps, heated swimming pools and domestic hot water systems in non-freezing locations. Coverplate is single-glaze, low-iron glass over a copper absorber plate/serpentine-tube system; insulation is isocyanurate foamed in place, 1 in. on the sides and 1½ in. on bottom; and the aluminum case is 5¼ in. thick (.032-in. aluminum under baked polyester enamel). Silicone adhesive is used. Recommended antifreeze fluid should be potable 60/40 propylene glycerol/water. Temperature limiter system will maintain internal collector temperatures below 270° F regardless of ambient conditions.

Dimensions and net absorber area: gross, 80¾ × 44½ × 5¼ in; net, 21 sq ft effective aperture. **Features:** single glaze. **Installation requirements/considerations:** tilt angle must exceed 15 degrees above horizontal. **Guarantee/warranty:** 5-yr. **Maintnenace requirements:** bi-annual check. **Manufacturer's technical services:** system design and technical assistance. **Suggested retail price:** $433.

SUNNY DAYSTAR COLLECTOR, MODEL 21-C

This collector is designed and engineered for applications where output temperature requirements will be less than 60° F above average ambient condition and is used for solar assisted water-to-air heat pumps, heated swimming pools and domestic hot water systems in non-freezing locations. Coverplate is single glaze, low-iron glass over a copper parallel tube; insulation is isocyanurate foamed in plate, 1 in. on sides and 1½ in. on bottom; and the aluminum case is 5¼ in. thick (.032-in. aluminum under baked polyester enamel). Silicone adhesive is used. Recommended antifreeze fluid should be potable 60/40 glycerol/water or 60/40 propylene glycerol/water.

Dimensions and net absorber area: gross, 72¾ × 44½ × 5¼ ; net, 21 sq ft effective aperture. **Features:** single glaze. **Installation requirements/considerations:** tilt angle must exceed 15 degrees above horizontal. **Guarantee/warranty:** 5-yr. **Maintenance requirements:** bi-annual check. **Manufacturer's technical services:** system design and technical assistance. **Suggested retail price:** $292.

SUNPANEL, MODEL 120, 121, 220, 221

Libby Owens Ford Co.
1701 East Broadway
Toldeo, OH 43605
Lloyd E. Bastian (419) 247–4350

An all-copper absorber collector for use with water, aqueous glycol solutions or silicone fluids. The extruded aluminum case has welded corners, no screws or bolts, and is weather sealed; 3 in. of fiberglass insulation.

Dimensions and net absorber area: gross, 36 × 84 × 4¾ in. **Features:** integral handles on case function as hold-down brackets when collector is installed. **Options:** clear float or low-iron glass, selective or non-selective absorber coating. **Guarantee/warranty:** 1-yr limited warranty. **Availability:** 6 to 8 wk. **Suggested retail price:** $12 to $14/sq ft.

SUN-SERT AIR COLLECTOR PANEL

Northern Solar Power Co.
311 South Elm St.
Moorhead, MN 56560
Bruce Hilde (218) 233–2515

The Sun-Sert is an air-collector module installed between stud walls or roof joists (2 × 6 in., 16 in. o.c.). Panels are double glazed with Teflon and Kalwall SunliteII. The flat black absorber has air circulation on the back side between the fiberglass batt and the aluminum absorber.

Dimensions and net absorber area: gross, 14 × 93 × 3 in; net, 9 sq ft. **Features:** lightweight and easy to install between stud walls or roof joists. **Options:** flat plate or porous fiberglass absorber. **Installation requirements/considerations:** should be installed by building contractor on new construction. **Guarantee/warranty:** 5-yr. **Manufacturer's technical services:** installation and engineering of air controller. **Regional applicability:** northern United States. **Availability:** 2-wk between order and delivery. **Suggested retail price:** $60.

SUNSTREAM™ SOLAR COLLECTOR, MODEL 60A

Grumman Energy Systems, Inc.
4175 Veterans Memorial Highway
Ronkonkoma, NY 11779
Arthur L. Barry (516) 575–2549

Model 60A aluminum collectors are suitable for residential and commercial applications requiring heating or preheating temperatures below 140° F; designed to satisfy low/moderate temperature requirements for domestic hot water, solar-assisted heat-pump space heating and swimming pools.

Dimensions and net absorber area: gross, 40.75 × 110.5 × 9 in; net, 24.46 sq ft. **Features:** lightweight aluminum frame, acrylic covers; low shutdown temperature. **Options:** aluminum or copper fluid passages. **Installation requirements/considerations:** can be mounted on roof or on separate support structures. **Guarantee/warranty:** 5-yr on components and parts, including labor, shipping, removal and reinstallation costs. **Maintenance requirements:** replace antifreeze every 2 yr if used. **Manufacturer's technical services:** thermodynamic sizing analysis and design support. **Availability:** stock. **Suggested retail price:** $281.

SUNSTREAM™ SOLAR COLLECTOR, MODEL 60F

For heating and preheating temperatures below 140° F, the Sunstream Model 60F liquid collectors feature acrylic glazing, copper Finplank absorber plates, fiberglass insulation, and aluminum frames coated with white enamel. Gross collector size is 31.27 sq ft.

Dimensions and net absorber area: gross, 110½ × 40¾ × 108 in. **Features:** low shutdown temperature. **Installation requirements/considerations:** separate support structure and mounting provided by dealer. **Guarantee/warranty:** 5-yr warranty. **Maintenance requirements:** replace Sunstream antifreeze every 2-yr (if used). **Manufacturer's technical services:** thermodynamic sizing analysis and design support. **Availability:** stock. **Suggested retail price:** $322.

SUNSTREAM™ SOLAR COLLECTOR, MODEL 100F

Model 100F series solar collectors are used for domestic hot water, space, process and swimming pool heating. Glazing is tempered glass; absorber plate is copper or copper passages (Finplank); insulation is fiberglass; case material is aluminum extrusions and aluminum sheet. Collector frame can be integrated into waterproof walls and roofs; aperture size is 23.82 sq ft.

Dimensions and net absorber area: gross, 108.75 × 36.75 × 4.44 in; net, 23.82 sq ft aperture. **Features:** welded frame construction with mounting flanges; copper water passages. **Options:** single or double-glazed, painted or black chrome surface; special sizes and plumbing configurations available. **Installation requirements/considerations:** method of installation to roof optional, not supplied. **Guarantee/warranty:** 5-yr on components and parts. **Maintenance requirements:** antifreeze solution must be changed every 2-yr (if used). **Manufacturer's technical services:** thermodynamic sizing analysis and design support. **Availability:** stock from distributors. **Suggested retail price:** 100 F, $305.

SUNSTREAM™ SOLAR COLLECTOR, MODEL 100-2

Model 100-2 series solar collectors are used for domestic hot water, space, process and swimming pool heating. Glazing is tempered glass; absorber plate is copper Roll-bond; insulation is fiberglass; case material is aluminum extrusions and aluminum sheet. Collector frame can be integrated into waterproof walls and roofs; aperture size is 23.82 sq ft.

Dimensions and net absorber area: gross, 108.75 × 36.75 × 4.44 in; net, 23.82 sq ft aperture. **Features:** welded frame construction with mounting flanges; copper water passages. **Options:** single or double glazed. Painted or black chrome surfaces. Special sizes and plumbing configurations available. **Installation requirements/considerations:** method of installation to roof optional, not supplied. **Guarantee/warranty:** 5-yr on components and parts. **Maintenance requirements:** antifreeze solution must be changed every 2-yr (if used). **Manufacturer's technical services:** thermodynamic sizing analysis and design support. **Availability:** stock from distributors. **Suggested retail price:** consult manufacturer.

SUNSTREAM™ SOLAR COLLECTOR, MODEL 200F

The Model 200F solar collector is used for domestic hot water, space, process and swimming pool heating. Glazing is tempered glass; absorber plate is Finplank copper passages; insulation is fiberglass; case material is aluminum extrusion and aluminum sheet. Collector frame can be integrated into waterproof walls and roofs; aperture size is 23.82 sq ft.

Dimensions and net absorber area: gross, 108.75 × 36.75 × 4.44 in; net, 23.82 sq ft. **Features:** welded-frame construction with mounting flanges, copper water passages. **Options:** single or double glazed; painted or black chrome surface; special sizes and plumbing configurations available. **Installation requirements/considerations:** method of installation to roof optional, not supplied. **Guarantee/warranty:** 5-yr on components and parts. **Maintenance requirements:** antifreeze solution must be changed every 2-yr if used. **Manufacturer's technical services:** thermodynamic sizing analysis and design support. **Availability:** stock from distributors. **Suggested retail price:** $390.

SUN*TRAC

Future Systems, Inc.
12500 W. Cedar Dr.
Lakewood, CO 80229
Bill Thompson (303) 989–0431

This extended-surface-area type collector has a blackened aluminum vertical fin absorber plate and $^{3}/_{16}$-in. tempered glass glazing. Insulation is 2½-in. polyurethane in back and 2-in. spun glass sides and back. The case is ⅛-in. fiberglass-reinforced plastic. Absorber area is approximately 165 sq ft. One to three collectors are commonly used in a series; applications include space heating, drying, and if coupled with a hydronic coil, domestic hot water preheat.

Dimensions and net absorber area: gross, 98 × 50 × 12 in. **Features:** vertical fin extended-surface-area collector. **Options:** single or double glazed. **Installation requirements/considerations:** mounting position provided at back side of collector. **Guarantee/warranty:** 5-yr on materials and workmanship. **Maintenance requirements:** annual check of seals; repairs with silicone caulk if needed. **Availability:** 3 to 4 wk between order and delivery. **Suggested retail price:** $630.

TELLURIDE SOLARWORKS COLLECTOR

Telluride Solarworks
Box 700
Telluride, CO 81435
Dean Randle (303) 728–3303

A hot-air collector designed for integral installation on 24-in. centers, in roof, walls or greenhouses.

Dimensions and net absorber area: gross, 24 in × any length; net, 1.8 sq ft × linear ft. **Options:** configurations available for new or retrofit installations, domestic to commercial scale. Greenhouse unit has no absorber duct. **Guarantee/warranty:** 5-yr. **Maintenance requirements:** washing in dusty climates, occasional refinishing or Tedlar coating. **Availability:** 4 to 6 wk. **Suggested retail Price:** $7/sq ft.

SUNUP COLLECTOR

Solar Technology Corp.
2160 Clay St.
Denver, CO 80211
Richard Speed (303) 455–3309

An air-cooled flat plate collector available in dimensions of 3 × 6.5 ft with double glazing (Thermoglass) or 4 × 8 ft with single-acrylic glazing.

Dimensions and net absorber area: 3 × 6.5 ft with double glass; 4 × 8 ft with acrylic; net area same less approximately 2 in. **Features:** model with acrylic glazing is lightweight, less expensive, and highly resistant to breakage. **Guarantee/warranty:** 5-yr. **Manufacturer's technical services:** engineering design and backup, installation. **Regional applicability:** Colorado and adjacent states. **Availability:** 4 to 8 wk. **Suggested retail price:** $11/sq ft with acrylic glazing, $14/sq ft with glass.

THERMO TREK 36-120UE

TechniTrek Corp.
1999 Pike Ave.
San Leandro, CA 94577
Vernon Butorovich (415) 352–0535

The heart of this circulating-water solar system is an insulated, copper absorber plate with soldered copper tubing and manifolds. Panels have aluminum housing and end plates. Standard panel is 3 × 10 ft.

Dimensions and net absorber area: gross, 38 × 122 × 3 in; net, 30 sq ft. **Options:** 8-ft panels, glazed or unglazed. **Guarantee/warranty:** 5-yr full warranty; 5-yr warranty against absorber system corrosion. **Manufacturer's technical services:** design engineering. **Availability:** small orders from stock, large orders 4 to 6 wk. **Suggested retail price:** $260.

VACU-FLOW COLLECTOR

Pleiad Industries, Inc.
Springdale Road
West Branch, IA 52358
Donald Laughlin (319) 643–5650

Distributed flow, flat plate collector operates under vacuum; absorber plate is stainless steel and glazing is reinforced fiberglass. The collector has fiberglass and polyurethane insulation and 24-ga. steel casing. Slip joint header coupling allows roof installation without plumbing.

Dimensions and net absorber area: gross, 4 × 10 ft; 37.7 sq ft. **Features:** black headers are part of the absorber surface. **Options:** increments of 40 sq ft available. **Installation requirements/considerations:** outdoor feeder pipes require heat tapes. **Guarantee/warranty:** 5-yr limited warranty. **Maintenance requirements:** periodical inspection and cleaning of glazing. **Manufacturer's technical services:** solar systems design with the Vacu-Flow. **Avail-**

ability: stock. **Suggested retail price:** $11/sq ft (approximate).

YANKEE

Dixon Energy Systems, Inc.
47 East St.
Hadley, MA 01035
Jane Nevin (413) 584–8831

A solar panel for new construction or retrofit which may be surface mounted or recessed; all-copper collector, soldered tubes and plate, double glazed.

Dimensions and net absorber area: gross, 34½ × 76½ in (18.3 sq ft); net, 17.5 sq ft. **Features:** 2-in. Spinglas insulating board has an R=8.33 value. **Options:** aluminum mounting extrusions for surface or recessed mounting. **Guarantee/warranty:** 5-yr warranty (HUD). **Availability:** 3 wk. **Suggested retail price:** $390.

THRUFLO™ SOLAR COLLECTOR

Park Energy Co.
Star Route, Box 9
Jackson, WY 83001
Frank D. Werner (307) 733–4950

A site-assembled, hot-air solar collector that may be assembled in variable lengths and widths, construction is not unitized.

Features: cover is mounted from the underside, eliminating external frames and/or batten strips. **Options:** individual components may be purchased separately. **Guarantee/warranty:** 5-yr. **Maintenance requirements:** air must be filtered. **Availability:** 2 to 4 wk. **Suggested retail price:** typically $8.95/sq ft.

YING SP 4120 COLLECTOR

Ying Manufacturing Corp.
1957 W. 144th St.
Gardena, CA 90249
George E. Home, III (213) 770–1756

This rigid collector utilizes welded steel waterways, 3-in. foamed-in-place insulation and polycarbonate glazing.

Dimensions and net absorber area: gross, 49.5 × 145.5 × 4.1 in (50 sq ft); net, 45.5 sq ft. **Options:** electrodeposition-painted exterior; custom fittings. **Installation requirements/considerations:** full-perimeter mounting flange integral with case. **Guarantee/warranty:** 3-yr full guarantee in California, 5-yr limited warranty elsewhere. **Maintenance requirements:** semiannual washdown of glazing. **Availability:** 2 to 4 wk. **Suggested retail price:** $534.50.

COLLECTORS—CONCENTRATING

ACUREX 3001 COLLECTOR

Acurex Corp.
485 Clyde Ave.
Mt. View, CA 94042
Edward L. Rossiter (415) 964–3200, Ext. 3341

The Model 3001 Concentrating Collector is a reflecting parabolic-trough solar collector designed to heat fluids to temperatures between 140° and 600° F. Typical applications include industrial process hot water and steam, space cooling, and organic Rankine-cycle systems.

Dimensions and net aperture area: each module, 6-ft aperture, 10-ft length; standard assembly is a row of eight modules. **Options:** OVER-TEMPERATURE/NO-FLOW fault detection; assemblies with eight or more modules in a row. **Guarantee/warranty:** 1-yr on workmanship and materials. **Manufacturer's technical services:** warranty service and service contract. **Suggested retail price:** consult manufacturer.

HELIODYNE TRACKING CONCENTRATOR

OMNIUM-G
1815 Orangethorpe Park
Anaheim, CA 92801
Stanley Zelinger (714) 879–8421

This model HTC-25 is a two-dimensional solar imagery parabolic reflector. The unit tracks the sun and focuses the sun rays from a 6-m diameter reflector with a concentration ratio of greater than 10,000 to 1. It provides 25 kw of energy for high-temperature experimentation. (See Electrical Generating Plant under Miscellaneous).

Dimensions and net aperture area: mirror: 6-m diameter, focal length 4 m. **Features:** the unit is servo controlled and is pointed by both sun-seeking and open-loop data. There is a time-sequencing control system for open-loop pointing and retrace control. The unit can withstand hurricane force winds in the stow position and will operate at winds up to 80 km/hr. Experiments ranging up to 500 lb can be placed on the inner gimbal. Additional 750 lb can be placed on the outer gimbal. The 10-ft diameter, walk-in enclosure provides weatherproof housing for instrumentation and other equipment. **Options:** focal-plane experiment mounting hardware. **Installation requirements/considerations:** price includes installation labor provided by company; adequate foundation is required. **Maintenance requirements:** Preventative maintenance - minor cleaning, depending on atmospheric conditions, approximately once a month; regular maintenance - lubrication of gimbals and drive mechanism once a year. **Manufacturer's technical services:** installation services, engineering services, special adaptation for different experimental situations. **Regional applicability:** requires full sunlight for meaningful experimentation. **Availability:** delivery is 90 days ARO. **Suggested retail price:** $26,500.

HEXCEL CONCENTRATOR

Hexcel Corp.
11711 Dublin Blvd.
Dublin, CA 94566
George Branch (415) 828–4200

A parabolic-trough, tracking concentrator, this collector uses air, water or oil as a heat transfer medium. The concentration ratio is 60:1; temperature to 600° F, efficiency is 64% at 300° F; modules from 9 × 20 to 9 × 80 ft.

Dimensions and net aperture area: 9 × 20 to 9 × 80 ft/row; modular to 662 sq ft aperture. **Features:** large modules may be oriented N-S or E-W. **Options:** absorber configuration, material, self-contained control system. **Installation requirements/considerations:** foundation or supports required only every 20 ft. **Guarantee/warranty:** 5-yr. **Maintenance requirements:** minimal; lubrication and cleaning. **Manufacturer's technical services:** computer sizing of field requirements. **Availability:** 4 to 12 wk. **Suggested retail price:** consult manufacturer.

HOT LINE™ COLLECTOR HL 8, HL 10, CUSTOM

Hot Line Solar, Inc.
1811 Hillcrest Drive
Bellevue, NE 68005
Dan Lightfoot (402) 291–3888

Tracking and focusing mechanisms are eliminated from this optical tracking concentrator, which operates from a stationary position. Receives direct and diffuse energy from entire winter sky.

Dimensions and net aperture area: HL 8: 26 × 96 × 10 in. (typical). **Features:** close orientation not necessary due to patented optical tracking. **Installation requirements/considerations:** HVAC skills and methods. **Guarantee/warranty:** 1-yr full, 5-yr limited. **Maintenance requirements:** lubricate motor, keep glazing clean. **Availability:** 4 to 6 wk. **Suggested retail price:** $205 to $305.

MARK IV SOLAR HEATER

Solar Kinetics Corp.
P.O. Box 17308
West Hartford, CT 06117
James A. Pohlman (203) 233–4461

Linear concentrating solar collector, U.S. Patent 3954097 and Foreign Patents. Two-pass absorber tube comprised of two concentric spiral copper pipes. Inner tube preheats fluid and spiral outer shell in final heating stage. Parabolic reflector is highly polished aluminum with a bonded oxide surface to protect it from corrosion backed with a bonded marine fiberglass for rigidity.

Dimensions and net aperture area: 40 × 60 × 12 in. **Features:** weight, 32 lb. **Options:** sun-tracking device. **Installation requirements/considerations:** installation by authorized and trained dealer. **Guarantee/warranty:** 5-yr limited warranty including freight. **Maintenance requirements:** periodic cleaning. **Manufacturer's technical sergices:** training, application/engineering assistance. **Availability:** 2 to 4 wk. **Suggested retail price:** $299.

MEGA® SUNTRAC 40X

Mega Engineering
1717 Elton Rd.
Silver Spring, MD 20903
Richard E. Dame (301) 445–1110

Designed for hot water and hot fluid heating, this concentrating collector is aluminum coated with 3M FEK material and has a thermal shielded copper nickel receiver.

Dimensions and net aperture area: 42-in. wide × variable length in 2-ft increments. **Features:** receiver tube supported through the bearing to minimize joints. **Options:** length variable; color of support frame variable. **Installation requirements/considerations:** HUD minimum property standards, NBSIR-77-1272. **Guarantee/warranty:** 1-yr with optional extension coverage. **Maintenance requirements:** test pH of fluid and correct; inspect and replace seals and test focus. **Availability:** 6 to 8 wk between order and delivery. **Suggested retail price:** consult manufacturer.

QUICK SILVER CONCENTRATING COLLECTOR

Thermal Dynamics, Inc.
2285 Emerald Hts. Court
Reston, VA 22091
M.D. Shell (703) 620–3014

A non-tracking concentrating collector employing reinforced-plastic resin body, first surface silver mirrors with glass resin coating, evacuated tubular absorbers, and glazings (choice of plastic films or glass). There are no metal parts in the glazing system, mirror substrate, or housing. Mounting systems designed and fabricated for each application.

Dimensions and net aperture area: 16 × 54 × 108 in. **Features:** non-corrosive, stationary, durable. **Options:** choice of color and glazing materials. **Installation requirements/considerations:** snap-fit parts; reflective geometry can be modified for maximum concentration at any latitude or orientation angle. **Guarantee/warranty:** 5-yr limited warranty. **Maintenance requirements:** periodic cleaning of glazing. **Manufacturer's technical services:** load and system engineering, computer analysis of system, installation supervision. **Regional applicability:** 50 degrees latitude to equator. **Availability:** 6 mo between order and delivery. **Suggested retail price:** consult manufacturer.

SOLAR COLLECTOR LTC-367

General Solar Systems Division,
General Extrusions, Inc.
4040 Lake Park Road
Youngstown, OH 44507
William Kenney (216) 783–0270

A solar collector designed specifically for the industrial or commercial application which requires process temperatures of 70 to 90° C (158 to 203° F). It is a low concentration ratio (3.67:1), all-aluminum module with 3.04 sq m (32.7 sq ft) capture aperture. Simple mounting requirements, hardware included.

Dimensions and net aperture area: gross, 54 × 120 in. (1.37 × 3.04m); capture aperture, 32.7 sq ft (3.04 sq m). **Features:** requires only infrequent manual re-aiming. **Installation requirements/considerations:** less than 5 lb/sq ft roof loading. **Guarantee/warranty:** 1-yr, materials and workmanship. **Maintenance requirements:** re-aiming of collectors 12 times a year; actual measured time for 100-unit array - 2 men, ½ hour. No other periodic maintenance required. **Manufacturer's technical services:** engineering design assistance included; installation and/or supervision available. **Availability:** 6 to 8 wk. **Suggested retail price:** $625 per module.

SOLAR RANKINE POWER SYSTEM

S.R.D. Corp.
6625 4th St. South
St. Petersburg, FL 33705
Daniel R. White (813) 866–1346

A Rankine concentrating solar energy system primarily for the generation of electricity, 5 kw and up. Waste heat can help provide domestic hot water, space heating or power absorption air-conditioning.

Dimensions and net aperture area: 10 × 36 ft (approximate; custom sized). **Features:** tube tracking, direct evaporation cycle, custom-designed, high-efficiency turbine. Non-tracking collector, adjusted seasonally. **Options:** absorption air-conditioning and other high-temperature waste heat applications. **Guarantee/warranty:** 3-yr. **Maintenance requirements:** adjustment of collector seven times a year, minimal lubrication schedule. **Manufacturer's technical services:** complete design, engineering, manufacturing, installing and servicing. **Regional applicability:** recommended for areas with over 70 percent sunshine. **Suggested retail price:** $3000/kw.

SOLARCO 7 SUN TREK

Alpha Solarco
1014 Vine St., Suite 2230
Cincinnati, OH 45202
M. Uroshevich (513) 621–1243

Sun Trek 7 is a mobile, self-contained, full-tracking concentrator, solar energy research laboratory. Its receiver is enclosed in a variable vacuum chamber, and the unit has a variable flow pump, multipoint automatic temperature recording, a build-in heat exchanger and accumulator. Concentrator mirror is aluminum and plastic; receiver is black chrome. The Sun Trek 7 can be used with heat transfer fluids such as water, silicone, Therminols, ethylene glycol, etc. Solar rate tracking is automatic, and the Sun Trek 7 fully instrumented. Suitable for research, education, and data acquisition. Options and variations available.

Dimensions and net aperture area: 1-sq-m concentrator mirror. **Features:** can be ordered to customer specifications. **Installation requirements/considerations:** completely portable system. **Guarantee/warranty:** 90 days parts, 5-yr limited warranty. **Maintenance requirements:** minor oiling, liquid system flushing. **Manufacturer's technical services:** full consulting, project suggestions, curriculum suggestions. **Availability:** 8 to 12 wk between order and delivery on standard unit; 12 to 16 wk between order and delivery on custom unit. **Suggested retail price:** $3,995 to $8,995.

SOLARCO SUN TREK ATH

Alpha Solarco
1014 Vine St., Suite 2230
Cincinnati, OH 45202
M. Uroshevich (513) 621–1243

This concentrating collector has six parabolic-trough tracking mirrors in array and is designed for power, air conditioning, process heating, heat pumps, turbines and high-temperature applications. The mirrors are protected aluminum and plastics; the receiver is black chrome on steel. Overall mirror size is 10-sq-m (107.6 sq ft), and each array is self-contained. Therminol-60 fluid is used and passes through the concentrators in a cascade manner. The array has automatic diurnal tracking and annual adjustment.

Dimensions and net aperture area: 10-sq-m of total mirror surface. **Features:** automatic acquisition and tracking, specially shaped receiver, and high-temperature synthetic bearings. **Options:** receiver shapes, pumps, receiver materials, glazing, vacuum jackets, 2-axis tracking, photovoltaic-powered drive. **Installation requirements/considerations:** requires flat space, 21 by 10½ ft, to install each array. Welding or hard soldering might be required in installation. **Guarantee/warranty:** 90 days on parts, 5-yr limited warranty. **Maintenance requirements:** periodic inspection of moving parts. In some climates, mirror material might have to be replaced after 5 to 10 yr; consult manufacturer. **Manufacturer's technical services:** full installation instructions. **Availability:** 6 to 8 wk between order and delivery. Suggested retail price: $2300 to $4995.

SOLARCO SUN TREK JR.

This high-temperature parabolic concentrating collector is fully instrumented, has manual tracking, and works with water or Therminol fluids in a "boiler" (sic) mode. It has a built-in inclinometer and its receiver is mounted on removable ceramic rods. The concentrator mirror of aluminum and plastic is 8.8 sq ft. This collector can be used for experimental and educational purposes.

Dimensions and net aperture area: 8.8-sq-ft concentrator mirror. **Features:** 87 percent reflectance mirror, built-in pressure and temperature gauges, automatic pressure relief valve, selective and non-selective receivers. **Options:** tempered glass glazing. **Installation requirements/considerations:** fully portable. **Guarantee/warranty:** 90 days parts. **Manufacturer's technical services:** consulting service; software. **Availability:** 2-wk between order and delivery. **Suggested retail price:** $159 to $199.

SOLARTRON VACUUM TUBE COLLECTOR

General Electric Company
Advanced Energy Programs
P.O. Box 13601
Philadelphia, PA 19101
William F. Moore (215) 962–2112

The Solartron can be used for daytime cooling, process heat, and high-temperature heating. Evacuated tube design uses GE fluorescent lamp tubes fabricated into a "thermos bottle" unit. Thermal energy is removed from glass tubes by independent copper hydronic system. Collector fluid is ethylene glycol/deionized water.

Dimensions and net aperture area: 47¾ × 52½ × 3¼ in. **Options:** installation accessories. **Guarantee/warranty:** 1-yr materials and workmanship. **Manufacturer's technical services:** application and installation assistance. **Availability:** 90 days between order and delivery. **Suggested retail price:** depends on quantity.

SOL-R-BEAM SB-205

Beam Engineering, Inc.
732 N. Pastoria Ave.
Sunnyvale, CA 94086
Benjamin H. Beam (408) 738–4573

Package includes preassembled array of five SB-201 concentrating collectors, complete with insulated header piping, solar tracking linkage, and prefabricated support hardware for mounting at 45° elevation angle on a level surface. Each collector is an aluminum parabolic cylinder with copper focus tube and polyvinyl fluoride cover. Shipping crate is 9 by 15 by 1-ft.

Dimensions and net aperture area: gross, 9 × 15 × 1 ft; aperture total, 68.5 sq ft. **Options:** mounting hardware for other elevation angles or surface contours. **Installation requirements/considerations:** components can be handled by two persons. **Guarantee/warranty:** 3-yr warranty on repair or replacement of defective components. **Maintenance requirements:** none. **Manufacturer's technical

services: systems engineering and consultation; repair and replacement of materials. **Availability:** 6 wk. **Suggested retail price:** $1400.

SOL-R-BEAM CONCENTRATING COLLECTOR SB 201

Packaged kit includes one preassembled 22 × 96-in. aluminum, parabolic cylinder concentrating collector, copper water-piping focus tube, polyvinyl fluoride cover, support frame for mounting at 45° elevation angle on a level surface, 24-vac tracker, 50-ft of electrical cable, relay box for wiring tracker into 110-vac circuit and complete instructions for assembly.

Dimensions and net aperture area: gross, 22 in. × 8 ft 10 in; aperture, 13.7 sq ft (1.27 sq m). **Options:** 220-vac or other supply voltages available. Accessory kits for water heater and hot tub connections available. **Installation requirements/considerations:** kit can be assembled by one person. **Guarantee/warranty:** 3-yr warranty on repair or replacement of defective kit materials. **Maintenance requirements:** none. **Manufacturer's technical services:** systems engineering and consultation, repair and replacement of defective kit materials. **Availability:** 4 wk. **Suggested retail price:** $550.

SUNPOWER LINEAR PARABOLIC CONCENTRATOR, SERIES 200

Sunpower Systems Corporation
510 South 52nd Street
Tempe, AZ 85281
Dan Ikeler (602) 894–2331

The Sunpower Linear Parabolic Concentrator Series 200 is used on applications requiring temperatures under 250° F. Concentrating surface is polished aluminum with copper absorber tube.

Dimensions and net aperture area: nominal, 2 × 10 ft. **Features:** installation and banking flexible; framing system allows as many as 80 collectors to be driven with one drive motor and one tracking system. **Options:** black chrome absorber piping, dual-walled insulated convection shield, elevated racking. **Installation requirements/considerations:** southern-sloped roof (altitude or orbital tracking), flat roof (altitude or orbit), ground mount (orbit). **Guarantee/warranty:** 5-yr limited warranty. **Maintenance requirements:** periodic lubrication of jackscrew. **Manufacturer's technical services:** architectural engineering design consultation. **Regional applicability:** southwestern United States. **Availability:** 2 to 6 wk between order and delivery. **Suggested retail price:** $14 sq ft, including framing and tracking.

SUNPOWER PARABOLIC CONCENTRATOR, SERIES 400

Used on applications requiring temperatures to 400° F, the Series 400 collector has a polished aluminum surface and copper absorber tube. Sunpower Model 3010 tracking device is supplied with concentrator.

Dimensions and net aperture area: 4 × 10 ft. **Features:** flexibility in banking of collectors. **Options:** black chrome absorber, insulated convection shields, elevated racking. **Installation requirements/considerations:** southern-sloped roof (altitude or orbit tracking), flat roof (altitude or orbit), ground mount (orbit). **Guarantee/warranty:** 5-yr limited warranty; 1-yr limited warranty on tracking device. **Maintenance requirements:** periodic lubrication of jackscrew. **Manufacturer's technical services:** architectural engineering, design consultation. **Regional applicability:** southwestern United States. **Availability:** 2 to 6 wk between order and delivery. **Suggested retail price:** $14 sq ft, including framing and tracking.

SUNPOWER SOLAR CAROUSEL

Sunpower Systems Corp.
510 S. 52nd St.
Tempe, AZ 85281
Patricia Matlock, Dan Ikeler, Duane Magni (602) 894–2331

Series 400 collectors are arranged on a framework which rotates 210 degrees daily (for 33.5° N latitude) on a cement or metal ring to track the sun's orbit while the collectors simultaneously track its altitude; allows for nearly continuous solar collection throughout the day with performance in the morning and evening approaching that available during peak collection hours; adaptable to photovoltaic applications with the advantage of continuous exposure of the surface of the cells at 90-degree angle of incidence.

Dimensions and net aperture area: nominal 4 × 10 ft; framework. **Features:** dual-axis tracking; central distribution manifold (patent pending). **Options:** elevated racking to eliminate shade for use with photovoltaic cells. **Installation requirements/considerations:** large flat area (can be ground- or roof-mounted). **Guarantee/warranty:** 5-yr limited parts and labor, 1-yr electronic controls. **Maintenance requirements:** periodic lubrication of jackscrew. **Manufacturer's technical services:** architectural engineering; design consultation. **Regional applicability:** for use primarily in direct-sunshine areas. **Availability:** 2 to 6 wk. **Suggested retail price:** $20/sq ft (including framing and tracking).

SUNPUMP SCM-200 & SCM-201

Entropy Limited
5735 Arapahoe Ave.
Boulder, CO 80303
Hank Valentine (303) 443–5103

Non-tracking, focusing collector (patented); water vaporized within absorber tube and self-transported to storage tank/heat exchanger. Passive system, one square meter (10.8 sq ft) collection aperture; 1.3:1 concentration ratio, double glazing. No auxiliary electrical equipment needed for operation.

Dimensions and net aperture area: 80.5 in. long × 29.7 in. wide × 15.7 in. deep (204 × 75.4 × 39.9 cm). **Features:** system output always has temperature of 200° F minimum. System is self-pumping from collector to storage. **Options:** distilled water may be recirculated from storage to collectors. **Installation requirements/considerations:** collectors may be mounted on horizontal, vertical, or sloping surfaces; on building or on ground-racks. **Guarantee/warranty:** 1-yr limited warranty on materials and workmanship. (Design life: 20-yr with periodic maintenance.) **Maintenance requirements:** periodic inspection and cleaning of external glazing surfaces. **Manufacturer's technical services:** total system design and sizing. **Regional applicability:** not practical for parts of Pacific Northwest and Alaska. **Availability:** 30 to 90 days, depending upon quantity. **Suggested retail price:** contact manufacturer.

SUNSHINE CONCENTRATOR ARRAY

Sunshine Manufacturing Co.
4870 S.W. Main, No. 4
Beaverton, OR 97005
David Carson (503) 642–1123, (503) 643–6172

This concentrator is designed for high-temperature, total energy systems and is used to charge Solar Power Cells by sunshine. A selective surface absorber tube within a Pyrex-glass vacuum tube is located at the focus of the reflector trough; the collector array is supplied with an integral differential gas pressure generator that has a double-acting piston for automatic sun tracking. The concentrator is designed to operate between 250 and 550° F and on an average clear-sky day will produce 15,000 to 25,000 Btu's at 550° F.

Dimensions and net aperture area: collector trough: 96 × 46 × 15 in. **Features:** electronic output temperature control, collector trough cover for mirror protection, automatic sun tracking. **Options:** special absorber tubes with high-temperature silicon cell as absorption surface (consult manufacturer for availability and prices). **Installation requirements/considerations:** level concrete footings onto which to bolt array with insulated main supply and return piping. **Guarantee/warranty:** 5-yr warranty against normal operating conditions. **Maintenance requirements:** cover cleaning or replacement as needed. **Regional applicability:** areas with over 200 days of direct sunshine per year. **Availability:** 90 days. **Suggested retail price:** $2500/array.

WHITELINE CONCENTRATING SOLAR COLLECTOR

Whiteline, Inc.
P.O. Box 3071
Asheville, NC 28802
George T. White (704) 258–8405

This parabolic trough has an aluminum reflector, chemically-blackened copper pipe, and acrylic cover plate. Overall size is 18½ in. by 8¼ ft. Panels weigh approximately 22-lb each; frame(s) should be designed to support each panel.

Dimensions and net aperture area: 18½ in. × 8¼ ft; 11 sq ft effective collector area. **Features:** wash-out hole for cleaning in place. **Options:** glass cover plate. **Installation requirements/considerations:** panels to face south. **Guarantee/warranty:** 5-yr limited warranty. **Maintenance requirements:** monthly focusing. Manufacturer's technical services: sizing and installation advice. **Availability: 4 to 6 wk between order and delivery. Suggested retail price:** $14/sq ft.

WINSTON CPC (COMPOUND PARABOLIC CONCENTRATOR)

G.N.S. Company
79 Magazine Street
Boston, MA 02119
David Schwartz (617) 442–1000

A non-tracking, focusing-type collector which relies on the

Winston principle to reflect the solar energy onto an absorber tube. The concentration achieved is approximately two times the flux available.

Dimensions and net aperture area: 48 or 51 in. width; 4 to 120 in. length in modular increments of 4 in. **Features:** non-tracking parabolic design produces higher operating temperature than flat plate collectors. **Options:** vacuum shielded absorber tubes (Mk 3); available as single-trough modules or complete collector arrays. **Installation requirements/considerations:** same mounting techniques as flat plate collectors; piping and accessories must be rated for low-pressure steam. **Guarangee/warranty:** 5-yr unconditional except for glazing; 20-yr limited on absorber coil only. **Maintenance requirements:** none. **Manufacturer's technical services:** will supply schematic drawings and specifications for complete systems using GNS Winston units. **Availability:** 6 wk. **Suggested retail price:** Mk 2: $11/sq ft; Mk 3: $15/sq ft.

CONTROLS AND INSTRUMENTATION

AIR HANDLER/ELECTRONIC CONTROLS 800, 900, 1900

Helio Thermics, Inc.
110 Laurens Rd.
Greenville, SC 29607
William Haas (803) 235–8529

Control computer is mounted within air handler and automatically operates solar collection of heat, storage and distribution, natural cooling, and back-up heating and cooling.

Differentials available: ON, 27° F Delta T; OFF, 15° F Delta T; optional settings available. **Features:** no separate blower system needed; individually motorized dampers power driven in both directions; mode readout LED panel; LED temperature readout. **Installation requirements/considerations:** connect to existing ductwork. 110-vac, installation manual provided. **Guarantee/warranty:** 1-yr limited warranty on handler and associated mechanisms; 5-yr limited warranty on computer control. **Maintenance requirements:** oil motor annually, change filter as needed. **Manufacturer's technical services:** can vary rpm and cfm as required. **Availability:** 60 days between order and delivery. **Suggested retail price:** $2695 to $3655.

BTU METER S45

Natural Power, Inc.
Francestown Turnpike
New Boston, NH 03070
Charles Puckette (603) 487–5512

Temperature difference across an element or elements of a system is measured by means of linear RTD temperature sensors feeding linear analog circuitry. Flow is measured by a commercial paddle-wheel flow sensor having excellent sensitivity and linearity combined with very low resistance to flow, capable of operation to temperatures of 220° F. Signals from these sensors are combined to measure flow rate; the rate is integrated electronically to accumulate total energy transferred. The instantaneous values of the two temperatures, their difference, the flow rate and the rate of energy transferred, are displayed on a front panel mounted meter by means of a selector switch; the total energy transferred is displayed on a separate electro-mechanical counter which provides a non-volatile readout of the information.

Features: sensitivity, linearity, low impedance to flow, high-temperature flow meter, non-volatile BTU readout. **Options:** additional sensor input, calibration in watt-hours, calories or other units. **Guarantee/warranty:** 1-yr limited warranty. **Availability:** 4 to 6 wk. **Suggested retail price:** $880.

C-60 DIFFERENTIAL TEMPERATURE CONTROL

Independent Energy, Inc.
P.O. Box 732
42 Ladd St.
East Greenwich, RI 02818
R. Dowdell (401) 884–6990

The C-60 is a proven, solid-state differential temperature control designed specifically for solar domestic water, space, and pool heating. The C-60 is reliable, economical, and applicable to air or hydronic systems. The C-60 senses collector and storage temperature and automatically activates the pump or fan to initiate collection of energy in the presence of sufficient sunlight.

Differentials available: 20° F ON/5° F OFF; 8° F ON/5° F OFF. **Features:** all C-60's are factory computer tested, easy to install, quiet (no chattering relays), and use interchangeable sensors. **Options:** storage high-temperature limit (140° F), freeze protection (recirculate or drain down), ac line cord and outlets, auxiliary output; C-60SP. **Guarantee/warranty:** 1-yr limited warranty. **Availability:** 2-wk. **Suggested retail price:** consult manufacturer.

C-100 DIFFERENTIAL TEMPERATURE CONTROL

Independent Energy, Inc.
P.O. Box 732
42 Ladd St.
East Greenwich, RI 02818
R. Dowdell (401) 884–6990

The C-100 is a differential temperature control with performance monitoring in one unit; microcomputer based; for domestic water heating. Performance monitoring allows operational testing at system startup and continuous fine tuning of performance. Helps spot malfunctions; controls up to three outputs; monitors up to six sensors; three-digit-LED temperature display (°C or °F).

Differentials available: 20° F ON/5° F OFF, 8° F ON/3° F OFF, custom thresholds on volume orders. **Features:** all C-100's are factory computer tested, easy to install, compact, quiet, safe, have interchangeable sensors. **Options:** proportional control storage high-temperature limit 140° F, freeze protection (recirculate or drain down 38° F) ac line cord and outlets, 230-vac input power; °F or °C readout. **Guarantee/warranty:** 1-yr limited warranty. **Availability:** 2-wk. **Suggested retail price:** consult manufacturer.

C-110 DIFFERENTIAL TEMPERATURE CONTROL

C-110 is a temperature control and performance monitor in one, microcomputer based; for solar pool heating systems; monitors up to five sensors; allows operational testing at system startup and continues fine tuning of system performance; helps spot malfunctions; makes maximum energy efficiency possible so less conventional energy is consumed.

Differentials available: 8° F ON/4° F OFF; custom threshold on volume orders. **Features:** unit construction, raintight enclosure, adjustable high-temperature control avoids overheating, interchangeable sensors, safe, compact, immune to noise or lightning surges. **Options:** nocturnal cooling turns off at 3° F cooler, ac line cord and outlets, 230-vac input power, display readout in °C or °F. **Guarantee/warranty:** 1-yr limited warranty. **Availability:** 2 to 4 wk. **Suggested retail price:** consult manufacturer.

C-120 DIFFERENTIAL TEMPERATURE CONTROL

The C-120 Differential Temperature Control and performance monitor is used in space and domestic water heating; adds control of energy distribution to collector function; microcomputer based; monitors up to four sensors and two thermostats; controls up to four outputs; offers indirect heating and continuous fine tuning of system performance; helps spot malfunctions; maximum energy efficiency.

Differentials available: ON/OFF 20° F/5° F; PREP 16° F/3° F. **Features:** interchangeable sensor, fast installation, multiple outputs available, monitors up to four sensors and two thermostats, three-digit display, compact, quiet, brownout protected. **Options:** proportional control, high-temperature limit 140° F, freeze protection, ac line cord, 230-vac, display readout °C or °F, direct or indirect solar heating. **Guarantee/warranty:** 2 to 4 wk. **Suggested retail price:** consult manufacturer.

C-200 DIFFERENTIAL TEMPERATURE CONTROL

Controls direct/indirect space heating; C-200 is designed to maximize energy conservation; monitors up to 6 temperature sensors, 10 inputs, up to 16 solid-state outputs; formulates control strategy via its microcomputer by matching internal environmental demand to a most economical energy source; computer performance tested; applicable to air or hydronic systems; controls single or dual heat pumps, the boiler furnace or radiant heat source, auxiliary source and domestic hot water.

Differentials available: not specified. **Features:** unit construction, accurate to 1° F, safe, low-voltage sensor, factory calibrated, compact, quiet, brownout protection, interchangeable sensors. **Options:** S-200 system monitor custom control operation; storage high-temperature limit 140° F; proportional control of outputs No. 1 and 2, 230-vac input power. **Guarantee/warranty:** 1-yr limited warranty. **Availability:** 2 to 4 wk. **Suggested retail price:** consult manufacturer.

DEKO-LABS TC-3 TEMPERATURE COMPARATOR CONTROL

Deko-Labs of Friberg Dekold Corp.
Route 4, Box 256
Gainesville, FL 32601
Donald F. Dekold (904) 372–6009

The TC-3 is a differential thermostat for the control of a circulating pump or blower. Two sensors employed with the unit monitor temperatures of solar collector and storage reservoir; when the collector temperature exceeds the storage temperature by the pre-determined differential, the circulator is turned on.

Differentials available: ON, 10° F; OFF, 5° F. **Options:** 1-hp relay, custom differentials, fixed set-point operation, 240-vac, 12 vdc. **Installation requirements/considerations:** wiring for 120-vac input and output must meet local codes. **Guarantee/warranty:** 1-yr parts and labor. **Availability:** from stock. **Suggested retail price:** $59.

DELTA-T DIFFERENTIAL TEMPERATURE THERMOSTAT (SERIES DTT-80 AND DTT-90)

Heliotrope General
3733 Kenora Drive
Spring Valley, CA 92077
Al Abernathy (714) 460–3930

This differential-temperature thermostat is an automatic-motor-control device that turns a circulating pump or blower on when the collector is warmer than the storage. It can be used to regulate temperatures for domestic hot water, space heating, and nocturnal cooling.

Differentials available: four sets of differentials available from 15° ON to 1.5° OFF. **Features:** UL listed. Plug receptacles and line cords included. **Options:** 240 v and normally closed (NC) relay contacts. **Guarantee/warranty:** 5-yr limited warranty. **Manufacturer's technical services:** engineering support available, including system design review. **Availability:** stock. **Suggested retail price:** $38.

DELTA-T DIFFERENTIAL TEMPERATURE THERMOSTAT (SERIES DTT-100)

This differential thermostat turns the circulating pump or blower on when heat can be distributed. The DTT-100 series has an overriding feature which prohibits pump/blower from switching on until the collector is above 80° F.

Differentials available: four sets of differentials available from 15° ON to 1.5° OFF. **Features:** UL listed. OFF position maintained when collector remains below 80° F. **Options:** 240 v and normally closed relay contacts. **Guarantee/warranty:** 5-yr limited warranty. **Manufacturer's technical services:** engineering support available, including system design review. **Availability:** stock. **Suggested retail price:** $43.

DELTA-T DIFFERENTIAL TEMPERATURE THERMOSTAT (SERIES DTT-290)

This differential-temperature thermostat is designed to regulate pump activity; allows circulation when heat can be picked up from the collector and delivered to storage. The DTT-290 Series has a recycle freeze protection sensor that turns the pump to ON when temperatures drop to 36° F. The thermostat is designed for use in areas where freezing occurs several times a year.

Differentials available: four sets of differentials available from 15° ON to 1.5° OFF. **Features:** UL listed, with plug receptacles and line cord. **Options:** 240 v and normally closed relay contacts. **Guarantee/warranty:** 5-yr limited warranty. **Manufacturer's technical services:** engineering support available, including system design review. **Availability:** stock. **Suggested retail price:** $43.

DELTA-T DIFFERENTIAL TEMPERATURE THERMOSTAT (SERIES DTT-400)

This differential-temperature thermostat is an automatic-motor-control device that turns a circulating pump on when the collector is warmer than the storage. The thermostat also turns the pump off when the high limit setting of water storage has been reached.

Differentials available: four sets of differentials available from 15° ON to 1.5° OFF. **Features:** UL listed. OFF when storage capacity reaches the high-limit setting. **Options:** 240 v and normally closed relay contacts. **Guarantee/warranty:** 5-yr limited warranty. **Manufacturer's technical services:** engineering support available, including system design review. **Availability:** stock. **Suggested retail price:** $43.

DELTA-T DIFFERENTIAL TEMPERATURE THERMOSTAT (SERIES DTT-690)

In addition to controlling the circulation pump for domestic hot water, the Delta-T, Differential-Temperature Thermostat 690 also controls the solenoid valves which will cause the collector to drain upon electrical failure or freezing conditions.

Differentials available: four sets of differentials available from 15° ON to 1.5° OFF. **Features:** UL listed, with plug receptacles and line cord. **Guarantee/warranty:** 5-yr limited warranty. **Manufacturer's technical services:** engineering support available, including system design review. **Availability:** stock. **Suggested retail price:** $43.

DELTA-T DIFFERENTIAL TEMPERATURE THERMOSTAT (SERIES DTT-700 AND DTT-790)

This differential-temperature control can be adjusted to vary the "ON" differentials from 3° F to 24° F; the control is fixed at 1.5° F "OFF". The differentials have freeze and high-limit features.

Differentials available: adjustable from 3° to 24° F. **Features:** UL listed, with line cord. **Options:** 240 v and normally closed (NC) relay contacts. **Guarantee/warranty:** 5-yr limited warranty. **Manufacturer's technical services:** engineering support available, including system design review. **Availability:** stock. **Suggested retail price:** $48.

DELTA-T DIFFERENTIAL TEMPERATURE THERMOSTAT (SERIES DTT-3410)

The Delta-T Series, 3410, differential-temperature thermostat not only controls the circulating pump/blower when the heat can be transferred to storage, but it also allows adjustments to a high-level OFF condition.

Differentials available: four sets of differentials available from 15° ON to 1.5° OFF. **Features:** UL listed. **Options:** 240 v and normally closed (NC) relay contacts. **Guarantee/warranty:** 5-yr limited warranty. **Manufacturer's technical services:** engineering support available, including system design review. **Availability:** stock. **Suggested retail price:** $53.

DIFFERENTIAL TEMPERATURE CONTROLLER

TA Controls
652 Glenbrook Rd.
Stamford, CT 06906
Leif Eiderberg (203) 324–0106

A solid-state controller working with two NI-1000-ohm sensors. Control output is ON-OFF or time proportioning. Adjustments are: Differential Temperature 0 to 30° C; Integral Action 1 to 11° C; proportional bank 0 to 13° C; sensitivity 1 to 27° C.

Differentials available: 0 to 30° C. **Features:** Triac output. **Guarantee/warranty:** 18-mo after sale; 12-mo after installation. **Availability:** from stock. **Suggested retail price:** $400.

DIFFERENTIAL THERMOSTAT S25 & S26

Natural Power, Inc.
Francestown Turnpike
New Boston, NH 03070
Charles Puckette (603) 487–5512

Model S25 is a two-sensor differential thermostat featuring adjustable turn-on and turn-off temperature settings, built-in metering and self-test capability. It is available with various options to protect against freeze-up and overheating. Model S26 has the same capability and features as the S25, and in addition is provided with a third temperature input which may be used for supplementary control logic.

Differentials available: 0-25° F adjustable standard, 0-10° or 0-50° F on special order. **Features:** linear stable sensors. Built-in metering and test. Fully adjustable operating points. Rugged housing. **Guarantee/warranty:** 1-yr limited warranty. **Availability:** stock. **Suggested retail price:** $120 up, depending upon model and options.

DIFFERENTIAL THERMOSTAT, MODELS S, SF, SFH

Thanor Enterprises, Inc.
817 West St.
Wilmington, DE 19801
Frank W. Arnoth (302) 571–9515

This electronic, solid-state differential thermostat operates motors that have up to ½ hp. Lower and upper differential limits are independently adjustable, and settings are linear to within plus or minus 10 percent over full operating range.

Differentials available: upper and lower differential limits independently adjustable. **Features:** MAN-AUTO switch, pump indicator light, two glass probe thermistors. **Options:** stainless steel clad thermistors; Model SF with freeze protection, Model SFH with freeze and high-temperature protection. **Installation requirements/considerations:** 120-vac, single phase. **Guarantee/warranty:** 1-yr. **Manufacturer's technical services:** repair after warranty expires. **Availability:** 3-wk between order and delivery. **Suggested retail price:** $79.95, $94.94, $109.95.

DIFFERENTIAL THERMOSTAT

Solarsystems Industries, Ltd.
5511 128th St.
Surrey, B.C. V3W 4B5
Erich W. Hoffmann (604) 596–2665

Electronic differential thermostats with remote sensors. ON and OFF differential separately adjustable, with output rating up to 2 hp at 230 v. Five different models.

Differentials available: ON, 10° C; OFF, 1° C. **Options:** special ranges and output relays available. **Installation requirements/ considerations:** to be installed in dry locations only. **Guarantee/warranty:** 1-yr. **Manufacturer's technical services:** system design available. **Availability:** from stock. **Suggested retail price:** $69 to $89.

DIFFERENTIAL THERMOSTAT 1100 AND 1200 WITH SUNMETER

American Solar Heat Corp.
7 National Place
Danbury, CT 06810
Joseph Heyman (203) 792–0077

A differential thermostat for either closed-loop (1100) or drain-down (1200) space heating or domestic hot water systems with exclusive "Sunmeter," which gives a constant readout of the solar panel and storage tank temperatures.

Differentials available: optional. **Features:** Sunmeter may be mounted anywhere in home. **Options:** any length cable from Sunmeter; drain-down or closed-loop, 24-vac or 110-vac pump control. **Installation requirements/ considerations:** grounded 110-vac outlet nearby. **Guarantee/warranty:** 1-yr. **Maintenance requirements:** none. **Manufacturer's technical services:** installation diagram backed up by engineering staff. PC board can be unplugged and returned for repair or replacement. **Availability:** 10-days. **Suggested retail price:** 1100, $148.75; 1200, $168.75.

FIXFLO CONTROLS H-1503-A, H-1503-B, H-1504-A, H-1505-A, H-1506-A

Hawthorne Industries, Inc.
Solar Energy Division
1501 S. Dixie Highway
West Palm Beach, FL 33401
Ray Lewis, J.B. Carr, Dale Kline (305) 659–5400

ON/OFF-type, differential-thermostat controls with various features. H-1503-A has recirculation freeze protection and stopped-circulation, upper-limit protection; H-1505-A has dual outlets; H-1503-B is printed-circuit-board-only configuration; H-1504-A and H-1506-A are designed for drain-down freeze and upper-limit protection.

Differentials available: differentials available, 16° F ON, 3° F OFF. **Features:** all solid-state construction. **Options:** consult manufacturer. **Installation requirements/ considerations:** 117-vac supply. **Guarantee/warranty:** 1-yr limited warranty. **Manufacturer's technical services:** limited custom controls design; advice; private labeling. **Availability:** stock. **Suggested retail price:** consult manufacturer.

HELIO-MATIC SPA/PUMP CONTROLLER

Heliotrope General
3733 Kenora Drive
Spring Valley, CA 92077
Al Abernathy (714) 460–3930

The Helio-Matic Spa/Pump Controller for spas and swimming pools turns on a circulating pump when the pool temperature is below the collector temperature. Adjustment for maximum water temperature.

Differentials available: 3° ON, 1.5° OFF. **Features:** UL listed. Adjustable high-limit temperature. **Options:** available in 120 and 240 v. **Guarantee/warranty:** 5-yr limited warranty. **Manufacturer's technical services:** engineering support available, including system design review. **Availability:** stock. **Suggested retail price:** $74.

HELIO-MATIC VALVE CONTROL

The Helio-Matic Valve Control is a swimming pool valve control system. The Helio-Matic senses pool temperature and collector temperature and allows flow through the collector when the pool filtration pump is on. Twelve-volt solenoid valves regulate water flow when heat is available and prevents heat loss through the collector during cloud cover. Maximum pool temperature is restricted to dial setting.

Differentials available: 3° ON, 1.5° OFF. **Features:** UL listed. Adjustable, high-temperature limit prevents pool water from becoming too hot. **Options:** available in 120 and 240 v. Also available with one or two valves. **Guarantee/warranty:** 5-yr limited warranty. **Manufacturer's technical services:** engineering support available, including system design review. **Availability:** stock. **Suggested retail price:** $120.

HOMEMASTER 77-620 CONTROLLER

Solar Control Corp.
5721 Arapahoe Rd.
Boulder, CO 80303
Larry Trudell (303) 449–9180

The Homemaster is designed for installation in either air or hydronic transport space heating systems. It operates in five modes, handling solar heat storage, direct solar heating, solar storage heating, auxiliary heating and solar domestic water pre-heating.

Differentials available: collection ON, 40° F/20° F; collection OFF, 20° F/4° F. **Options:** many custom options available. **Guarantee/warranty:** limited. **Maintenance requirements:** none. **Availability:** 1-wk. **Suggested retail price:** $249.

HVAC SOLAR CONTROLS

Virginia Solar Components, Inc.
Highway 29 South
Rustburg, VA 24588
Robert Savage (804) 239–9523

The HVAC Solar Control System consists of a pump manifold assembly and control head. These two components interface collector panels, a water circulating fireplace grate, hot-water coil, and a water-source heat pump to a solar storage tank and cooling tower. All functions are controlled by the home's wall thermostat.

Differentials available: not applicable. **Features:** plate on control head has a complete diagram of the system with a built-in electronic thermometer to measure the operation of the system. **Options:** fireplace grate control per installation manual. **Guarantee/warranty:** limited 5-yr warranty. **Maintenance requirements:** possible debris in solenoid valves if strainers are not used; no other maintenance required. **Manufacturer's technical services:** provides installation instructions; will advise or consult by phone or actual field inspection if necessary. **Availability:** 4 to 5 wk. **Suggested retail price:** $1012.32.

INTEGRATING PYRANOMETER S43

Natural Power, Inc.
Francestown Turnpike
New Boston, NH 03070
Charles Puckette (603) 487–5512

This instrument senses the intensity of the radiation incident upon the pyranometer head and integrates this information over time, resulting in a measure of the total available radiation. The pyranometer head has the same cosine angular response characteristic as that of a flat plate collector, enabling its use in evaluating the majority of solar collection systems. The display, which reads to 99,999 shows either instantaneous or total radiation values, and is available calibrated in millivolt-hours, cal/cm^2, Watt-hours/M^2 or BTU/ft^2.

Differentials available: maximum accumulation: 99,999 units. **Features:** power failure protection, small size. **Guarantee/warranty:** 1-yr limited warranty. **Availability:** stock to 2-wk. **Suggested retail price:** $835.

LOGIC CONTROL UNIT, LCU-110

Contemporary Systems, Inc.
68 Charlonne St.
Jaffrey, NH 03452
John C. Christopher (603) 532–7972

A flexible, fully automatic logic controller for air-circulating solar systems that regulates collector and storage operating temperatures, collector/storage differential, manual-automatic back-up system, summer storage temperature-limiting and system-status display in an easy-to-install, maintenance-free package.

Differentials available: variable collector/storage differential. **Features:** manual switching provision, easily accessible component boards to facilitate servicing. **Options:** custom-tailored logic to meet varied system parameters. **Installation requirements/considerations:** any indoor location. **Guarantee/warranty:** 1-yr materials and workmanship. **Manufacturer's technical services:** custom engineering to meet design parameters of specific systems. **Availability:** 4-wk for standard units, 4 to 8 wk for custom units. **Suggested retail price:** $584, including probes.

MAXWELL'S DEMON DIFFERENTIAL THERMOSTAT

Energy Design, Inc.
1925 Curry Rd.
Schenectady, NY 12303
Gary Poukish (518) 355–3322

A differential-control thermostat with LED temperature display for sensor points in °F and °C. Outside air temperature is also monitored. Delta-T and upper-limit settings are adjustable in this all solid-state unit.

Differentials available: adjustable 3° F to 25° F. **Features:** each unit is pretested for 100 hr. **Options:** outside air temperature or other additional monitoring points than three supplied. **Guarantee/warranty:** 5-yr; replacement if unit sent back unopened. **Availability:** 2-wk. **Suggested retail price:** $95.

MULTI-MODE S10 SERIES

Natural Power, Inc.
Francestown Turnpike
New Boston, NH 03070
Charles Puckette (603) 487–5512

The NPI S10 consists of a main frame upon which can be built any control configuration, however complex, to automatically control the active elements of a multi-mode solar or solar-augmented system. It incorporates priorities, interlocks, logic and control parameters required by a specific application and is furnished on a proposal/custom order basis.

Differentials available: adjustable 0 - 10, 0 - 25, or 0 - 50°, as many as required. **Features:** Metering of all temperatures, fully adjustable differentials and set points built-in self-test capability, recorder outputs. **Options:** as required. **Guarantee/warranty:** 1-yr limited warranty. **Availability:** 6 to 8 wk ARO. **Suggested retail price:** quotation only.

PROPORTIONAL DIFFERENTIAL THERMOSTAT S27 & 28

Model S27 is a two-sensor differential thermostat featuring adjustable proportional control of output, built-in metering and self-test capability. It is available with various options to protect against freeze-up and overheating. Model S28 has the same capability and features as the S27, and in addition is provided with a third temperature input which may be used for supplementary control logic.

Differentials available: 0-25° F standard, 0-10° F and 0-50° F available special order. **Features:** linear stable sensors. Built-in metering and test. Fully adjustable operating points. Rugged housing. **Guarantee/warranty:** 1-yr limited warranty. **Availability:** stock. **Suggested retail price:** price on request.

RECORDING PYRANOMETER S41 & S42

The S41 is an instrument which senses the intensity of incident radiation and records it on a strip chart recorder. The radiation sensor or pyranometer has the same angular response as a flat-plate collector, making the instrument extremely useful in evaluating the energy available for a solar DHW or heating system. The S42 adds to the above features an integrating function, marking discrete energy accumulations on the edge of the chart.

Differentials available: 0 - 2 cal/cm^2/min. standard, others on special order. **Features:** cosine angular response, accuracy, stability, linearity. **Options:** extended temperature range. **Guarantee/ warranty:** 1-yr limited warranty. **Availability:** stock to 2 wk. **Suggested retail price:** S41, $525; S42, $635.

RHO SIGMA SOLAR CONTROLS/ DIFFERENTIAL THERMOSTAT RS 104

Rho Sigma, Inc.
11922 Valerio St.
North Hollywood, CA 91605
Preston Welch (213) 982–6800

This differential thermostat is a dual-output relay closure control for use in drain-down (antifreeze) systems. The relays energize and de-energize corresponding to the differences in temperature between the collector and storage tanks.

Differentials available: available to customer's specifications. **Features:** PUMP-ON, AUTOMATIC, ANTI-FREEZE ON switch. **Options:** DPDT relays; 3 PDT relays. **Installation requirements/considerations:** local electrical codes. **Guarantee/warranty:** 1-yr limited warranty against defects in workmanship and materials. **Manufacturer's technical services:** application advice. **Availability:** allow 2-wk ARO. **Suggested retail price:** $151.10.

RHO SIGMA SOLAR CONTROLS/ DIFFERENTIAL THERMOSTAT RS 106

This differential thermostat energizes when the storage temperature is 20° F less than the temperature of the collector and de-energizes when the temperature difference reaches 3° F. This single-output relay closure control is housed in the standard NEMA box and is used for space and hot water heating systems.

Differentials available: to customer's specifications. **Features:** adequate space in high-voltage compartment to house a booster relay to handle horsepower-rated fans and pumps. **Options:** 2PDT relay; 3PDT relay. **Installation requirements/considerations:** local electrical codes. **Guarantee/warranty:** 1-yr guarantee workmanship and materials. **Manufacturer's technical services:** recommendations on interfacing RS 106 and available options. **Availability:** allow 2-wk ARO. **Suggested retail price:** $132.19.

RHO SIGMA SOLAR CONTROLS/ DIFFERENTIAL THERMOSTAT RS 260

This differential thermostat is used to control water flow through NO or NL solenoid valves used with pools and/ spas and will circulate the water through the collector loop when the collector becomes 6° F hotter than pool water temperature. The relay out-put has normal-open and normal-closed contacts for the valves.

Differentials available: temperature ON, 7° F; temperature OFF, 3° F; adjustable high-temperature cut-off. **Features:** rain-tight enclosure; ON/OFF automatic switch; adjustable high-temperature cut-off. **Options:** 240-vac configuration. **Installation requirements/ considerations:** local electrical codes. **Guarantee/ warranty:** 1-yr warranty against defects in workmanship and materials. **Availability:** allow 2-wk ARO. **Suggested retail price:** $114.61.

RHO SIGMA SOLAR CONTROLS/ DIFFERENTIAL THERMOSTAT RS 280

This differential thermostat is designed to control "solar-assist" (sic) pumps used for pool and spa heating. The control circulates water through the collector loop when the collector temperature reaches 7° F above that of the pool water. Standard DPDT relay is rated at 10 amp.

Differentials available: temperature ON, 7° F; temperature OFF, 3° F, adjustable, high-temperature cut-off. **Features:** ON-OFF-AUTO switch; adjustable, high-temperature cut-off. **Options:** 240-vac configuration available. **Installation requirements/considerations:** local electrical codes. **Guarantee/warranty:** 1-yr limited warranty. **Availability:** allow 2-wk ARO. **Suggested retail price:** $132.05.

RHO SIGMA SOLAR CONTROLS/ INTEGRATED DIFFERENTIAL THERMOSTAT RS 360

This integrated differential thermostat is designed for collector-to-storage heating, storage-to-domestic water heating, and auxiliary space heating. It is a four-input, three-output control system with a fourth output available. The three SPDT relays are rated at 10 amp. The unit can be pre-programmed to customer specifications and controls two differentials and up to four fixed-value temperatures.

Differentials available: to customer's specifications. **Features:** SPDT relays rated at 10 amp; high-temperature cut-off capability; freeze protection. **Options:** fourth SPDT relay, high-temperature cut-off circuitry; freeze protection circuits; field adjustable settings. **Installation requirements/considerations:** local electrical codes. **Guarantee/warranty:** 1-yr limited warranty. **Manufacturer's technical services:** custom programming and application advice. **Availability:** allow 4-wk ARO. **Suggested retail price:** (base) $289.62

RHO SIGMA SOLAR CONTROLS/ DIFFERENTIAL THERMOSTAT RS 500-1S-2

This solid-state, dual-output control for domestic water systems with drain-down freeze protection controls the ON/OFF pump control (first output) and the drain-down valve or secondary pump (second output). The second output can also be used for high-temperature drain-down.

Differentials available: to customer's specifications: **Features:** ON-OFF-AUTO switch; indicating light; 6 × 6 × 3-in. steel enclosure. **Options:** either H high-temperature drain-down, or L low-temperature drain-down. **Installation requirements/considerations:** local electrical codes. **Guarantee/warranty:** 1-yr limited warranty. **Suggested retail price:** $112.87.

RHO SIGMA SOLAR CONTROLS/ DIFFERENTIAL THERMOSTAT RS-500-1-P

This solid-state, single-output thermostat for proportional pump control in domestic water systems provides variable pump speed based on the temperature difference between the collector and storage tank.

Differentials available: Delta-T on minimal flow = 3.5° F; Delta-T on full flow = 12° F. **Features:** ON-OFF-AUTO switch; indicating light; 6 × 6 × 3-in. steel enclosure. **Options:** H high temperature shut off (to prevent tank from overheating); L freeze-protection override. **Installation requirements/considerations:** local electrical codes. **Guarantee/warranty:** 1-yr limited warranty. **Availability:** allow 2-wk ARO. **Suggested retail price:** (base) $107.21.

RHO SIGMA SOLAR CONTROLS/ DIFFERENTIAL THERMOSTAT RS 500-1P-2

This solid-state, dual-output control thermostat is for

proportional pump control (first output) and drain-down freeze protection control of a secondary pump (second output). The second output can also be used for high-temperature, drain-down protection.

Differentials available: T on minimum flow = 3.5° F; full flow = 12° F. **Features:** ON-OFF-AUTO switch; 6 × 6 × 3 in. steel enclosure. **Options:** either H high-temperature drain-down or L low-temperaurte drain-down. **Installation requirements/considerations:** local electrical codes. **Guarantee/warranty:** 1-yr limited warranty. **Suggested retail price:** $128.30.

RHO SIGMA SOLAR CONTROLS/ DIFFERENTIAL THERMOSTAT RS 500-1-S

This differential thermostat is a solid-state, single-output control for ON/OFF pump control in domestic water systems. The control is determined by the temperature difference between the collector and storage tank.

Differentials available: to customer's specifications. **Features:** ON-OFF-AUTO switch; indicating light; 6 × 6 × 3-in. steel enclosure. **Options:** H - high-temperature shut-off (to prevent tank from overheating), and L - freeze-protection override to prevent collector freeze. **Installation requirements/considerations:** local electrical codes. **Guarantee/warranty:** 1-yr limited warranty. **Availability:** allow 2-wk ARO. **Suggested retail price:** (base) $91.87.

SOLAR COMMANDER DIFFERENTIAL THERMOSTAT SD-10

Robertshaw Controls Co.
Temperature Controls Marketing Group
100 W. Victoria St.
Long Beach, CA 90805
K. LaGrand, P.S. Johnson (213) 638–6111

SD-10 Solar Differential Thermostat features solid-state load-switching and interchangeable thermistor temperature sensors.

Differentials available: 5° F to 25° F; optional 8° F to 20° F adjustable. **Options:** fixed or adjustable cut-in differential; recycling high limit; pump ON freeze protection; power indicating lights. **Installation requirements/considerations:** available in steel case or as encapsulated circuit board assembly only. **Guarantee/warranty:** 18-mo limited warranty. **Availability:** 4 to 8 wk between order and delivery. **Suggested retail price:** consult manufacturer.

SOLAR CONTROL SYSTEM

Pelcasp Solar Designs
4817 Sheboygan Ave.
Madison, WI 53705
Michael Pellett (608) 274–1752

A control package with simple services. It uses a Hawthorne differential controller, AMF relays, and Honeywell secondary sensors. Pelcasp will custom design a control system to meet your mode requirements.

Differentials available: not specified. **Features:** custom designed to meet every conceivable mode set-up. **Options:** heating only, heating-and-cooling modes, and heating, cooling-and-hot-water-heating modes. **Installation requirements/considerations:** standard wiring practices must be followed for safety and reliability. **Maintenance requirements:** locate in a dust free, clean, dry and heated area. **Manufacturer's technical services:** design of systems to meet mode requirements for design of a heating system to meet and fit the basic control system limitations. **Regional applicability:** Pelscap will sell a system and include one inspection to anyone within 200 mi of Madison. Sales outside this area are at buyer's risk. **Availability:** 60-days. **Suggested retail price:** $475 and up.

SOLAR HOT WATER CONTROLLER

Solar Control Corp.
5721 Arapahoe Rd.
Boulder, CO 80303
Larry F. Trudell (303) 449–9180

A solid-state differential thermostat capable of fully controlling a solar hot water system. The unit is designed for a lifetime of maintenance free service and incorporates freeze and boil protection circuitry.

Differentials available: differentials: ON 20° F; OFF 4° F. **Options:** ambient probe for temperature control. **Installation requirements/considerations:** mounts on 4 × 4 in. J box. **Guarantee/warranty:** limited. **Maintenance requirements:** none. **Availability:** 1-wk. **Suggested retail price:** $65.

SOLAR STAT SSA

C & M Systems, Inc.
Saybrook Industrial Park
Elm St., P.O. Box 475
Old Saybrook, CT 06475
Nelson Messier (203) 388–3429

A differential thermostat designed to control the operation

of a heat transfer fluid circulator, the Solar Stat consists of two sensor elements and a control box. The unit has an external switch for off, automatic, or manual circulator operations, indicator lamps for power and circulator-on.

Differentials available: 12°(±3°)F IN; 5°(±2°)F OUT. **Features:** high-temperature cut-out will not allow circulator to run if storage temperature is greater than 170° F. **Guarantee/warranty:** 1-yr limited warranty against defects in materials or workmanship. **Availability:** from stock. **Suggested retail price:** $129.

SOLAR SYSTEM MONITOR 77-180

Solar Control Corp.
5721 Arapahoe Rd.
Boulder, CO 80303
Larry F. Trudell (303) 449–9180

Provides the homeowner with accurate outdoor, indoor, collector and storage temperature readings. The solar system's mode is indicated by three light-emitting diodes to indicate collecting, solar distribution or back-up heating function.

Guarantee/warranty: limited. **Maintenance requirements:** none. **Availability:** 1-wk. **Suggested retail price:** $280.

SOLARICS DIFFERENTIAL CONTROLLER SPC-1000, SPC-2000, SPC-3000

Solarics Energy Control Systems, Inc.
P.O. Box 15183
Plantation, FL 33318
Ron Stein (305) 971–0391

A solid-state differential controller with two outputs (6 amp standard, 16 amp available). Two matched sensors are included with each unit.

Differentials available: factory set at 16° F ON, 5° F OFF. **Features:** solid-state with hysteresis; power consumption less than 3 w. **Options:** high-temperature or freeze-protection controls available, lightning protection, manual override, external ac outlet. **Installation requirements/considerations:** mounts to wall. Professional installation not necessary with external ac outlet; internal terminal strip is standard. **Guarantee/warranty:** 1-yr replacement or repair warranty. **Manufacturer's technical services:** custom engineering services available for special control requirements. **Availability:** 3-wk. **Suggested retail price:** $29.50 to $34.50.

SOLARLOGIC AUTOMATIC TEMPERATURE CONTROL

Solar Energy Systems, Inc.
One Olney Ave.
Cherry Hill, NJ 08003
Nathan E. Brussels (609) 424–4446

A differential thermostat that combines solid-state temperature sensing circuitry with a mechnaical switching relay in order to provide a precise and reliable temperature control.

Differentials available: adjustable differential setpoints. **Features:** adjustable on-off setpoints, solid-state linear sensor output, immersion-type sensors. Both NO and NC 10-amp contacts. **Installation requirements/considerations:** power input: 115-vac, 60 Hz, single phase. **Guarantee/warranty:** 1-yr. **Availability:** 2-wk. **Suggested retail price:** consult manufacturer.

SOLTROLLER PROPORTIONAL CONTROL, SPC SERIES

Ecotronics, Inc.
8502 E. Cactus Wren Rd.
Scottsdale, AZ 85253
Alden Stevenson (602) 948–8003

This high-sensitivity, proportional controller is designed to maximize heat collection from various collector systems, especially drain-down systems, by providing a variable flow rate at a small differential temperature. This solid-state control provides two set points triac and relay outputs and is most effectively used in high-flow-rate collector systems.

Differentials available: differentials factory or user set (1° C ± .5° C normal). **Features:** fill cycle for drain-down systems; indicator lights; set points for Solar A/C. **Options:** SPC-1 without relay; SPC-2 without fill features and programming switches. **Installation requirements/considerations:** 110-vac input; two sensors; pump motor outputs; relay contacts output (optional). **Guarantee/warranty:** 1-yr limited warranty. **Manufacturer's technical services:** consulting and custom design. **Availability:** stock. **Suggested retail price:** $159 to $199.

SUN BUCKET

Mann Russell Electronics Inc.
1401 Thorne Rd.
Tacoma, WA 98421
George F. Russell (206) 383–1591

This direct-tracking heliostat platform is mounted on a

carriage structure that can track the sun in dual axes of freedom of rotation. It has automatic cloud-cover solar tracking and returns to sunrise position after sundown. It can mount flat plate collectors, heat pipes or photovoltaic cells. The platform has an internal power supply, umbilical swiveling input and exit liquid collector lines.

Differentials available: adjustable, and depends on atomsphere; less than ¼ of 1%. **Features:** custom platform size and remote control. **Options:** adjustable counterbalancing available. **Installation requirements/recommendations:** can be bolted for wind factors. **Guarantee/warranty:** 1-yr, workmanship and materials. **Maintenance requirements:** reduction gears must be kept full of oil. **Manufacturer's technical services:** installation instructions and phone consultation. **Availability:** 30 to 60 days between order and delivery. **Suggested retail price:** $1000 to $25,000.

SUN-LOC 1

Delavan Electronics
14605 N. 73rd St.
Scottsdale, AZ 85260
Hardy K. Landskov (602) 948–6350

A tracking device used to control a motor that drives a parabolic trough collector to follow the sun.

Features: small in size, with low power consumption. **Installation requirements/considerations:** accurate alignment of sun sensor head. **Guarantee/warranty:** 2-yr warranty on parts and labor. **Maintenance requirements:** remove dust and dirt from sun sensor head weekly. **Manufacturer's technical services:** repair service available, no charge in warranty; $29.50 flat fee out of warranty, excluding relays. **Availability:** stock item, ship same day. **Suggested retail price:** $155 (in quantities of 1 to 3).

SUN-LOC 2

Totally automatic solar tracking device with over temperature detection, day/night detection, wind velocity detection, and four other programmable fault circuits. Designed to run off of two 12-v automobile batteries connected in series. Will power up to ¾-hp motor.

Features: first-order, linear feedback system to drive permanent magnet dc motor; transistor switching; battery charger included. **Options:** relay switching instead of transistor switching. **Installation requirements/considerations:** accurate alignment of sun sensor head. **Guarantee/ warranty:** 2-yr warranty on parts and labor. **Maintenance requirements:** remove dust and dirt from sun sensor head weekly. **Manufacturer's technical services:** repair service available; no charge in warranty; $29.50 flat fee out of warranty, excluding relays. **Suggested retail price:** $615 (in quantities from 1 to 3).

SUN-TRACK SYSTEMS INTERFACE MODULE

Energy Applications
Route 5, Box 383A
Rutherfordton, NC 28139
Napoleon P. Salvail (704) 287–2195

This Interface Module provides complete interface between the user motor drive system and the Sun-Track Sensor system. It has a regulated 15-vac, 120-ma power supply and two, 3-amp, solid-state relays for directional control of the drive motor. It is also equipped with an ON-OFF switch, fuse, and 6-ft grounded cord for 120-v, 60-Hz power. Terminal strips provided for sensor-unit and drive-motor connections.

Features: provides complete interface. **Installation requirements/considerations:** must be mounted in weather-proof enclosure. **Guarantee/warranty:** 90 days parts and labor; satisfaction guaranteed. **Manufacturer's technical services:** full support services and consulting. **Availibility:** 1 to 2 wk or 4 to 6 wk, depending on size of order. **Suggested retail price:** $60.

SUN-TRACK SYSTEMS SENSOR UNIT, ST-100, ST-100A

Energy Applications
Route 5, Box 383A
Rutherfordton, NC 28139
Napoleon P. Salvail (704) 287–2195

This sun-tracking sensor unit is designed to control north-south oriented collectors. It is equipped with isolated reed relay outputs with 100 ma capability and provides fully automatic pointing east to west tracking sensor and day-night sensor to return collector east each evening.

Features: electronic, solid-state design. **Options:** ST-100A lower-cost version for OEM applications has transistor output with 20 ma capability instead of reed relays. **Installation requirements/considerations:** proper orientation in unshaded location. **Guarantee/warranty:** 90 days parts and labor; satisfaction guarantee. **Manufacturer's technical services:** full support services and consulting. **Availability:** 1 to 2 wk or 4 to 6 wk, depending on size of order. **Suggested retail price:** $50 (ST-100A); $60 (ST-100).

SUN-TRACK SYSTEMS SENSOR UNIT, ST-110, ST-110WA

These sun-tracking sensing units for seasonal and daily tracking of east-west oriented collectors include two sensor units for north-south elevation movement and south-north elevation movement. Sensor unit is equipped with reed relay outputs, 100 ma capability.

Features: fully automatic operation; solid-state design; 5-hr tracking field of view. **Options:** ST-110WA has two additional sensor units to expand tracking field of view to 160 degrees or 11 hrs. **Installation requirements/considerations:** proper orientation in unshaded location. **Guarantee/warranty:** 90 days parts and labor; satisfaction guaranteed. **Manufacturer's technical services:** full support services and consulting. **Availability:** 1 to 2 wk or 4 to 6 wk, depending on size of order. **Suggested retail price:** $75 (ST-110); $150 (ST-110WA).

SUN-TRACK SYSTEMS SENSOR UNIT, ST-200

This sun-tracking sensing unit is designed for collectors and devices that require both azimuth and elevation tracking. It has three sensor units, two for north-south elevation tracking and one for east-west azimuth tracking. Equipped with reed relay outputs, 100 ma capability.

Features: electronic, solid-state design; dual-axis control. **Installation requirements/considerations:** proper orientation in unshaded location. **Guarantee/warranty:** 90 days parts and labor; satisfaction guaranteed. **Manufacturer's technical services:** full support services and consulting. **Availability:** 1 to 2 wk or 4 to 6 wk, depending on size of order. **Suggested retail price:** $135.

TEMPERATURE DIFFERENTIAL SWITCH, SERIES 101AC, 101DC

West Wind
Box 542
Durango, CO 81301
Geoffrey Gerhard

This dual-probe, temperature-differential switch activates the collector pump or blower when collector is hotter than storage. Temperature difference between probes is user adjustable.

Differentials available: ON, 10° C user adjustable; OFF, less than 1° C, factory adjustable. **Features:** integrated circuitry for precise temperature sensing. Probes matched to keep differential constant over wide range (0° C to 100° C). **Options:** HEAT-COOL switch; freeze or overheat turn ON or OFF; thermostatic operation. **Guarantee/warranty:** 2-yr limited warranty. **Manufacturer's technical services:** applications advice; modifications for special applications. **Availability:** 2-wk, AC; 4-wk DC. **Suggested retail price:** $35; quantity discounts available.

TEMPERATURE MONITOR S35

Natural Power, Inc.
Francestown Turnpike
New Boston, NH 03070
Charles Puckette (603) 487–5512

Electronic thermometer reading temperatures and temperature differences on a 3½ in. rectangular meter calibrated in both °F and °C. Housed in rugged enclosure suitable for permanent wall mounting. Switch-selectable metering. Two direct and one differential monitor outputs provided.

Differentials available: 0-250° F direct, 0-25° F differential standard, others on special order. **Features:** 3½ in rectangular meter, linear scale, recorder/monitor outputs. **Options:** additional switch-selector inputs, digital meter. **Guarantee/warranty:** 1-yr limited warranty. **Availability:** stock. **Suggested retail price:** $135 up, depending upon options.

TEMPERATURE MONITOR S36

Electronic thermometer reading temperatures and temperature differences on a 3½ in rectangular meter calibrated in both °F and °C. Housed in rugged enclosure suitable for permanent wall mounting. Switch-selectable metering. Two direct and one differential monitor outputs provided. Recorder/remote monitor output provided for each temperature input and differential.

Differentials available: 0 - 250° F direct, 0 - 25° F differential standard, others available special order. **Features:** 4½ in. rectangular meter, linear scale, recorder/monitor outputs. **Options:** digital meter. **Guarantee/warranty:** 1-yr limited warranty. **Availability:** stock to 6 wk. **Suggested retail price:** depends upon number of inputs and options.

THERMOSOL™ HVAC CONTROL PANEL

Pioneer Energy Products
Rt. 1, Box 189
Forest, VA 24551
Timothy M. Hayes

A solar-control package including electronic components,

pumps and valves to regulate any hydronic system. A pump grid installs on a wall and the all-weather cable plugs into a control panel with function indicator lights. Pre-wired, pre-plumbed and pre-tested.

Differentials available: various differentials available. **Options:** additional pumps and/or valves available for special requirements. **Installation requirements/considerations:** pump grid must be installed 2-ft below storage tank water level. **Guarantee/warranty:** 18-mo on pumps, 1-yr on all other components. **Manufacturer's technical services:** will size to each job. **Availability:** 2-wk. **Suggested retail price:** consult manufacturer.

VARIFLO CONTROLS H-1510-A, H-1510-B, H-1511-A, H-1512-A

Hawthorne Industries, Inc.
Solar Energy Division
1501 S. Dixie Highway
West Palm Beach, FL 33401
Ray Lewis, J.B. Carr, Dale Kline (305) 659–5400

These are proportional-output, differential-thermostat controls. H-1510-A has recirculation freeze protection and stopped-circulation upper limit. H-1512-A has dual outlets; H-1510-B is printed-circuit-board-only configuration. H-1511-A is designed for drain-down freeze and upper-limit protection.

Differentials available: 16° F full on; 3° F initial slow flow. **Features:** all solid-state construction. **Options:** consult manufacturer. **Installation requirements/considerations:** 117-vac supply. **Guarantee/warranty:** 1-yr limited warranty. **Manufacturer's technical services:** limited custom controls design; advice; private labeling. **Availability:** stock. **Suggested retail price:** consult manufacturer.

GLAZINGS

FILON^{R}/TEDLAR^{R} GLAZING, TYPE 548 AND 558

Filon Div.
Vistron Corp.
12333 Van Ness Ave.
Hawthorne, CA 90250
James E. Whitridge (213) 757–5141

These acrylic-enriched polyester panels are reinforced with 25 to 27 percent random-strand fiberglass and parallel nylon strands. Exposed side is surfaced with DuPont's Tedlar^{R} polyvinyl fluoride.

Maximum operating temperature: 200° F. **Features:** high transmissivity, low reflectance, shatterproof; Tedlar^{R} surface resists degradation, erosion, and corrosive atmospheres. **Options:** Type 548 (4 oz psf) available in corrugated and flat surfaces; Type 558 (5 oz psf) in various sizes. **Maintenance requirements:** periodic water rinse to remove accumulated dirt. **Availability:** stock. **Suggested retail price:** $.81 to $.98/sq ft.

GLAZING A2-3 THROUGH A2-9

Park Energy Co.
Star Route, Box 9
Jackson, WY 83001
Frank D. Werner (307) 733–4950

Polycarbonate extruded into a double glazing (1-in. space) in a 1-ft-width module. The glazing is mounted from its underside, using C1 brackets, and solvent bonded, resulting in a monolithic cover plate of nearly any desired length and width.

Maximum operating temperature: 200° F continuous, 300° F short time. **Longevity:** greater than 10-yr. **Features:** high-impact strength, impervious to thermal shock. The plastic formulation includes UV inhibitor and the extrusion has UV screen coating. Can be field cut to various sizes and shapes. **Installation requirements:** requires C1 brackets, solvent bonding and edge fittings, per installation manual. **Guarantee/warranty:** 2 to 4 wk. **Suggested retail price:** typically $3.30/sq ft.

HELIO THERMICS COLLECTOR COVER

Helio Thermics, Inc.
110 Laurens Rd.
Greenville, SC 29607
Willaim Haas (803) 235–8529

Glazing consists of two layers of cross-corrugated fiberglass. Top layer is Tedlar coated.

Maximum operating temperature: 212° F. **Longevity:** in excess of 20-yr. **Features:** replaces conventional roofing material on south facing roof; available in 4 × 8 ft to 4 × 16 ft panels in 2 ft increments. **Options:** redwood strips, screws, washer, adhesive and flashing plus double glazing available at $1.80/sq ft. **Installation requirements/considerations:** easily installed by builder on site. **Guarantee/warranty:** 20-yr, top layer. **Maintenance requirements:** occasional washing. **Manufacturer's technical**

services: instructions and sizing assessment. **Availability:** 2 to 3 wk between order and delivery. **Suggested retail price:** top layer, $0.68/sq ft; bottom, $0.52/sq ft.

LO-IRON™

ASG Industries, Inc.
P.O. Box 929
Kingsport, TN 37662
M.L. Lilly (615) 245–0211, (800) 251–0266

Fully transparent, tempered sheet glass with not more than 0.05% iron-oxide content provides total solar transmittance of 87.8% to 89.1%, relative to thickness, at a substantially lower cost than iron-free glass, as Sunadex®. **Maximum operating temperature:** 400° F. **Longevity:** Indefinitely.

LUMASITE FIBERGLASS ACRYLIC SHEETS

American Acrylic Corp.
173 Marine St.
Farmingdale, NY 11735
M. Ziegler (516) 249–1129

These cast acrylic sheets range in thickness from .045 in. to .125 in. and are reinforced with a light loading of glass fibers to improve rigidity and strength. Light transmission approaches that of plate glass, and the sheets are UV stabilized to reduce long-term yellowing.

Maximum operating temperature: 200° F continuous; 225° F intermittent. **Longevity:** panels should have useful life in excess of 7-yr. **Options:** available in sizes up to 4 × 10 ft. **Guarantee/warranty:** in case of failure to maintain 85% of light transmission over a 7-yr period, replacement material provided on **pro rata** basis. **Availability:** 4-wk between order and delivery for orders not exceeding 10,000 sq ft. **Suggested retail price:** consult manufacturer.

POLY-GLAZ

Sheffield Plastics, Inc.
P.O. Box 248, Salisbury Rd.
Sheffield, MA 01257
Thomas Kradel (413) 229–8711

This glazing material is a clear, light stable polycarbonate available in sizes up to 6 × 8 ft; thickness ranges from 0.030 to 0.5 in.

Maximum operating temperature: 250° F. **Longevity:** 10 to 15 yr. **Suggested retail price:** consult manufacturer.

SUNADEX®

ASG Industries, Inc.
P.O. Box 929
Kingsport, TN 37662
M.L. Lilly (615) 245–0211, (800) 251–0266, (800) 251–0461

This low-iron crystal glass contains virtually no iron and has a solar transmittance of 91.6%. The lightly stippled surface of this rolled glazing reduces specular reflectance.

Maximum operating temperature: 400° F continuous. **Longevity:** indefinitely.

SUN-LITE®

Kalwall Corp., Solar Components Div.
P.O. Box 237
Manchester, NH 03105
Scott F. Keller (603) 668–8186

Sun-Lite is a fiberglass-reinforced, polymer glazing material specifically developed for solar collector cover applications. This partially translucent material offers 85 to 90 percent solar transmittance and high-impact strength and durability:

Maximum operating temperature: 300° F. **Longevity:** 20-yr min. **Features:** Kalwall Weatherable Surface applied at factory for additional surface erosion protection. **Installation requirements/considerations:** typical unsupported span of 30 in.; minimum distance to mechanical fastener of ¾ in. **Guarantee/warranty:** none expressed or implied. **Manufacturer's technical services:** customer requirements discussed. Maintenance requirements: cleaning and field-applied weatherable surface coating at 10-yr intervals suggested. **Availability:** approximately 2-wk between order and delivery. **Suggested retail price:** $.45 to $.65/sq ft.

SUN-LITE® GLAZING PANELS

These insulated covers for individual solar collectors or large, site-built arrays are fiberglass-reinforced polymer sheets sandwiched with aluminum and laminated to an aluminum frame. The panels have a maximum operating temperature of 300° F and are available in three standard sizes.

Maximum operating temperature: 300° F. **Longevity:** 20-yr min. **Features:** fiberglass/aluminum bonding system. **Options:** ½-in., 1½-in. or 2¾-in. thick panels in custom sizes. Also available with inner layer of Teflon.

Installation requirements/considerations: maximum unsupported span is 2-ft for the ½-in. panel and 4-ft for the 1½-in. panel. **Guarantee/warranty:** none expressed or implied. **Maintenance requirements:** cleaning and field applied weatherable surface coating at 10-yr intervals suggested. **Availability:** 2-wk between order and delivery for standard panels; 4 to 6 wk for custom panels. **Suggested retail price:** $2.00 to $2.50/sq ft.

SUNWALL[R]

Sunwall[R] is the Kalwall[R] panel system composed of fiberglass-reinforced polymer Sun-Lite[R] face sheets permanently bonded to a supporting aluminum grid core. These sandwiched panels have a maximum operating temperature of 225° F.

Maximum operating temperature: 225° F. **Longevity:** 20-yr min. **Features:** Clamp-tite[R] aluminum installation system. **Options:** intermediate glazing layers; partially translucent (standard) or transparent. **Installation requirements/considerations:** complete instructions available. **Guarantee/warranty:** consult manufacturer. **Maintenance requirements:** cleaning and field-applied weatherable surface coating at 10-yr intervals suggested. **Manufacturer's technical services:** quotation service when inquiry accompanied by architectural plans. Technical services to discuss requirements. **Availability:** 4 to 6 wk between order and delivery. **Suggested retail price:** $4.50 to $6.50/sq ft.

HEAT EXCHANGERS

BRYANT SOLAR HOT WATER COIL

BDP Co.
7310 West Morris St.
Indianapolis, IN 46231
Robert J. Johnson (317) 243–0851

This plate-finned coil transfers energy from solar heated air to water circulating through tubes in the coil. Coil has aluminum fins, copper tubes, and fiberglass insulation.

Features: low pressure drip. **Installation requirements/considerations:** consult manufacturer. **Guarantee/warranty:** 1-yr limited warranty. **Availability:** 8-wk between order and delivery. **Suggested retail price:** consult manufacturer.

CHILL CHASER™ ROOMWARMER, MODEL WD-10-TF

Turbonics, Inc.
11200 Madison Ave.
Cleveland, OH 44102
David R. Essen, Mgr. Sales (216) 228–9663

The Chill Chaser Roomwarmer is a unit heater containing a motor, fan, water coil, water pump, and temperature and speed control; economical heat extraction from any hot water source; ideal for solar hydronic space heating.

Features: water pump is sealed and magnetically driven and speed coordinated with fan motor. **Options:** wall-mounted room thermostat. **Installation requirements/considerations:** pipe-connections to water source (can be static, not pumped) and two-wire ac. **Guarantee/warranty:** standard limited, 1-yr from shipment. **Maintenance requirements:** oil motor, clean filter. **Manufacturer's technical services:** system application engineering suggestions available. **Availability:** from stock, 1-wk ARO. **Suggested retail price:** $298, FOB Cleveland, $5 higher in west.

HEAT EXCHANGER 5840, 5841

Refrigeration Research, Inc., Solar Research Div.
525 N. Fifth St.
Brighton, MI 48116
Jerry Kay (313) 227–1151

May be used to provide heat interchange between heat transfer fluid and domestic hot water.

Features: continuous copper coil with no internal joints. **Installation requirements/considerations:** should be installed for counterflow operation. Depending upon application, a circulating pump will be required in one or both of the water circuits. **Guarantee/warranty:** 5-yr, pro rata, limited. **Maintenance requirements:** freeze and corrosion protection required for collector circuit. **Manufacturer's technical services:** limited assistance is available by mail or telephone. **Availability:** 4 to 6 wk. **Suggested retail price:** consult manufacturer.

HEAT EXCHANGERS

Heat exchangers for heat pumps, commercial refrigeration, air-conditioning, etc. Installed in discharge line between compressor and condenser. Designed to pick up not over 25% of the rated load of the condenser.

Features: designed for interchange between super-heated

refrigerant gas and water. **Installation requirements/considerations:** should be installed on systems using automatic or thermostatic expansion valves. **Guarantee/warranty:** 5-yr pro rata, limited warranty. **Availability:** 4 to 6 wk.

HEAT RECLAIMING HEAT EXCHANGER 5832

Where there is more or less steady or frequent use of hot water, this heat exchanger can be used to reclaim much of the heat that normally leaves the building as warm waste water. Primarily for commercial applications.

Installation requirements/considerations: installed horizontally, inclined slightly downward with a short, upward inclined riser at the outlet. Should be insulated and connected for counterflow operation. **Guarantee/warranty:** 5-yr pro rata, limited warranty. **Maintenance requirements:** nothing should project inside the drain line to interfere with the flow. **Availability:** 4 to 6 wk. **Suggested retail price:** consult manufacturer.

ROM-AIRE HX-10

Solar Energy Products Co.
121 Miller Road
Avon Lake, OH 44012
William L. Maag (216) 933–5000

A highly effective, air-to-water heat exchanger fabricated with a continuous copper tube expanded onto aluminum fins. It is designed for 10,000 Btu/hr at nominal air and water temperatures.

Features: integral 16-ga. mounting flange, top and bottom. **Installation requirements/considerations:** to be used in conjunction with a 300 to 500 cfm blower. **Guarantee/warranty:** 5-yr limited warranty. **Availability:** from stock. **Suggested retail price:** $75.

SOLAR SYSTEMIZER HYDRONIC MODULE 1100, 1200, 1300, 1400

Solar Energy Systems, Inc.
One Olney Ave.
Cherry Hill, NJ 08003
Nathan E. Brussels (609) 424–4446

A heat exchanger/pump/control package, pre-wired and preassembled. Models are available for domestic hot water and for space heating systems in single-wall and double-wall versions.

Features: differential thermostat offers adjustable set-points and new sensor design. **Options:** double- or single-wall construction. **Installation requirements/considerations:** must be mounted on vertical wall. Capacities are for 50% solution of propylene glycol. **Guarantee/warranty:** 5-yr on heat exchanger, 1-yr on other components. **Availability:** 2-wk. **Suggested retail price:** consult manufacturer.

SOLATHERM FIREPLACE WATER-HEATERS (Nine models)

Solatherm Corp.
1255 Timber Lake Drive
Lynchburg, VA 24502
W.W. Hays, III (804) 237–3249

A water cooled fireplace heat absorber/fuel grate that extracts heat for use as solar supplement or as sole heat source, for domestic hot water or space heating.

Features: multiple sizes to fit standard masonry or steel fireplaces, for new or existing construction. **Installation requirements/considerations:** standard hydronic plumbing techniques. **Guarantee/warranty:** 25-yr guarantee against burnout, limited to replacement of unit; 5-yr material and workmanship. **Maintenance requirements:** none, if operated according to recommendations. **Availability:** 2 to 3 wk. **Suggested retail price:** $160 to $420.

SUN SPONGE HEAT EXCHANGER

A-1 Prototype, Inc.
1288 Fayette St.
El Cajon, CA 92020
Jerry Hull (714) 449–6726

100-gal stainless steel, non-pressurized tank, 26-in. sq × 62½ in. high. Single-wall heat exchanger, 120 ft of ¾-in. copper. 3-in. styrene insulation, R = 15.

Features: automatic drain-down for freeze protection, sensors included. **Options:** pump and controller. **Installation requirements/considerations:** make sure tank is vented while filling (non-pressurized system). **Guarantee/warranty:** 5-yr. **Maintenance requirements:** drain tank every 6 mo. **Manufacturer's technical services:** design assistance. **Availability:** 2-wk. **Suggested retail price:** $850.

HEAT TRANSFER FLUIDS

BRAYCO 888 & BRAYCO 888 HF

Bray Oil Company, Inc.
1925 North Marianna Ave.
Los Angeles, CA 90032
Eugene R. Slaby (213) 268–6171

A heat transfer fluid based on a synthetic oil and formulated to provide long life in the presence of copper and aluminum. Designed for use in most solar collectors, it is noncorrosive and nonpoisonous.

Features: exhibits excellent thermal and oxidative stability and has an unusually high flash point for a hydrocarbon fluid of this type. **Guarantee/warranty:** 1-yr from date of shipment. **Manufacturer's technical services:** quality control and research labs; technical representatives. **Availability:** immediate delivery. **Suggested retail price:** 888, $3.80/gal; 888HF, $4.85/gal.

SOLAR WINTER BAN

Solar Alternative, Inc.
30 Clark Street
Brattleboro, VT 05301

This heat transfer fluid consists of propylene glycol, dipotassium phosphate and water with food coloring; it can be used with single or double walled heat exchanger in closed loop systems. Acceptability is conditional upon acceptance of the state, conformance to local health codes, and future determination of potability by government agencies.

Features: food grade certified red dye for easy identification. **Options:** gal jugs or 55 gal drum. **Guarantee/warranty:** 1-yr limited warranty. **Maintenance requirements:** pH must be checked annually. **Availability:** 3-wk between order and delivery. **Suggested retail price:** $4.90/gal.

SUNSOL 60

Sunworks Div., Enthone, Inc.
P.O. Box 1004
New Haven, CT 06508
Floyd C. Perry (203) 934–6301

A non-toxic, non-flammable, heat transfer medium for solar collectors, Sunsol 60 contains corrosion inhibitors to prolong the life of copper and steel components of solar heating installations.

Special features: contains a certified non-toxic dye to aid in identifying leaks that may occur in systems. **Options:** 1, 5 and 55 gal containers. **Installation requirements/considerations:** undiluted, Sunsol 60 freezes at -55° F. In areas where the minimum winter temperature is higher, water may be added. **Guarantee/warranty:** none. **Maintenance requirements:** without periodic pH and hydrome-

ter testing, replace every 3-yr; with testing, replace only if required. **Availability:** 3-5 days. **Suggested retail price:** consult manufacturer.

INSULATION

BOARD URETHANE INSULATION

Bally Case and Cooler, Inc.
Bally, PA 19503
Leon Prince (215) 845–2311

Board insulation manufactured of 97%, closed-cell urethane, foamed in place under high temperature and pressure in steel molds.

Features: high insulating value: R = 8.475/in.; 25 flame spread rating; easy to cut and shape. **Options:** available in widths of 24 and 48 in., lengths of 51½ and 103 in., thickness of 1 to 4 in. in 1 in. increments. **Installation requirements/considerations:** requires warm side vapor barrier and compliance with local building code requirements. **Availability:** 6-wk. **Suggested retail price:** consult manufacturer.

DYRELITE

Dyrelite Corp.
63 David St.
P.O. Box B-947
New Bedford, MA 02741
Randall T. Weeks (617) 993–9955

Expanded polystyrene insulation, produced as small beads, expanded under controlled conditions to form a lightweight prefoam which is then fused together to form large blocks, subsequently cut into sheets of required size and thickness. R = 4.0 to 5.0/in. depending upon density of formulation.

Features: lightweight; rot, mildew, vermin and rodent proof. **Options:** densities from 1 to 2½ lb/ft^3; sizes to 4 × 8 ft; ¼ to 20 in. thick. Available in self-extinguishing formulation. **Guarantee/warranty:** none. **Availability:** 2 to 14 days. **Suggested retail price:** depends on quantity and type. Consult manufacturer.

HEAT SAVER INSULATED SWIMMING POOL BLANKET

MacBall Industries, Inc.
5765 Lowell St.
Oakland, CA 94608
W.A. McKirdy (415) 658–1124

A reinforced foam floating insulating blanket designed to reduce evaporation and heat loss and to help keep pool clean. Winding and storage devices available for large pools.

Options: comes in one piece or sections. **Guarantee/warranty:** 1-yr, optional second-year service contract. **Maintenance requirements:** occasional hosing off. **Availability:** 2 to 4 wk. **Suggested retail price:** 45¢ to 55¢/sq ft.

INSTA-FOAM & INSTA-SEAL FROTH PAKS

Insta-Foam Midwest
8000 47th St.
Lyons, IL 60534
Kenneth H. Mettam

Pressurized one- and two-component polyurethane foam paks for insulating and caulking ducts, board stock, collectors, tanks and pipe chases.

Features: no external power source needed to operate; portable and self-contained. **Manufacturer's technical services:** technical data and application bulletins. **Availability:** in stock. **Suggested retail price:** $5.75 to $432.

INSULJACK

Urethane Molding Inc.
RFD No. 3, Route 11
Laconia, NH 03246
Randy H. Annis (603) 524–7577

Pipe insulation designed to slide over fluid carriers as the lines are being installed. Ridged PVC exterior with an interior of urethane foam, molded to a specific diameter.

Features: insulates both supply and return lines in one, waterproof, exterior jacket. **Guarantee/warranty:** limited to the purchase price of material found to be of defective manufacture. **Availability:** 7 to 10 days, on orders of 500 ft or less. **Suggested retail price:** average $3.75/ft.

IS HIGH "R"™ SHADE

Insulating Shade Limited Partnership
17 Water St.
Guilford, CT 06437
Thomas P. Hopper (203) 453–9334

This high-resistance insulation shade is made of multiple layers of metallized plastic films with unique spacers attached at intervals to separate the layers when pulled down. A three-layer shade develops a resistance of about R = 6; a five-layer shade develops a resistance of about R = 10. Shade is effective for both heating and air conditioning. Side and top seals prevent air circulation at the perimeter. Shades are 56 × 96 in.

Features: compact size (a five-layer shade compacts to diameter of 2½ in.). **Options:** number of layers (3 or 5). **Installation requirements/considerations:** maximum efficiency gained if window is insulated. Instructions accompany purchase. **Guarantee/warranty:** limited warranty. **Suggested retail price:** $1.75/sq ft for the three-layer shade; $2.75/sq ft for the five-layer shade.

KLEGE-CELL INSULATION/TYPE 33

American Klegecell Corp.
204 N. Dooley St.
Grapevine, TX 76051
Philip Wilkens (817) 481–3547

Rigid, closed-cell, polyvinyl-chloride foam for insulation and structural sandwich construction; to be used on tanks, piping, and components where temperatures do not exceed 200° F.

Features: rigid sheets or blocks, heat formable, easy to cut and shape, unaffected by water, direct sunlight or cold temperatures. **Options:** ten types that vary in density. **Installation requirements/considerations:** not for use where temperatures exceed 200° F. **Manufacturer's technical services:** assistance available. **Availability:** 2 wks. **Suggested retail price:** consult manufacturer.

THERMACOR A, C, D, AND THERMAFAB

Thermacor Process, Inc.
P.O. Box 4529
Forth Worth, TX 76106
Richard B. Bender II (817) 624–1181

A patented thermal-insulation and corrosion-control process for use on nearly any type or size of pipe. The process provides an inner and outer layer of urethane molecules several mils thick, yielding excellent compressive strength and low K factor.

Features: excellent insulation properties, initial value per inch R = 7.7. **Installation requirements/considerations:** suitable for applications from -325° to +225° F. **Guarantee/warranty:** 1-yr (requires notification of fault within 30 days of detection). **Availability:** 2-wk. **Suggested retail price:** consult manufacturer.

THERMACOR CTI

Thermacor Process, Inc.
P.O. Box 4529
Ft. Worth, TX 76106
Richard B. Bender II (817) 624–1181

Closed-cell urethane insulation that is machined from foam that has been manufactured as pre-cured slab stock to provide Cut Thermal Insulation for pipe with an all-service jacket of aluminum, fiberglass and Kraft paper or cut to desired size and shape.

Features: excellent insulating value, more than R = 6.66/in. **Options:** various outer jackets. **Installation requirements/considerations:** suitable for applications from -65° F to +250° F. **Guarantee/warranty:** 1-yr (requires notification of fault within 30 days of detection). **Availability:** 2-wk. **Suggested retail price:** consult manufacturer.

MEASUREMENT AND DATA COLLECTION DEVICES

AEOLIAN KINETICS DATA LOGGER

Aeolian Kinetics
P.O. Box 100
Providence, RI 02901
Ralph Beckman (401) 274–3690

This solid-state data logger with CMOS components is used for remote data collection of solar, wind, or any other analog or digital data. Data is stored in semiconductor memory for recall at any time.

Features: operates to -40° C for 6-yr on four D cells. **Options:** available as accessories: up to 64K memory, A-D converter, input scaling. **Guarantee/warranty:** 1-yr limited warranty. **Availability:** stock. **Suggested retail price:** $725 with clock and 3k memory.

ALPHATOMETER

Devices & Services Co.
3501-A Milton
Dallas, TX 75205
Charles Moore (214) 368–5749

The Alphatometer is a miniaturized, thermopile-type pyranometer that makes possible routine and rapid measurements of solar radiation, reflectivity, absorptivity, and transmissivity. The detector comes with a mounting board, which provides a support for absorbing and transparent materials, and a solar angle of incidence indicator.

Features: small size, easy to use. **Options:** portable mounting board for collector testing, tripod for mounting standard board. **Installation requirements/considerations:** detector mounts on board which mounts on standard photographic tripods. **Guarantee/warranty:** 1-yr warranty. **Maintenance requirements:** a desiccant is provided to eliminate condensation inside of bubble that covers front of detector. **Manufacturer's technical services:** available for applications and maintenance. **Availability:** 2-wk. **Suggested retail price:** $550.

ANGSTROM PYRHELIOMETER ANG

The Eppley Laboratory, Inc.
12 Sheffield Ave.
Newport, RI 02840
George L. Kirk (401) 847–1020

A standard pyrheliometer for the accurate measurement of direct solar radiation. A calibrating instrument.

Features: used to maintain International Pyrheliometric Society (1956) scale; accurate to 1.5%. **Options:** control unit with digital readout. **Guarantee/warranty:** 1-yr. **Manufacturer's technical services:** complete repair and calibration facilities. **Availability:** 8-wk. **Suggested retail price:** $3,100.

BLACK AND WHITE PYRANOMETER 8-48

An instrument for the measurement of global (total sun and sky) radiation. This pyranometer is suitable for long-term monitoring of solar collection systems or determination of available solar energy.

Features: accuracy and precision of 2%. **Options:** readout devices. **Installation requirements/considerations:** clear field of view either horizontally or in the plane of the collector. **Guarantee/warranty:** 1-yr. **Maintenance requirements:** periodic inspection and cleaning of dome. **Manufacturer's technical services:** complete repair and calibration facilities. **Availability:** 3 wk. **Suggested retail price:** $590.

CHAMPION SOLAR MONITOR

Champion Home Builders Co.
Solar Products Div.
118 Walnut St.
Waynesboro, PA 17268
Al Cool (717) 762–3113

Solar Monitor has five temperature sensors, probe select, ambient temperature scale, and collection, distribution and auxiliary furnace indicator lights that report the system's mode of operation.

Availability: consult manufacturer. **Suggested retail price:** consult manufacturer.

DIFFERENTIAL THERMOSTAT TESTER

Heliotrope General
3733 Kenora Drive
Spring Valley, CA 92077
Al Abernathy (714) 460–3930

This differential thermostat tester verifies the performance and accuracy of the Delta-T Differential Temperature Thermostats. The tester simulates temperature. It has no recording capability; no power input is required to use the tester.

Guarantee/warranty: 90 days. **Availability:** stock. **Suggested retail price:** $13.50.

DIGITAL ELECTRONIC THERMOMETER (ITS-600 and ITS-650)

Sun Spot Research
1070 South Leyden
Denver, CO 80224
Harold L. Moses (303) 321–7323

This hand-held, self-calibrating, digital electronic thermometer is equipped with an internal NiCad energy cell.

The model has one or two temperature range selections and reads out in Fahrenheit or Centigrade. An AC adaptor is featured. Size is 3 × 6 × 1¼ in.

Installation requirements/considerations: interchangeable probes not included. **Guarantee/warranty:** 1-yr warranty. **Maintenance requirements:** can not be submersed in liquids. **Availability:** 5 to 7 days between order and delivery. **Suggested retail price:** $139.95 (ITS-600); $149.95 (ITS-650).

DIGITAL ELECTRONIC THERMOMETER (PM-100 and PM-150)

Sun Spot Research
1070 South Leyden
Denver, CO 80224
Harold L. Moses (303) 321–7323

This panel-mounted, digital-readout electronic thermometer uses a 12 vdc power source and is equipped with a metal encapsulated thermistor. It fits a standard 2½-in.-cutout and reads Fahreheit or Centigrade.

Features: Fahrenheit or Centigrade readout on either model. **Options:** 200 plus feet of probe wire from monitor. **Installation requirements/considerations:** 12 vdc power source (unregulated). **Guarantee/warranty:** 1-yr warranty. **Availability:** 1-wk between order and delivery. **Suggested retail price:** $79.95.

DESK-TOP TEMPERATURE MONITOR

Sun Spot Research
1070 South Leyden
Denver, CO 80224
Harold L. Moses (303) 321–7323

This desk-top or console-mounted, electronic temperature monitor comes with a four-channel input with optional additional inputs. The monitor is 8½ × 6¼ × 3¼ in. and weighs 2 lb. It reads out in Centigrade or Fahrenheit.

Features: four-channel input. **Options:** two temperature range selections. **Installation requirements/considerations:** AC adaptor for internal NiCad energy cell. **Guarantee/warranty:** 1-yr warranty. **Maintenance requirements:** cannot be submersed in fluids. **Availability:** 7 to 10 days between order and delivery. **Suggested retail price:** $265.95 plus probes.

DIGITAL THERMOMETER/MODEL 100

Iowa Solar, Inc.
Box 246
North Liberty, IA 52317
Tom Miller (319) 626–2342

Solid-state, continuous readout temperature monitor with twelve channels. Uses thermistor probes and measures from 30 to 212° F (also reads out in Centigrade). Accuracy is plus or minus 1°.

Options: chart recorder output and recorder; can be built for other ranges. **Guarantee/warranty:** 1-yr on parts and labor. **Manufacturer's technical services:** repair and calibration. **Availability:** 1 to 2 wk. **Suggested retail price:** 265 (probes are $18).

ELECTRONIC THERMOMETER

Heliotrope General
3733 Kenora Drive
Spring Valley, CA 92077
Al Abernathy (714) 460–3930

This electronic thermometer, with strip chart capability, monitors temperatures from 0 to 350° F at distances up to 1000 ft from the instrument. Eleven different locations can be monitored.

Options: assorted designs of sensors available. **Guarantee/warranty:** 90 days. **Availability:** stock. **Suggested retail price:** $74.50.

ELECTROTHERM DIGITAL ELECTRONIC THERMOMETER (M-99 and TC-100)

Sun Spot Research
1070 South Leyden
Denver, CO 80224
Harold L. Moses (303) 321–7323

This digital, electronic temperature measuring device gives a Fahrenheit or Centigrade readout on either the M-99 or TC-100 model. The M-99 uses a 9-v disposable battery; the TC-100 uses a rechargeable internal NiCad. Temperature range is 32° F to 230° F.

Features: 2½ in. wide by 5 in. long by 2 in. high. Weight: 5½ oz. **Guarantee/warranty:** 1-yr warranty. **Availability:** approximately 1-wk between order and delivery. **Suggested retail price:** $59.95 (M-99) and $69.95 (TC-100).

EMISSOMETER MODEL AE

Devices and Services Co.
3501-A Milton
Dallas, TX 75205
Charles Moore (214) 368–5749

An instrument for measuring emissivity. No temperature measurements are needed. With high and low emissivity standards provided, the Emissometer can be periodically recalibrated by the user by means of a set-screw adjustment.

Features: rapid measurements and easy to operate. **Installation requirements/considerations:** voltmeter with at least 0.01-mv sensitivity is required to measure output of Emissometer. **Guarantee/warranty:** 1-yr. **Maintenance requirements:** periodic calibration, which is done by user with supplied standards. **Manufacturer's technical services:** for applications and maintenance. **Availability:** 2-wk. **Suggested retail price:** $550.

NORMAL INCIDENCE PYRHELIOMETER

The Eppley Laboratory, Inc.
12 Sheffield Ave.
Newport, RI 02840
George L. Kirk (401) 847–1020

An instrument for the measurement of either total or spectral direct solar intensity.

Features: accuracy and precision of 1%. **Options:** colored glass filters to isolate different wavelength regions; readout devices. **Installation requirements/considerations:** generally employed with a solar tracker. **Guarantee/warranty:** 1-yr. **Maintenance requirements:** periodic inspection to insure alignment on sun and that cable is not twisted and window clear. **Manufacturer's technical services:** complete repair and calibration facilities. **Availability:** 6-wk. **Suggested retail price:** $890.

PRECISION INFRARED RADIOMETER PIR

The Eppley Laboratory, Inc.
12 Sheffield Ave.
Newport, RI 02840
George L. Kirk (401) 847–1020

An instrument for the measurement of unidirectional, terrestrial, long-wave radiation, 4 to 50 micrometers.

Features: new silicon hemisphere for improved performance. **Options:** readout devices. **Installation requirements/considerations:** clear field of view, either horizontally or in the plane of the collector; often used inverted. **Guarantee/warranty:** 1-yr. **Maintenance requirements:** periodic inspection and cleaning of dome. **Manufacturer's technical services:** complete repair and calibration services. **Availability:** 6-wk. **Suggested retail price:** $1,190.

PRECISION SPECTRAL PYRANOMETER PSP

The Eppley Laboratory, Inc.
12 Sheffield Ave.
Newport, RI 02840
George L. Kirk (401) 847–1020

An instrument for the precise measurement of sun and sky radiation, totally or in defined wave-length bands.

Features: accuracy and precision of 1%. **Options:** colored glass hemispheres to isolate different wave-length regions; readout devices. **Installation requirements/considerations:** clear field of view, either horizontally or in the plane of the collector. **Guarantee/warranty:** 1-yr. **Manufacturer's technical services:** complete repair and calibration facilities. **Availability:** 6-wk. **Suggested retail price:** $990.

LINI-TEMP TEMPERATURE SENSOR

Texas Controls, Inc.
P.O. Box 59469
Dallas, TX 75229
F.W. Schempf (214) 386–5000

Lini-Temp is a -50° F to 250° F temperature sensor with linear output (typically 3.218 v at 100° F) that changes 7.16 mv/°F and is inversely proportional to temperatures. Solid-state components are used for detection, linerization, and regulation. Sensors are interchangeable without the need for calibration.

Features: linear output across entire -50° F to 250° F range, reference junction not required. **Options:** enclosure options: ambient-air type, surface, liquid immersion, air duct. **Installation requirements/considerations:** twisted shielded pair (AWG 20), 24 vdc power supply regulated to 0.10 percent, draws 100 microamps/sensor. **Guarantee/warranty:** 90 days against defects in material and workmanship. **Manufacturer's technical services:** consultation from Dallas office. **Availability:** 30 days. **Suggested retail price:** $26 in single quantity.

RECORDING PYRANOMETER MR-5A

Hollis Observatory
One Pine St.
Nashua, NH 03060
Joseph F. Litwin (603) 882–5017

A silicon-based pyranometer system for the continuous recording of total sun and sky radiation. Recording is on dry-writing, pressure-sensitive paper.

Features: temperature and cosine response corrected. **Options:** MR-5, used with potentiometric readout equipment. Also available, LM-100 digital integrator. **Guarantee/warranty:** 1-yr. **Maintenance requirements:** periodic inspection for cleanliness and replacement of chart paper. **Manufacturer's technical services:** repair and recalibration; engineering services for specialized instrumentation development. **Availability:** 2-wk. **Suggested retail price:** consult manufacturer.

SCALING DIGITAL VOLTMETER

Devices and Services Co.
3501-A Milton
Dallas, TX 75205
Charles Moore (214) 368–5749

This readout enables the Alphatometer and Emissometer user to read reflectivity, transmissivity, and emissivity directly. With fixed scale the Alphatometer (or other pyranometer) output can be displayed in engineering units. The liquid crystal display is easily read outdoors.

Features: variable scaling, liquid crystal display, portable, single 9-v transistor, battery powered. **Installation requirements/considerations:** comes complete with battery, ready to use. **Guarantee/warranty:** 1-yr. **Maintenance requirements:** battery, carbon zinc lasts over 120 hr of use, alkaline over 200 hr. **Availability:** 2-wk. **Suggested retail price:** $295.

SHADOW MAPPER

Campbell Engineering
1302 Toney Drive
Huntsville, AL 35802
Dr. R. A. Campbell (205) 883–9866

A portable solar shadow mapping instrument which provides immediate and accurate data for determining the time, date, and duration of annual shadows at proposed solar collector sites. Can be operated by non-technical personnel.

Features: aluminum case coverts into the instrument's mounting platform. **Guarantee/warranty:** 1-yr guarantee against defects in materials or workmanship. **Availability:** 4 to 5 wk after receipt of order. **Suggested retail price:** $376.50

SOL-A-METER MK 1-G RADIOMETER

Matrix, Inc.
537 S. 31st St.
Mesa, AZ 85204
Don Pershing (602) 832–1380

A weatherproof pyranometer for measuring total radiation from sun and sky. Fast response.

Features: calibrated in both English and metric scales. **Options:** meter or recorder readout available. **Guarantee/warranty:** 1-yr. **Maintenance requirements:** dome must be kept clean. **Availability:** from stock. **Suggested retail price:** $195.

SOL-A-METER MK 3 RADIOMETER

A weatherproof pyrheliometer for measurment of direct solar radiation at normal incidence. Constructed with an internally baffled, blackened, 10-in. collimating tube with a 5.7 degree aperture, in compliance with W.M.O. recommendation.

Features: hand-held measurements can be made because of rapid response time. Calibrated in both English and metric scales. **Guarantee/warranty:** 1-yr. **Maintenance requirements:** keep window clean. **Availability:** from stock. **Suggested retail price:** $295.

SOL-A-METER MK 14-E RADIOMETER

This weatherproof pyranometer gives total integrated insolation on a digital display, instantaneous insolation on a meter. Self-contained, deriving all necessary power from batteries and a solar-powered battery charger built in.

Installation requirements/considerations: instrument must be level. **Maintenance requirements:** keep clean; check air-drying crystals to make sure internal humidity is not too high. **Guarantee/warranty:** 1-yr. **Availability:** from stock. **Suggested retail price:** $695.

SOL-A-METER SOLAR RADIOMETER MK 6

Hand-held solar radiometer and transmittance meter

measures solar energy directly in Btu/hr/ft^2 and solar transmission and reflectance of materials in percent. Self-powered.

Features: highly portable: 5½ by 5½ by 2½ in. under 2lb. **Options:** galvanometric, inkless recorder available. **Guarantee/warranty:** 1-yr. **Availability:** from stock. **Suggested retail price:** $195.

SOLAR INSOLATION METER SIM-300

Entropy Limited
5735 Arapahoe Ave.
Boulder, CO 80303
Hank Valentine (303) 443–5103

Measures and displays real time solar radiation in Langleys/min plus integrated insolation in Langleys. Solar powered, with battery for night time integrated readouts. Packaged in industrial suitcase with 50-ft cable. May be removed from case for permanent installation.

Features: provides automatic digital display of incident insolation in Langleys/min, will also display integrated insolation readout to 99,999 Langleys before resetting to zero. **Installation requirements/considerations:** normal operating range: -20° C to +45°C. **Guarantee/warranty:** 90 days, materials and workmanship. **Suggested retail price:** $1200.

SOLAR TRACKER ST-1, ST-3

The Eppley Laboratory, Inc.
12 Sheffield Ave.
Newport, RI 02840
George L. Kirk (401) 847–1020

A device that accomodates from one to three Normal Incidence Pyrheliometers, and points them directly at the sun to permit continuous measurements of direct solar radiation.

Features: tracking accuracy of ±0.25% daily. **Options:** 120 vac/60 Hz standard, 220 vac/50 Hz available. **Installation requirements/considerations:** same as Normal Incidence Pyrheliometers. **Guarantee/warranty:** 1-yr. **Maintenance requirements:** periodic inspection to remove twist from pyrheliometer cable and to check alignment. **Manufacturer's technical services:** complete repair and calibration facilities. **Availability:** 6-wk. **Suggested retail price:** $740 and $1,190.

SOLAR POWER METER

Liconix
1400 Stierlin Road
Mountain View, CA 94043
Mark W. Dowley (415) 964–3062

Designed for direct measurement of solar radiation, the Solar Power Meter is battery powered and based on a calibrated silicon solar cell. It has dial reading and chart recorder output; reads in kw/m^2.

Features: direct reading of solar power; includes ambient suppression circuitry. **Options:** laser wavelength calibrating; consult manufacturer. **Guarantee/warranty:** 1-yr limited warranty. **Maintenance requirements:** battery replacement. **Availability:** 2-wk between order and delivery. **Suggested retail price:** $800.

SOLAR SITE SELECTOR

Don Lewis Associates
Box L
Sutter Creek, CA 95685
Sheri Lewis (209) 296–4943

The Solar Site Selector calculates total solar day/hours and determines shading patterns occuring during winter for correct site of collector orientation. Baseplate includes bubble level, compensating compass, 180° distortion optic at viewpoint. Grids silk-screened in 2-degree even latitutes on transparent mylar.

Features: when viewed through the 180° distortion optic, the predictable sunpaths and hour segments are superimposed on the site being studied. **Options:** threaded insert

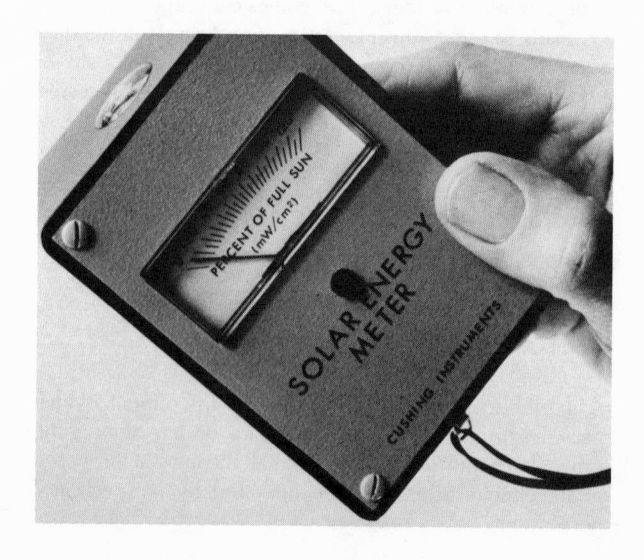

in baseplate fits photographic tripods. **Guarantee/warranty:** 90 days for defects in materials and workmanship. **Availability:** 4-wk max. **Suggested retail price:** $29.50 Plus $2 postage and handling.

TEMPERATURE ALARM MONITOR

Sun Spot Research
1070 South Leyden
Denver, CO 80224
Harold L. Moses (303) 321–7323

This desk-top or console-mounted electronic thermometer has a high and low alarm system and an internal NiCad energy cell. Alarm can be audio or visual.

Features: 8½ × 3¼ × 6¼ in. Centigrade or Fahrenheit readouts; interchangeable probes not included. **Installation requirements/considerations:** ac adaptor for internal NiCad energy cell. **Guarantee/warranty:** 1-yr warranty. **Maintenance requirements:** can not be submersed in liquids. **Availability:** 7 to 10 days between order and delivery. **Suggested retail price:** $279.95 plus probes.

PASSIVE PRODUCTS

BEADWALL[R]

Zomeworks Corp.
P.O. Box 712
Albuquerque, NM 87103
Steve Baer (505) 242–5354

Double-glazed windows when empty, insulated walls when filled, Beadwalls are designed to be filled (with Styrofoam™ beads) and emptied by means of vacuum and blower-motor storage units. Beadwalls may be used at angles up to 45° and are 2½ in. thick (standard). Insulation value is R-8 when filled, R-1.5 when empty. Typical dimensions are 38⅛ or 50⅛ × 80⅛ in. for single panels and 80⅛ or 104⅛ × 80⅛ in. in pairs. Each motor/storage unit can fill and empty a pair of panels up to 40 ft distant.

Features: Zomeworks provides all components except glazing and the electrical and bead-transport connections between panels and motor storage units. **Options:** automatic solar control assembly for emptying and filling. Plans, including license, for building your own Beadwall available at $15. **Installation requirements/considerations:** wall should be south wall, not more than 45° from vertical. **Guarantee/warranty:** 1-yr on materials and workmanship. **Manufacturer's technical services:** advice on feasibility of installations. **Availability:** 2-wk. **Suggested retail price:** $400 to $600.

EGGE RESEARCH BEADWALL[R]

Egge Research
Box 394B RFD
Kingston, NY 12401
Hank Starr (914) 336–5597

Egge Research Beadwalls are clear, double-glazed windows with a 4-in. space between glazing into which Styrofoam™ beads are introduced by blowers that distribute the beads through a standard plastic pipe. The system is automatically controlled, using a light sensor, and has a manual override. When beads are not in use they are stored in a 16-ga galvanized steel tank by means of suction. Units available (prefabricated): 3 × 6 ft or 4 × 8 ft; custom sizes also available. Window frames are clear 5/4 pine; glazing bowed for rigidity; insulation is white Styrofoam™ beads. U-factor when window is empty, 0.5; when full, 0.078; glass, acrylic, and mar-resistant polycarbonate sheets available. Standard unit is .100-in.-thick acrylic.

Features: Beadwall allows for large areas of glazing without the customary large heat losses. **Options:** windows are available for walls or roofs with a minimum slope of 45 degrees. Custom sizes and various glazing materials also available. Supply pipe can be clear PVC and run exposed to view moving beads. **Installation requirements/consideration:** allowance for standard 1½-in. plastic plumbing to windows from tank. **Guarantee/warranty:** 1-yr parts and labor. **Maintenance requirements:** beads must be treated periodically with antistatic solution and window frames re-stained every 7 to 10 yr. **Manufacturer's technical services:** installation, service, and custom design. **Availability:** 1 to 6 wk between order and delivery. **Suggested retail price:** 3 × 6 ft, $300, 4 × 8 ft, $535. Custom windows are $10 to $20/sq ft, depending on size.

HORIZONTAL INSULATING CURTAIN, MODEL 102

Thermal Technology Corp.
P.O. Box 130
Snowmass, CO 81654
Ron Shore (303) 963–3185

An insulating curtain designed to reduce the losses from

large glazing areas (greenhouse, large skylights). Travels horizontally and effectively reduces the ceiling height; fabric of low emissivity; layers provide air spaces for the high insulation value.

Options: totally automated (solar sensor and override); manual motorized operation; manual non-motorized operation; finished fabric of customer can be incorporated into unit for visible inner layer applications, sinsulate view window. **Installation requirements/considerations:** stacking space (min. 70-ft-long curtain, 4-ft stack). **Guarantee/warranty:** unconditional guarantee for 5-yr of motor and limit switch. **Manufacturer's technical services:** full shop drawing and assistance for custom applications. **Availability:** 30 days. **Distributors:** Thermal Technology Corporation only. **Suggested retail price:** $5/sq ft.

KALWALL[R] SOLAR FURNACE (PAT. PENDING)

Kalwall Corp., Solar Components Div.
P.O. Box 237
Manchester, NH 03105
Scott F. Keller (603) 668–8186

Kalwall's combined collection and storage tubes are placed directly behind Sunwall. Set vertically and filled with water or other material, these tubes act as both the solar absorber and storage unit.

Features: collection and storage are combined into one unit. **Options:** multi-layed solar windows and movable insulation. **Installation requirements/considerations:** should be installed on south-facing wall or roof. **Guarantee/warranty:** none expressed or implied. **Maintenance requirements:** manufacturer suggests cleaning and field refinishing of weatherable surface coating at 10-yr intervals. **Manufacturer's technical services:** sales and technical staff are available to discuss specific projects. **Availability:** 2-3 wk. **Suggested retail price:** consult manufacturer.

NIGHTWALL[R] CLIPS

Zomeworks Corp.
P.O. Box 712
Albuquerque, NM 87103
Steve Baer (505) 242–5354

Magnetic devices for mounting lightweight sheets of rigid insulation, such as 1-in. beadboard, to window areas as an alternative to double glazing. Each ½ × 6-in. clip comprises a steel strip lying on a magnetic strip, each with an adhesive backing. The magnet is adhered to the insulation board, the steel strip to the window, at intervals of 10 to 18 in. around the perimeter of the glazing. The temporary, removable insulation can cut night-time heat losses in winter and reduce unwanted insolation in summer.

Guarantee/warranty: Zomeworks will replace any clips that don't work. **Availability:** from stock. **Suggested retail price:** minimum order, 20 clips for $7.00; .29 to .35 each depending upon quantity ordered.

ONE DESIGN WATERWALL

One Design, Inc.
Mt. Falls Route
Winchester, VA 22601
Tim Maloney (703) 662–4898

These fiberglass waterwall modules measure 8 ft long, 2 ft tall, and 1 ft wide and hold approximately 900 lb of water per module. The horizontal configuration of the modules reduces water pressure, and the rectilinear foremat yields highest possible mass per sq ft of floor space. The units can be used for heating and cooling with no moving parts. Glazing is provided by others. Modules nest in transit and are self-stacking in installation. The molded fiberglass walls contain UV screen. Low water pressure and high volume result from the rectilinear shape of modules; fans not required to remove heat from storage.

Features: can be used in 2- to 12-ft-tall stacks with only .8 lb/sq in. water pressure. **Installation requirements/considerations:** to be located in relation to vertical south glass, greehouse glass, roof apertures, or clerestories. **Guarantee/warranty:** 5-yr unconditional guarantee on manufacturing. **Maintenance requirements:** tolerates same water conditions as a fiberglass bathtub. **Manufacturer's technical services:** installation information provided. **Regional applicability:** most cost-effective in regions requiring both heating and cooling. **Availability:** depends upon production run, 1 day to 8 wk between order and delivery. **Suggested retail price:** consult manufacturer.

SKYLID[R]

Zomeworks Corp.
P.O. Box 712
Albuquerque, NM 87103
Steve Baer (505) 242–5354

Skylids are shutter/louvers that open and close in response to the heat of the sun by a self-contained gravity-shifting system and are designed as ceiling installations for use under skylights. Paired refrigerant-charged canisters actuate louvers to control heat gain or loss. Pre-framed wooden box contains louvers of .040 aluminum curved over wooden ribs and filled with R-11 fiberglass insulation, sub-skylight mounting from horizontal to 75°.

Features: simple manual over-ride control. **Options:** special sizes, boxed units with double-glazed skylights are available at extra cost. **Installation requirements/considerations:** pre-formed, ready to install into a rough opening, much like a pre-hung door. **Guarantee/warranty:** 1-yr on materials and labor. **Maintenance requirements:** seals around edges of louvers may need replacement eventually. **Availability:** 2-wk. **Suggested retail price:** 4 × 4 ft, $266; 6 × 10 ft, $468.

SOLAR SAVER TUBE

U.S. Solar Pillow, Inc.
P.O. Box 88
Grand Junction, CO 81501
Dick Carmack (303) 242–2743

A three-wall plastic tube usually made of 6-mil polyethylene or PVC. It can be fabricated from different materials and in various weights for custom jobs. Applications include greenhouse, outdoor garden and pool heating. Collector and storage combined.

Features: lightweight, portable, versatile, inexpensive. **Options:** fittings for complete system. **Installation requirements/considerations:** minimal, site determined. **Guarantee/warranty:** guaranteed free of defects and workmanship at time of purchase. **Manufacturer's technical services:** performance sizing for systems; installation information and requirements. **Availability:** 6 to 8 wk on custom orders; small quantities available from stock. **Suggested retail price:** 49¢/linear foot.

PHOTOVOLTAICS

PHOTOVOLTAIC BATTERY CHARGERS, SERIES 125SL to 147S

Energy Saving Systems
Price and Pine Sts.
Holmes, PA 19043
Ray Speicher (215) 583–4780

Solar chargers designed for use in marine, aircraft, remote, and home facilities. Kits and educational items available. Panels designed for each specific application.

Features: marine panels designed for on-board hazards; aircraft panels are lightweight and drag-free. **Options:** marine, panels or built into hatch covers and kits; aircraft, panels or kits. **Installation requirements/considerations:** mounted by customer; instructions supplied. **Guarantee/warranty:** limited warranty. **Availability:** 2 to 6 wk between order and delivery. **Suggested retail price:** $79 to $299.

PHOTOVOLTAIC UNIPANELS

Solarex Corp.
1335 Piccard Dr.
Rockville, MD 20850
Bob Edgerton (301) 948–0202

A complete line of silicon photovoltaic cells and panels for direct conversion of sunlight to electricity. Power of panels range from 2 to 30 w/panel; 4, 6, and 12 v systems. **Options:** mounting hardware, voltage regulators, plastic covers, special voltages and currents. **Installation requirements/considerations:** need full view of sun. **Guarantee/warranty:** 1-yr on panels; up to 5-yr on complete systems of Solarex design. **Maintenance requirements:** none, except checking on battery water level. **Manufacturer's technical services:** sales and system engineering assistance. **Regional applicability:** used worldwide. **Availability:** 4 to 6 wk. **Suggested retail price:** varies.

STARBURST™ SOLAR ELECTRIC CELLS

Spire Corp.
Patriots Park
Bedford, MA 01730
Thomas E. Wilber (617) 275–6000

These ion-implanted, silicon photovoltaic cells yield an efficiency of greater than 12 percent AMI at 28° C. 3-in. diameter n on p cells have 4 solderable output pads. Minimum performance parameters are V_{oc} = 575 v; I_{sc} = 1.25 a, and I_{460} = 1.2 a. Typical performance parameters are: V_{oc} = 590 v, I_{sc} = 1.4 a and I_{460} = 1.3 a.

Options: higher efficiency cells and encapsulation are available upon request. **Installation requirements/considerations:** cell encapsulation is recommended. **Maintenance requirements:** none when encapsulated. **Guarantee/warranty:** none specified. **Availability:** 60 days. **Suggested retail price:** consult manufacturer.

PHOTOVOLTAIC SYSTEMS

Sensor Technology, Inc.
29416 Lassen St.
Chatsworth, CA 91311
Kees Van Der Pool (213) 882–4100

Photovoltaic power modules and supply systems.

Options: covering materials (glass, Lexan, silicone rubber). **Installation requirements/considerations:** determined per site. **Guarantee/warranty:** 5-yr on modules. **Maintenance requirements:** 1 or 2 inspections per year. **Manufacturer's technical services:** complete design of the whole system. **Availability:** depends on size of installation, between 2 and 26 wk. **Suggested retail price:** consult manufacturer.

PUMPS

CIRCULATOR 102B

Taco, Inc.
1160 Cranston St.
Cranston, RI 02920
John E. Jessup (401) 942–8000

This compact, bronze-bodied 1/12-hp seal-type pump will deliver 10 gpm at a maximum head of 10.3 ft.

Features: close-coupled, end-suction design; brass/plastic impeller. **Installation requirements/considerations:** may be used with water or aqueous solutions of ethylene or propylene glycol at a maximum temperature of 200° F. **Guarantee/warranty:** 1-yr limited warranty. **Availability:** from stock. **Suggested retail price:** consult manufacturer.

CIRCULATING PUMP, GPPS 45H SERIES

Hartell Div.
Milton Roy
70 Industrial Drive
Ivyland, PA 18974
Douglas Bingler (215) 322-0730

This circulating pump, available with various motors and fluid connections, is designed for high-pressure, low-flow systems where the fluid loop is vented for low system pressure. High discharge pressures make this pump ideal for multiple heat exchanger and solar drain back systems.

Features: rotary mechanical seal suitable for water or glycol solutions. **Options:** available with various motors and fluid connections; materials include bronze and plastic. **Installation requirements/considerations:** can be installed in any position to conserve space. **Guarantee/warranty:** 1 to 2 yr. Depends on application. **Manufacturer's technical services:** engineering and technical services from factory. **Availability:** stock to 20 wk. Depends on special features requested. **Suggested retail price:** starts at $60.

CIRCULATOR 110, 110B, 117B

Taco, Inc.
1160 Cranston St.
Cranston, RI 02920
John E. Jessup (401) 942–8000

These 1/12-hp pumps deliver up to 35 gpm at a maximum head of 7½ ft.

Features: the motor has built-in overload protection. **Options:** cast-iron or bronze bodies, both with plastic impeller; standard flanges: ½, ¾, 1, 1¼, 1½ in.; shut-off flanges: ¾, 1, 1¼ in. **Installation requirements/considerations:** may be used with water or aqueous solutions of ethylene or propylene glycol at a maximum of 240° F. **Guarantee/warranty:** 1-yr limited warranty. **Availability:** from stock. **Suggested retail price:** consult manufacturer.

CIRCULATOR 111

Taco, Inc.
1160 Cranston St.
Cranston, RI 02920
John E. Jessup (401) 942–8000

This ⅛-hp pump with a plastic impeller in a cast-iron housing will deliver 50 gpm at a maximum head of 12 ft.

Features: stainless-steel shaft, bronze-sleeve bearing. **Options:** standard cast-iron flanges, ½, ¾, 1, 1¼, 1½ in.; shut-off flanges, ¾, 1, 1¼ in. **Installation requirements/considerations:** may be used with water or aqueous solutions of ethylene or propylene glycol at a maximum temperature of 240° F. **Guarantee/warranty:** 1-yr limited warranty. **Availability:** from stock. **Suggested retail price:** consult manufacturer.

CIRCULATOR 008-V2

Taco, Inc.
1160 Cranston St.
Cranston, RI 02920
John E. Jessup (401) 942–8000

A compact $^{1}/_{25}$-hp ciruclator/air-separation package, this pump delivers 13 gpm at a maximum head of 16 ft.

Features: the cast-iron housing has a built-in air-separation chamber; all moving parts, including the plastic impeller, are housed in a patented, stainless-steel cartridge. **Installation requirements/considerations:** may be used with water or aqueous solutions of ethylene or propylene glycol at a maximum temperature of 240° F. **Guarantee/warranty:** 1-yr limited warranty. **Availability:** from stock. **Suggested retail price:** consult manufacturer.

Options: available with various motors and fluid connections; materials include bronze and plastic; choice of ball-bearing or sleeve-bearing motors in various configurations. **Installation requirements/considerations:** may be installed in any position to conserve space. **Guarantee/warranty:** 1 to 2 yr. Depends on application. **Manufacturer's technical services:** engineering and technical services from factory. **Availability:** stock to 20-wk. Depends on special features requested. **Suggested retail price:** starts at $60.

GRUNDFOS CIRCULATING PUMPS

Grundfos Pumps Corp.
2555 Clovis Ave.
Clovis, CA 93612
James M. Mueller (209) 299–9741

Grundfos Circulating Pumps are used in conjunction with ON/OFF or proportional-type differential controllers to circulate solar absorption fluids through thermal mass collection systems. Performance ranges to 21 ft head, 32 gpm. Maximum continuous operating temperature is 230° F.

Features: stainless-steel construction, two-speed motor, no mechanical seals, self-lubricating. **Options:** special models available for high heat applications and for use with hydro-carbon fluids. Pumps are available with flanged or union fittings (cast iron or bronze) and bronze isolation valves also available, as well as stainless-steel volutes. **Installation requirements/considerations:** must be sized properly for systems and proper volute materials utilized (cast iron for closed systems, stainless steel for open systems). **Guarantee/warranty:** 18 mo conditional warranty. **Manufacturer's technical services:** available upon request on all aspects of fluid handling. **Availability:** stock. **Suggested retail price:** varies.

MAGNETIC DRIVE PUMP, CDE SERIES

Hartell Div.
Milton Roy Co.
70 Industrial Drive
Ivyland, PA 18974
Douglas Bingler (215) 322–0730

This double-ended drive pump is designed to realize maximum energy input economy for dual fluid systems. The two pumping ends are driven by a single motor. Each end can be completely independent in terms of capacity, head, and materials of construction. Materials include bronze and plastic.

Features: sealless magnetic drive design. **Options:** various motors and fluid connections available. **Guarantee/warranty:** 1 to 2 yr, depends on use. **Manufacturer's technical services:** engineering and technical services from factory. **Availability:** stock to 20-wk, depends on special features requested. **Suggested retail price:** starts at $60.

MAGNETIC DRIVE PUMP, CP SERIES

Hartell Div.
Milton Roy Co.
70 Industrial Drive
Ivyland, PA 18974
Douglas Bingler (215) 322–0730

Designed for solar and energy reclaiming systems that require low power consumption, this pump can be used on vented or closed-loop systems. Service is to 200° F and 150 psi on continuous applications. The seal-less magnetic drive design eliminates seal failures.

MAGNETIC DRIVE PUMPS, SERIES CP-10B AND CP-12B

Hartell Div.
Milton Roy Co.
70 Industrial Drive
Ivyland, PA 18974
Douglas Bingler (215) 322–0730

This series of high-performance, magnetic-drive pumps

was designed for the high heads and capacities of the larger solar and waste heat systems. The pump can be used on vented or closed-loop systems with internal pressure up to 150 psi.

Options: available with various motors and fluid connections; materials include bronze and plastic. **Installation requirements/considerations:** can be installed in any position to conserve space. **Guarantee/warranty:** 1 to 2 yr, depending on application. **Manufacturer's technical services:** engineering and technical services from factory. **Availability:** stock to 20 wk. Depends on special features requested. **Suggested retail price:** starts at $60.

REFLECTIVE SURFACES

ACRYLIC MIRROR MATERIAL

Solar Usage Now, Inc.
Box 306
Bascom, OH 44809
Joe Deahl (419) 937–2226

0.125-in. acrylic material functions as rear-surface mirror. Light-weight, weather- and fade-resistant, this flexible reflector is said to be 20 percent more brilliant than silvered glass.

Options: 3 to 4 ft width; length to order. **Guarantee/warranty:** 5-yr. **Availability:** from stock. **Suggested retail price:** $3.20/sq ft.

BERRY SOLAR REFLECTORS

Berry Solar Products
P.O. Box 327
Edison, NJ 08817
Calvin C. Beatty (201) 549–3800

Berry Solar Products manufactures a variety of reflective materials, including stainless steel, metallized mylar and glass mirrors that can be laminated or mechanically bonded to metal and non-metal substrates to form flat and curved reflector panels designed to customer specifications.

Features: custom made with choice of reflective material. **Installation requirements/considerations:** might require frame or support. **Guarantee/warranty:** none. **Maintenance requirements:** occasional cleaning. **Manufacturer's technical services:** reflectivity analysis and design engineering. **Availability:** depends on order. **Suggested retail price:** on application.

REFLECTO-SHIELD

Madico
64 Industrial Parkway
Woburn, MA 01801
Roger Greene (617) 935–7850

A laminated reflecting film for application to existing windows, Reflecto-Shield provides solar energy protection up to 60 to 79 percent and is available in five architecturally desirable colors.

Features: 1-mil metallized surface is protected within a lamination. **Options:** light transmissions of 20, 40 or 60 percent in silver; 15 or 30 percent in bronze, gray, gold or copper green. **Installation requirements/considerations:** wet application requires qualified applicator or factory training. **Guarantee/warranty:** up to 5 yr against peeling, cracking, crazing or de-metallizing. **Availability:** 30 days. **Suggested retail price:** consult manufacturer.

SOLAR REFLECTIVE FOIL

Solar Usage Now, Inc.
Box 306
Bascom, OH 44809
Joseph Deahl (419) 937–2226

5-mil reflective material for use in all types of focusing or concentrating collectors, foil/mylar/foil composition.

Options: 12- or 48-in. wide rolls, self- or non-adhesive. **Availability:** from stock. **Suggested retail price:** $.40 to $1.30/sq ft.

STORAGE TANKS AND LINERS

FIBERGLAS STORAGE TANK

Owens-Corning Fiberglas Corp.
Fiberglas Tower
Toledo, OH 43659
George K. Hammond (419) 248–8063

Fiberglas reinforced plastic (FRP) underground storage

tanks range in capacities from 550 to 50,000 gal. Made of polyester resins reinforced with Fiberglas materials, the UL-labelled tanks will not corrode either internally or externally. The tanks can be used to store water to 180° F.

DOUBLE-WALLED FIBERGLASS SOLAR STORAGE TANK

Solar Systems
26046 Eden Landing Road
Suite 4
Hayward, CA 94545
Terry Elledge, Mark Perrin (415) 785–0711

Available in rectangular (350 gal) or cylindrical (350 to 3000 gal) shapes, these storage tanks have seamless reinforced fiberglass inner walls that provide structural integrity and prevent leaks. The 180° F Gelcoat fiberglass inner wall is resistant to corrosion; high temperature Gelcoat up to 210° F is also available. The outer fiberglass wall sandwiches 4 in. of flame-retardant urethane foam insulation into a sealed envelope that prevents the accumulation of moisture and protects the insulation properties of urethane.

Options: tank shape and size, from 350 gal rectangular to 350 to 3000 gal cylindrical. 180° F Gelcoat or 210° F Gelcoat available on the rectangular model. **Guarantee/warranty:** 1-yr. **Maintenance requirements:** protection from standing water when set below grade. **Manufacturer's technical services:** installation and engineering consulting. **Availability:** 10 days between order and delivery (plus shipping time). **Suggested retail price:** rectangular, $410 to $535; cylindrical, $560 to $2600.

Features: won't corrode, maintenance free. **Options:** wide range of accessories available. **Installation requirements/considerations:** must be installed according to manufacturer's requirements. **Guarantee/warranty:** 1-yr warranty. **Manufacturer's technical services:** available on request. **Suggested retail price:** available on request.

FORD STONE-LINED STORAGE TANK

Ford Products Corp.
Ford Products Road
Valley Cottage, NY 10989
Ernest L. Schoolfied (914) 358–8282

The Ford Stone-Lined Storage Tank is a stone-lined steel tank for water storage with an internal copper heat exchanger through which fluid from solar collectors is circulated. Tanks range in size from 40 to 120 gal; "E" models are equipped with a thermostatically controlled electric heating element for reliable standby water heating.

Options: single- or double-wall heat exchanger. **Installation requirements/considerations:** good workmanship only. **Guarantee/warranty:** limited 5-yr warranty. **Manufacturer's technical services:** as needed. **Availability:** 2-wk between order and delivery. **Suggested retail price:** $200 to $500; consult dealer.

HEAT KEEPER STORAGE TANKS, MODELS 250 THROUGH 1500

Carolina Solar Comfort, Inc.
1205 Kenilworth Ave.
Charlotte, NC 28204
James Syers (704) 372–6549

These seamless storage tanks are insulated with 3¾ in. of polyurethane sandwiched between two layers of polyester resin reinforced fiberglass. Outside fiberglass wall is pigmented to resist UV radiation; fittings of any size or number can be provided. Standby heat loss is less than 2° F in a 24-hr period.

Features: sizes range from 250 to 1500 gal. Lightweight (1000-gal tank weight 305 lb). Monolithic, seamless construction. **Options:** custom color. Any size, number or location of fiberglass pipe fittings. **Installation requirements/considerations:** designed for atmospheric pressure only. **Guarantee/warranty:** 5-yr guarantee. **Regional applicability:** North and Southeast. **Availability:** 3 to 6 wk between order and delivery. **Suggested retail price:** 1000-gal tank, $1400.

JACKSON SOLAR STORAGE TANK

W. L. Jackson Manufacturing Co., Inc.
Box 11168
Chattanooga, TN 37401
P. G. Para (615) 867–4700

An insulated, jacketed hot-water storage tank with two ¾-in. NPT fittings for connection to solar collector. The auxiliary heating element is controlled by an adjustable surface-mounted thermostat.

Options: element wattage from 1000 to 6000. **Installation requirements/considerations:** standard. **Guarantee/warranty:** 5-yr limited warranty on tank; 1-yr limited warranty on element and thermostat. **Maintenance requirements:** drain tank regularly to remove sediment. **Manufacturer's technical services:** installation and maintenance instructions. **Availability:** in stock in many areas; 2-wk for special order. **Suggested retail prices:** 80-gal $250, 100-gal $400, 120-gal $434.

SOLAR POWER CELL

Sunshine Manufacturing Co.
4870 S.W. Main, No. 4
Beaverton, OR 97005
David Carson (503) 642–1123, (503) 643–6172

This solid-state, dimorphic-crystal, thermal power storage cell has a constant temperature discharge rate of up to 1 million Btu's/hr at 400° F as well as a storage capacity of over 3 million Btu's. These hot fluid heat transfer storage cells can be connected in a series to increase total storage capacity and discharge rate. Each storage cell unit (with casing) is 8 cu ft.

Features: constant temperature discharge cycle. **Options:** variable discharge rates and temperatures available as options. Consult manufacturer. **Installation requirements/considerations:** foundation must be insulated and reinforced concrete pad. **Guarantee/warranty:** 5-yr warranty against normal operating conditions failure. **Manufacturer's technical services:** technical support available. **Regional applicability:** areas with over 200 days of direct sunshine per year. **Availability:** 90 days. **Suggested retail price:** $12,500.

SOLAR STORAGE TANK

American Appliance Manufacturing Corp.
2341 Michigan Ave.
Santa Monica, CA 90404
Paul Hegg (213) 870–8541

Solar storage tanks designed for open systems in three sizes, include electrical back-up element (240 vac, 4500 w). Insulation is Owens-Corning RA26-336 (R=12).

Features: taped sides and top, with anode and special, heavy insulation. **Guarantee/warranty:** 5-yr. **Availability:** 3-wk. **Suggested retail price:** consult manufacturer.

SOLAR STORAGE TANKS/1000 TO 2500 GAL.

Solatherm Corp.
1255 Timber Lake Drive
Lynchburg, VA 24502
W. W. Hays III (804) 237–3249

Precast internally insulated and water-proofed concrete storage tanks. Come with steel fitting plate, fittings, internal piping.

Features: ready for external hookup, no internal piping or insulation required. **Options:** internal piping, heat exchangers. **Installation requirements/considerations:** must be buried to a minimum of one half of tank depth. **Guarantee/warranty:** 1-yr on materials and workmanship, free repair of failure. **Maintenance requirements:** none if operated according to manufacturer's recommendations. **Manufacturer's technical services:** assistance in system-tank coordination. **Regional applicability:** United States. **Availability:** 4-wk. **Suggested retail price:** 80¢ to $1/gal.

STC ENERGY STORAGE TANKS, SERIES 221-S AND 221-HT

Solar Tanks and Components
Grand Prairie, TX 75050
Bruce Pratt (214) 255–8453, (214) 225–2765

These liquid storage tanks hold 221 to 1069 gal, depending on model. Standard model withstands liquid temperatures up to 180° F; high-temperature model to 450° F and above. Tanks have removable tops, fiberglass interior, urethane insulation, six outlets, and vent and overflow tubing. High-temperature models lose no more than 3°F/day; standard models, less than 5° F.

Features: non-corrosive interior and exterior. **Options:** additional outlets available at $25 each. **Installation requirements/considerations:** can be used only in non-pressurized liquid-type system. **Guarantee/warranty:** 5-yr warranty on standard model; 10-yr warranty on high-temperature model. **Suggested retail price:** $309.30 to $1870.75; depends on model (standard or high-temperature) and size.

STORAGE TANKS

Solarsystems Industries, Ltd.
5511 128th St.
Surrey, BC V3W 4B5
Erich W. Hoffmann (604) 596–2665

Water storage tanks made from fiberglass or polypropylene. Tanks are round, upright and range from 80 to 2900 gal capacity. Fiberglass tanks may be buried in the ground.

Options: heat exchangers are available for any size tank; custom sizes and shapes are available on special order. **Installation requirements/considerations:** tanks are for atomspheric pressure only. **Manufacturer's technical services:** system design available. **Guarantee/ warranty:** 1-yr. **Suggested retail price:** $149 to $2950.

SUNERATOR STORAGE TANKS

Solar Energy Systems, Inc.
One Olney Ave.
Cherry Hill, NJ 08033
Nathan E. Brussels (609) 424–4446

Highly insulated tanks for the storage of solar-heated hot water, available in domestic sizes of 66, 82 and 120 gal, as well as large, commercial sizes.

Features: tanks up to 120-gal are glass lined, larger tanks are Placite lined. **Options:** fiberglass or urethane insulation, ASME coded; large tanks in above- or below-grade design. **Guarantee/warranty:** 5-yr. **Availability:** 2-wk for tanks to 120-gal, 6-wk for larger. **Suggested retail price:** consult manufacturer.

SUN-LITE[R] STORAGE TUBES

Kalwall Corp., Solar Components Div.
P.O. Box 237
Manchester, NH 03105
Scott F. Keller (603) 668–8186

Manufactured from Sun-Lite fiberglass reinforced polymer, these cylindrical tubes and tanks are designed to contain storage materials for direct gain passive solar systems or for aquaculture experiments. The tubes and tanks are non-pressure containers that have a capacity of 22 to 725-gal, and they are available in standard 12, 18, or 58 in. diameters.

Features: although designed as non-pressurized containers, they can be manufactured with plumbing fittings for interconnections. **Options:** custom sizes available in minimum lots of five; factory applied black absorber coating available. **Installation requirements/considerations:** level surface required to maintain vertical tube orientation. **Guarantee/warranty:** none expressed or implied. **Manufacturer's technical services:** customer requirements discussed. **Availability:** approximately 2 to 3 wk between order and delivery. **Suggested retail price:** $37 to $152.

MISCELLANEOUS

CAST ACRYLIC FRESNEL LENSES

Swedlow, Inc.
12122 Western Ave.
Garden Grove, CA 92645
W.R. Lee

Fresnel lenses for photovoltaic and photothermal systems. 20-yr projected life.

Features: cast acrylic weatherability. Custom designed lens performance, large sizes. **Installation requirements/ considerations:** accurate solar tracking and mounting frames are required. **Maintenance requirements:** periodic cleaning. **Manufacturer's technical services:** system performance predictions, tracking and mounting recommendations, lens performance testing. **Availability:** approximately 6-mo to first article. **Suggested retail price:** depends on quantities.

COMMERCIAL WATER DISTILLATION SYSTEM

Aquarian Research
P.O. Box 378
Bedford, VA 24523
Jerry Rosenberg (703) 586–4850

This 32-sq ft basin distiller can be used as a single unit or in large arrays for desalination systems that process from 5 to 5 million gal/day. Completely automated, untended systems are available including design and installation services.

Features: collects rain water; automatic water feed is standard. **Options:** adjustable solar reflector for enhanced output; completely automated system controls. **Installa-**

tion requirements/considerations: per installation manual. **Guarantee/warranty:** 5-yr materials and workmanship. **Maintenance requirements:** per installation manual. **Manufacturer's technical services:** design, installation and maintenance. **Availability:** consult manufacturer. **Suggested retail price:** consult manufacturer.

HELIODYNE SOLAR-POWERED ELECTRICAL GENERATING SYSTEM, MODEL OG-7500

OMNIUM-G
1815 Orangethorpe Park
Anaheim, CA 92801
Stanley Zelinger (714) 879–8421

The OG-7500 is an energy system that provides electricity during daylight hours. The waste-heat from the system is useable and storage for this heat is provided. The system produces 75 kwhr of electricity during a 10-hr sunny day.

Features: lowest cost electrical generating system available. **Options:** various output voltage options in 50 and 60 Hz are available; ganged system for higher output available. **Installation requirements/considerations:** company installs. **Maintenance requirements:** periodic (typically once per month) cleaning and lubricating required; major overhaul once in 10-yr. **Manufacturer's techncial services:** application engineering, site engineering. **Regional applicability:** useful in areas with upwards of 300 sunny days/yr. **Availability:** 120 days ARO. **Suggested retail price:** $32,500.

HELIODYNE SOLAR-POWERED ELECTRICAL GENERATING PLANT, MODEL OG-7500S

The OG-7500S is a total energy system providing both electricity and useable heat. It is designed with a decentralized power plant for producing 75 kwhr of electricity and 60,000 Btu's of heat. It provides storage for both electricity and heat. Energy to be converted to electricity during non-sunlight hours is stored in the form of compressed air.

Features: compressed air storage for energy that is to be converted to electricity. **Options:** output voltage to match requirement in either 50 or 60 Hz; multiple systems can be ganged. Storage capacity can be increased. **Installation requirements/considerations:** company installs. **Maintenance requirements:** periodic cleaning and lubricating required on monthly basis, typically; major overhaul once in 10-yr. **Manufacturer's technical services:** application engineering, site engineering. **Regional applicability:** useful in areas with upwards to 300 sunlit days/yr. **Availability:** 180 days ARO. **Suggested retail price:** $70,000.

HELIODYNE SOLAR-POWER ELECTRICAL GENERATING PLANT, MODEL OG-7500UGI

OMNIUM-G
1815 Orangethorpe Park
Anaheim, CA 92801
Stanley Zelinger (714) 879–8421

An electricity-producing system designed to mesh with utility power grids. Puts energy into grid during sunny periods. Waste heat storage is provided to use in total energy environments.

Features: connects to power networks and provides power during sunny periods. **Options:** various 50- and 60-Hz voltage options available. **Installation requirements/considerations:** company installed. **Maintenance requirements:** periodic (typically, once per month) cleaning and lubricating. Major overhaul once in 10-yr. **Manufacturer's technical services:** application engineering, site engineering. **Regional applicability:** useful in areas where there are upwards of 300 sunny days per yr. **Availability:** 120 days ARO. **Suggested retail price:** $35,000.

HELIOSTILL

Approtech
770 Chestnut St.
San Jose, CA 95110
Kent Algred Dogey (408) 297–6527

Designed for distillation of liquids with vaporization temperatures below 250° F, Heliostill materials include tempered glass, stainless-steel basins, brass fittings, binderless insulation, and a collapsible aluminum reflector.

Features: collapsible design. **Options:** size to customer specifications and needs. **Installation requirements/considerations:** knowledge of chemistry of reactant; south facing. **Guarantee/warranty:** 1-yr limited warranty. **Maintenance requirements:** periodic removal of salts from basin; cleaning of glazing and reflector. **Manufacturer's technical services:** sizing/installation, microclimate analysis, payback assessments. **Availability:** 1 to 3 wk between order and delivery. **Suggested retail price:** $39.95.

HOUSEHOLD WATER DISTILLER, MODEL H

Aquarian Research
P.O. Box 378
Bedford, VA 24523
Jerry Rosenberg (703) 586–4850

This solar water purification unit produces over 1 gal/ sunny day from tap, sea or polluted water source. Storage trough is stainless-steel; absorber plate material is glass with fiberglass insulation and pressure-treated wood with weather-resistant coating; adhesive is silicone.

Features: stores 2-gal of distillate. **Options:** float valve; adjustable solar reflector. **Guarantee/warranty:** 5-yr warranty on materials and workmanship. **Maintenance requirements:** basin flush twice a year. **Manufacturer's technical services:** assistance available for unusual installations. **Availability:** stock to 4 wk. **Suggested retail price:** $125.

SOLAR AIR MOVER SAM-10, 20, 30

Solar Control Corp./Luxaire
5595 Arapahoe Rd.
Boulder, CO 80302
Larry Trudell (303) 449–9180

The SAM provides total system air flow and operational mode control, simplifying design and installation of hot air solar heating systems. SAM contains the blower, dampers and controller, eliminating complicated ductwork, dampers and costly field wiring.

Features: summer/winter mode. **Options:** 230-v operation; built-in, hot water heat exchanger. **Guarantee/ warranty:** limited. **Availability:** 4-wk. **Suggested retail price:** $1329, $1449, $1649.

SOLAROASTER '78

Approtech
770 Chestnut St.
San Jose, CA 95110
Kent Alfred Dogey (408) 297–6527

This concentrating air collector for cooking food and other materials to temperatures to 550° F is made of tempered glass, binderless fiberglass insulation, copper absorber, aluminum housing and reflector sheet, and has a thermometer to 600° F. Container size is 14 × 14 × 14 in. without the reflector; when reflector is attached the overall dimensions are 26 × 26 × 26 in. Manual tracking is required hourly, and internal rotating racks maintain heated objects in horizontal orientation.

Features: rotating oven rack; collapsible reflector. **Options:** (accessories) custom-designed iron cookware; eutectic thermal storage; collapsible oven for backpacks. **Installation requirements/considerations:** direct insolation. **Maintenance requirements:** cleaning of reflector and glazing. **Guarantee/warranty:** 1-yr limited warranty. **Manufacturer's technical services:** sizing/installation procedures, microclimate analysis, and payback assessment. **Regional applicability:** anywhere direct and diffuse insolation available. **Availability:** 1 to 3 wk between order and delivery. **Suggested retail price:** $49.94.

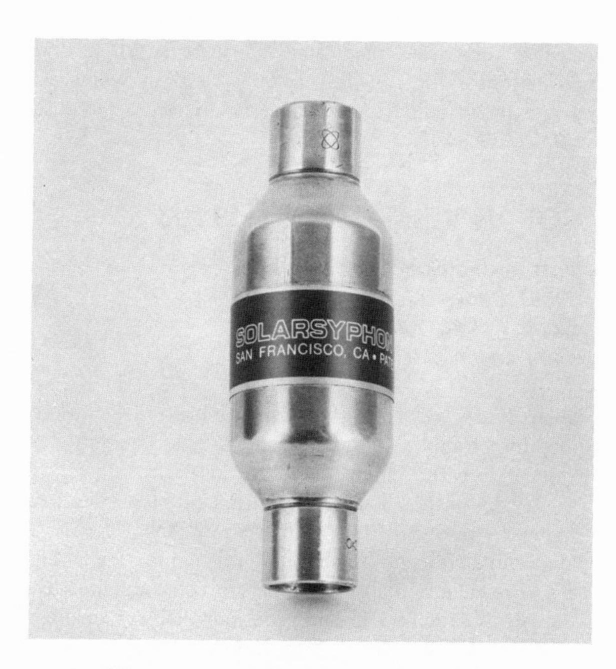

SOLARSYPHON DIODE™

Sun of Man Solar Systems
Drawer W
Bethel Island, CA 94511
Richard Brady (415) 684–3362

A sun-activated valve engineered to eliminate reverse thermosyphon action. The diode opens when the sun heats the fluid in a collector and closes automatically at the end of the solar cycle.

Features: required no external power. **Options:** customer's operational temperature, preset. **Installation requirements/considerations:** the valve should be placed as close to the collector as is possible, on the collector-to-tank (hot) line. All piping runs must be properly insulated to prevent thermal stratification. **Guarantee/warranty:** guaranteed against defects in manufacturing and materials, under normal use, for the life of the system in which it is originally installed. **Availability:** from stock. **Suggested retail price:** $97.

SOLAR VENT, MODEL 11706

Revere Chemical Corp.
30887 Carter St.
Solon, OH 44139
Mal Hansen (216) 248–0606

Solar Vent is a one-way roof vent that utilizes solar energy to pump trapped moisture from the roof assembly and eject it into the atomsphere. The one-way, elastomeric inlet valve inside Solar Vent's insulated stem sucks roof moisture into the vent's transparent solar dome. An aluminum collector plate turns the dome into a high-pressure chamber, absorbing the sun's heat and increasing the pressure within to force the moist air out through a one-way exhaust valve.

Features: one-way valves will not allow moisture to re-enter the roofing assembly. Removes moisture many times faster than ordinary vents. **Installation requirements/considerations:** area must get at least 6-hr of direct sunlight a day. Install one Solar Vent/1,000 sq ft. **Availability:** 3-wk. **Suggested retail price:** $49.76.

SOLTRAN™ SOLAR ONE HOLER, MODELS ST-100, 110, 200

Ecos, Inc.
21 Imrie Road
Boston, MA 02134
William M. Bell (617) 782–0002

This outdoor composting toilet facility uses acrylic glazing to facilitate the evaporation of liquid waste and the processing of solid wastes to compost. Indigenous rock is used as thermal mass; insulation is fiberglass batt; absorber coating is flat black paint. Structure is wood, and venting system is galvanized sheet metal. Overall dimensions are 6 × 7 × 13 feet high.

Features: no chemicals used. **Options:** finish, electric fan and light (accessories). **Installation requirements/considerations:** foundation location. **Maintenance requirements:** humus removal and local disposal every 2 to 6 mo. **Manufacturer's technical services:** architectural and engineering support. **Availability:** 9 to 12 wk between order and delivery. **Suggested retail price:** $2000 to $6000.

SUNPURE SOLAR STILL

Heliosystems Corp.
3407 Ross Ave.
Dallas, TX 75204
Gary deLarios (214) 824–5971

This water purification system consists of 3 × 7-ft stills of redwood construction with epoxy lining. Glazing is tempered glass. Time clock, solenoid valve and storage tank are provided. Unit purifies 1 to 2 gal of water daily.

Features: completely automatic, storage provided. **Options:** several units can be used to provide greater water supply. **Installation requirements/considerations:** southern exposure placement of distillation unit. **Guarantee/warranty:** 5-yr warranty against manufacturing defects. **Maintenance requirements:** annual cleaning. **Manufacturer's technical services:** sizing or design of large systems. **Regional applicability:** below 42 degrees North latitude unless protected from extreme freeze. **Availability:** 2 to 4 wk. **Suggested retail price:** $495.

Solar Manufacturers— Alphabetical Listing

For cross-referencing purposes, manufacturers of solar products mentioned in this book are listed below. Following each manufacturer are the product categories in which the manufacturer is represented in the Solar Product Buyer's Guide.

A-1 Prototype, Inc.
1288 Fayette St.
El Cajon, CA 92020
Jerry Hull (714) 449–6726
Heat exchangers, flat plate collectors

Acurex Corp.
485 Clyde Ave.
Mt. View, CA 94042
Edward L. Rossiter (415) 964–3200
Concentrating collectors

Aeolian Kinetics
P.O. Box 100
Providence, RI 02901
Ralph Beckman (401) 274–3690
Measurement & data devices

Alpha Solarco
1014 Vine St., Suite 2230
Cincinnati, OH 45202
M. Uroshevich (513) 621–1243
Concentrating & flat plate collectors

American Acrylic Corp.
173 Marine St.
Farmingdale, NY 11735
M. Ziegler (516) 249–1129
Glazings

American Appliance Manufacturing Corp.
2341 Michigan Ave.
Santa Monica, CA 90404
Paul Hegg (213) 870–8541
Hot water systems, storage tanks & liners

American Klegecell Corp.
204 N. Dooley St.
Grapevine, TX 76051
Philip Wilkens (817) 481–3547
Insulation

American Solar Heat Corp.
7 National Place
Danbury, CT 06810
Joseph Heyman (203) 792–0077
Controllers, flat plate collectors

American Solar Power Inc.
715 Swann Ave.
Tampa, FL 33606
Swimming pool heaters

Approtech
770 Chestnut St.
San Jose, CA 95110
Kent Alfred Dogey (408) 297–6527
Absorber plates, flat plate collectors

Aquarian Research
P.O. Box 378
Bedford, VA 24523
Jerry Rosenberg (703) 586–4850
Water distillation systems

Aquasolar, Inc.
1232 Zacchini Ave.
Sarasota, FL 33578
Gerald J. Zella (813) 366–7080
Swimming pool heaters

ASG Industries, Inc.
P.O. Box 929
Kingsport, TN 37662
M.L. Lilly (615) 245–0211 or (800) 251–0266
Glazings

Bally Case and Cooler, Inc.
Bally, PA 19503
Leon Prince (215) 845–2311
Insulation

BDP Co.
7310 West Morris St.
Indianapolis, IN 46231
Robert J. Johnson (317) 243–0851
Air handlers, flat plate collectors, heat exchangers, heating systems

Beam Engineering, Inc.
732 N. Pastoria Ave.
Sunnyvale, CA 94086
Benjamin H. Beam (408) 738–4573
Concentrating collectors

Berry Solar Products
P.O. Box 327
Edison, N.J. 08817
Calvin C. Beatty (201) 549–3800
Absorber plates, reflective surfaces

Bio-Energy Systems, Inc.
Box 489
Mountaindale Rd.
Spring Glen, NY 12483
Michael F. Zinn (914) 434–7858
Absorber plates, heating systems, hot water systems, swimming pool heaters

Bray Oil Company, Inc.
1925 North Marianna Ave.
Los Angeles, CA 90032
Eugene R. Slaby (213) 268–6171
Heat transfer fluids

C & M Systems, Inc.
Saybrook Industrial Park
Elm St. P.O. Box 475
Old Saybrook, CT 06475
Nelson Messier (203) 388–3429
Controllers

Calmac Manufacturing Corp.
150 S. Van Brunt St.
Englewood, NJ 07631
John Armstrong (201) 569–0420
Flat plate collectors, swimming pool heaters

Campbell Engineering
1302 Toney Drive
Huntsville, AL 35802
Dr. R. A. Campbell (205) 883–9866
Measurement & data devices

Carolina Solar Comfort, Inc.
1205 Kenilworth Ave.
Charlotte, NC 28204
James Syers (704) 372–6549
Storage tanks & liners

Chamberlain Manufacturing Corp.
845 Larch Ave.
Elmhurst, IL 60126
John E. Balzer (312) 279–3600
Flat plate collectors

Champion Home Builders Co.
Solar Div.
5573 E. North St.
Dryden, MI 48428
Henry Leck (313) 796–2211
Flat plate collectors, heating systems

Champion Home Builders Co.
Solar Products Div.
118 Walnut St.
Waynesboro, PA 17268
Al Cool (717) 762–3113
Measurement & data devices, hot water systems

Chicago Solar Corp.
1773 California St.
Rolling Meadows, IL 60008
T. Crombie (312) 358–1918
Flat plate collectors

Colt, Inc. of Southern California
71–590 San Jacinto Drive
Rancho Mirage, CA 92270
Charles Barsamiam (714) 346–8033
Flat plate collectors

Columbia Chase Corp.
Solar Energy Div.
55 High St.
Holbrook, MA 02243
Walter H. Barrett (617) 767–0513
Flat plate collectors

Conserdyne Corporation
4437 San Fernando Rd.
Glendale, CA 91204
Howard Kraye (213) 246–8404
Hot water systems, space heating systems, swimming pool heaters

Contemporary Systems, Inc.
68 Charlonne St.
Jaffrey, NH 03452
John C. Christopher (603) 532–7972
Air handlers, controllers, flat plate collectors

Daystar Corp.
90 Cambridge St.
Burlington, MA 01803
C. Greely (617) 272–8460
Flat plate collectors

Deko-Labs of Friberg Dekold Corp.
Route 4, Box 256
Gainesville, FL 32601
Donald F. Dekold (904) 372–6009
Controllers

Delavan Electronics
14605 N. 73rd St.
Scottsdale, AZ 85260
Hardy K. Landskov (602) 948–6350
Controllers

DeSoto, Inc.
1700 S. Mt. Prospect Rd.
Des Plaines, IL 60018
Kenneth Lawson (312) 391–9000
Absorber coatings

Devices and Services Co.
3501-A Milton
Dallas, TX 75205
Charles Moore (214) 368–5749
Measurement & data devices

Direct Energy Corp.
16221 Construction Circle West
Irvine, CA 92714
Karl E. Sterne (714) 552–6211
Flat plate collectors

Dixon Energy Systems, Inc.
47 East St.
Hadley, MA 01035
Jane Nevin (413) 584–8831
Flat plate collectors, hot water systems

Dumont Industries
Main St.
Monmouth, ME 04259
A. Douglas Scott (207) 933–4811
Hot water systems

Dyrelite Corp.
63 David St.
P.O. Box B–947
New Bedford, MA 02741
Randall T. Weeks (617) 993–9955
Insulation

Ecos, Inc.
21 Imrie Road
Boston, MA 02134
William M. Bell (617) 782–0002
Composting toilet (Misc.)

Ecotronics, Inc.
8502 E. Cactus Wren Rd.
Scottsdale, AZ 85253
Alden Stevenson (602) 948–8003
Controllers

Egge Research
Box 394B RFD
Kingston, NY 12401
Hank Starr (914) 336–5597
Greenhouses, passive solar devices

Energy Applications
Route 5, Box 383A
Rutherfordton, NC 28139
Napoleon P. Salvail (704) 287–2195
Controllers

Energy Design, Inc.
1925 Curry Rd.
Schenectady, NY 12303
Gary Poukish (518) 355–3322
Controllers

The Energy Factory
5622 East Westover
Suite 105
Fresno, CA 93727
Thomas W. Kristy (209) 292–6622
Greenhouses

Energy Saving Systems
Price and Pine Sts.
Holmes, PA 19043
Ray Speicher (215) 583–4780
Photovoltaics

Energy Shelters Inc.
2162 Hauptman Rd.
Saugerties, NY 12477
Morton Schiff (914) 246–3135
Greenhouses

Energy Systems, Inc.
4570 Alvarado Canyon Rd.
San Diego, CA 92120
Terrence R. Caster (714) 280–6660
Flat plate collectors, hot water systems

Entropy Limited
5735 Arapahoe Ave.
Boulder, CO 80303
Hank Valentine (303) 443–5103
Concentrating collectors, heating systems, measurement & data devices

The Eppley Laboratory, Inc.
12 Sheffield Ave.
Newport, RI 02840
George L. Kirk (401) 847–1020
Measurement & data devices

Fafco, Inc.
235 Constitution Drive
Menlo Park, CA 94025
Alex Battey (415) 321–3650
Swimming pool heaters

Filon Div.
Vistron Corp.
12333 Van Ness Ave.
Hawthorne, CA 90250
James E. Whitridge (213) 757–5141
Glazings

Ford Products Corp.
Ford Products Road
Valley Cottage, NY 10989
Ernest L. Schoolfied (914) 358–8282
Storage tanks & liners

Future Systems, Inc.
12500 W. Cedar Drive
Lakewood, CO 80228
Bill Thompson (303) 989–0431
Flat plate collectors, heating systems

Gem Manufacturing Co.
Star Route No. 18
Bascom, OH 44809
Joesph Deahl (419) 937–2225
Flat plate collectors

General Electric Company
Advanced Energy Programs
P.O. Box 13601
Philadelphia, PA 19101
William F. Moore (215) 962–2112
Concentrating collectors

General Energy Devices, Inc.
1751 Ensley Ave.
Clearwater, FL 33516
Clyde Bouse (800) 237–0137
Heating systems, hot water systems, swimming pool heaters

General Solar Systems Division,
General Extrusions, Inc.
4040 Lake Park Road
Youngstown, OH 44507
William Kenney (216) 783–0270
Concentrating collectors

G.N.S. Company
79 Magazine Street
Boston, MA 02119
David Schwartz (617) 442–1000
Concentrating collectors

Grumman Energy Systems, Inc.
4175 Veterans Memorial Highway
Ronkonkoma, NY 11779
Arthur L. Barry (516) 575–2549
Flat plate collectors

Grundfos Pumps Corp.
2555 Clovis Ave.
Clovis, CA 93612
James M. Mueller (209) 299–9741
Pumps

Gulf Thermal Corp.
629 17th Ave. W.
Bradenton, FL 33505
Dudley Slocum (813) 748–3433
Flat plate collectors

Halstead and Mitchell
P.O. Box 1110
Scottsboro, AL 35768
Troy Barkley (205) 259–1212
Flat plate collectors

Hartell Div.
Milton Roy
70 Industrial Drive
Ivyland, PA 18974
Douglas Bingler (215) 322–0730
Pumps

Hawthorne Industries, Inc., Solar Energy Division
1501 S. Dixie Highway
West Palm Beach, FL 33401
Ray Lewis, J. B. Carr, Dale Kline
(305) 659–5400
Controllers

Helio Thermics, Inc.
110 Laurens Rd.
Greenville, SC 29607
William Haas (803) 235–8529
Controllers, glazings

Heliosystems Corp.
3407 Ross Ave.
Dallas, TX 75204
Gary deLarios (214) 824–5971
Flat plate collectors, hot water systems, water purification systems

Heliotrope General
3733 Kenora Drive
Spring Valley, CA 92077
Al Abernathy (714) 460–3930
Controllers, measurement & data devices

Hexcel Corp.
11711 Dublin Blvd.
Dublin, CA 94566
George Branch (415) 828–4200
Concentrating collectors

Hollis Observatory
One Pine St.
Nashua, NH 03060
Joseph F. Litwin (603) 882–5017
Measurement & data devices

Horizon Enterprises, Inc.
P.O. Box V 1011 NW 6th St.
Homestead, FL 33030
Ed Glenn (305) 245–5145
Flat plate collectors, hot water systems

Hot Line Solar, Inc.
1811 Hillcrest Drive
Bellevue, NE 68005
Dan Lightfoot (402) 291–3888
Concentrating collectors

Hyperion, Inc.
7209 Valtec Ct.
Boulder, CO 80301
John Eatwell, Larry Brand (303) 449–9544
Flat plate collectors

Impac Corp.
312 Blondeau St.
P.O. Box 365
Keokuk, IA 52632
Duff Decker, Paul Hosemann (319) 524–3304
Flat plate collectors

Independent Energy, Inc.
P.O. Box 732
42 Ladd St.
East Greenwich, RI 02818
R. Dowdell (401) 884–6990
Controllers

Insta-Foam Midwest
8000 47th St.
Lyons, IL 60534
Kenneth H. Mettam
Insulation

Insulating Shade Limited Partnership
17 Water St.
Guilford, CT 06437
Thomas P. Hopper (203) 453–9334
Insulation

InterTechnology/Solar Corp.
100 Main St.
Warrenton, VA 22186
N. L. Beard (703) 347–9500
Flat plate collectors, hot water systems

Iowa Solar, Inc.
Box 246
North Liberty, IA 52317
Tom Miller (319) 626–2342
Measurement & data devices

W. L. Jackson Manufacturing Co., Inc.
Box 11168
Chattanooga, TN 37401
P. G. Para (615) 867–4700
Hot water systems, storage tanks & liners

Kalwall Corp., Solar Components Div.
P.O. Box 237
Manchester, NH 03105
Scott F. Keller (603) 668–8186
Flat plate collectors, glazings, passive products, storage tanks & liners

Lambda Selective Coatings
580 Alexander Road
Princeton, NJ 08540
Roger Mulock (609) 921–3330
Absorber coatings

Don Lewis Associates
Box L
Sutter Creek, CA 95685
Sheri Lewis (209) 296–4943
Measurement & data devices

Libby Owens Ford Co.
1701 East Broadway
Toledo, OH 43605
Lloyd E. Bastian (419) 247–4350
Flat plate collectors

Liconix
1400 Stierlin Road
Mountain View, CA 94043
Mark W. Dowley (415) 964–3062
Measurement & data devices

MacBall Industries, Inc.
5765 Lowell St.
Oakland, CA 94608
W.A. McKirdy (415) 658–1124
Swimming pool blankets

Madico
64 Industrial Parkway
Woburn, MA 01801
Roger Greene (617) 935–7850
Reflective surfaces

Mann Russell Electronics Inc.
1401 Thorne Rd.
Tacoma, WA 98421
George F. Russell (206) 383–1591
Controllers

Matrix, Inc.
537 S. 31st St.
Mesa, AZ 85204
Don Pershing (602) 832–1380
Measurement & data devices

McKim Solar Energy Systems, Inc.
1142 East 64th St.
Tulsa, OK 74132
Eric Paschall (918) 749–8896
Flat plate collectors

Mega Engineering
1717 Elton Rd.
Silver Spring, MD 20903
Richard E. Dame (301) 622–4030
Absorber plates, concentrating & flat plate collectors

Mid-West Technology
P.O. Box 26238
Dayton, OH 45426
Vern L. Huffines (513) 837–8551
Absorber plates

Miromit American Heliothermal Corp.
2625 S. Santa Fe Drive
Denver, CO 80223
Bill Phillips (303) 778–0650
Flat plate collectors

National Solar Supply
2331 Adams Drive N.W.
Atlanta, GA 30318
Sid Stansell (404) 352–3478
Flat plate collectors

Natural Power, Inc.
Francestown Turnpike
New Boston, NH 03070
Charles Puckette (603) 487–5512
Controllers, measurement & data devices

Northern Solar Power Co.
311 South Elm St.
Moorhead, MN 56560
Bruce Hilde (218) 233–2515
Flat plate collectors

Northrup
302 Nichols Dr.
Hutchins, TX 75141
(214) 225–4291
Flat plate collectors

Olin Brass
East Alton, IL 62024
J.I. Barton (618) 258–2443
Absorber plates

OMNIUM-G
1815 Orangethorpe Park
Anaheim, CA 92801
Stanley Zelinger (714) 879–8421
Concentrating collectors, generating systems

One Design, Inc.
Mt. Falls Route
Winchester, VA 22601
Tim Maloney (703) 662–4898
Passive solar devices

Owens-Corning Fiberglas Corp.
Fiberglas Tower
Toldeo, OH 43659
George K. Hammond (419) 248–8063
Storage tanks & liners

Park Energy Co.
Star Route, Box 9
Jackson, WY 83001
Frank D. Werner (307) 733–4950
Absorber plates, flat plate collectors, glazings

Pelcasp Solar Designs
4817 Sheboygan Ave.
Madison, WI 53705
Michael Pellett (608) 274–1752
Controllers

Permaloy Corporation
P.O. Box 1559
Ogden, UT 84402
Harry G. James (801) 731–4303
Absorber coatings

Pioneer Energy Products
Rt. 1 Box 189
Forest, VA 24551 Timothy M. Hayes
(804) 239–9020
Controllers, hot water systems

Pleiad Industries, Inc.
Springdale Road
West Branch, IA 52358
Donald Laughlin (319) 643–5650
Flat plate collectors

Prima Industries, Inc.
P.O. Box 141
Deer Park, NY 11729
A.L. Gruol (516) 242–6347
Flat plate collectors, hot water systems

Raypak, Inc.
31111 Agoura Rd.
Westlake Village, CA 91361
H. Byers (213) 889–1500
Flat plate collectors, hot water systems

Refrigeration Research, Inc., Solar Research Div.
525 N. Fifth St.
Brighton, MI 48116
Jerry Kay (313) 227–1151
Absorber coatings, absorber plates, flat plate collectors, heat exchangers, hot water systems

Revere Chemical Corp.
30887 Carter St.
Solon, OH 44139
Mal Hansen (216) 248–0606
Solar vent (Misc.)

Revere Solar and Architectural Products, Inc.
P.O. Box 151
Rome, NY 13440
(315) 328–2401
Flat plate collectors

Rho Sigma, Inc.
11922 Valerio St.
North Hollywood, CA 91605
Preston Welch (213) 982–6800
Controllers

Robertshaw Controls Co.
Temperature Controls Marketing Group
100 W. Victoria St.
Long Beach, CA 90805
K. LaGrand, P.S. Johnson (213) 638–6111
Controllers

Semco Corp.
1054 N.E. 43rd Street
Ft. Lauderdale, FL 33334
Jeff Prutsman (305) 565–2516
Flat plate collectors

Sensor Technology, Inc.
29416 Lassen St.
Chatsworth, CA 91311
Kees Van Der Pool (213) 882–4100
Photovoltaics

Sheffield Plastics, Inc.
P.O. Box 248, Salisbury Rd.
Sheffield, MA 01257
Thomas Kradel (413) 229–8711
Glazings

Shelley Radiant Ceiling Co., Inc.
456 W. Frontage Rd.
Northfield, IL 60093
William Shelley (312) 446–2800
Absorber plates

Solafern, Ltd.
536 MacArthur Blvd.
Bourne, MA 02532
Philip Levine (617) 563–7181
Flat plate collectors, heating systems, hot water systems

Solar Alternative, Inc.
30 Clark St.
Brattleboro, VT 05301
Jim Kirby (802) 254–6668
Flat plate collectors, heat transfer fluids, hot water systems

Solar American Corp.
106 Sherwood Drive
Williamsburg, VA 23185
R.J. Pegg (804) 874–0836
Hot water systems

Solar and Geophysical Engineering
P.O. Box 576
Sparta, NJ 07871
Gary Bubb (201) 729–7287
Hot water systems

Solar Control Corp.
5721 Arapahoe Rd.
Boulder, CO 80303
Larry Trudell (303) 449–9180
Controllers

Solar Control Corp./Luxaire
5595 Arapahoe Rd.
Boulder, CO 80302
Larry Trudell (303) 449–9180
Solar air mover (Misc.)

Solar Devices, Inc.
G.P.O. Box 3727
San Juan, Puerto Rico 00936
Frank Casa (809) 783–1775
Flat plate collector

Solar Energy Products Co.
121 Miller Road
Avon Lake, OH 44012
William L. Maag (216) 933–5000
Flat plate collectors, heat exchangers

Solar Energy Products, Inc.
Mountain Pass
Hopewell Junction, NY 12533
B. R. Kryzaniwsky (914) 226–8596
Hot water systems

Solar Energy Systems, Inc.
One Olney Ave.
Cherry Hill, NJ 08003
Nathan E. Brussels (609) 424–4446
Controllers, flat plate collectors, heat exchangers, storage tanks & liners

Solar Industries, Inc.
Monmouth Airport Industrial Park
Farmingdale, NJ 07727
Norman Reitman (201) 938–7000
Flat plate collectors, swimming pool heaters

Solar Kinetics Corp.
P.O. Box 17308
West Hartford, CT 06117
James A. Pohlman (203) 233–4461
Concentrating collectors

Solar Living, Inc.
P.O. Box 12
Netcong, NJ 07857
Richard Bonte (201) 691–8483
Flat plate collectors, hot water systems, swimming pool heaters

Solar Power West
709 Spruce St.
Aspen, CO 81611
Raymond N. Auger (303) 925–4698
Flat plate collectors

Solar Research Systems
3001 Redhill Ave. I–105
Costa Mesa, CA 92626
Dr. Joseph Farber (714) 545–4941
Swimming pool heaters

Solar Room Co.
Box 1377
Taos, NM 87571
Leah Alexander (505) 758–9344
Greenhouses

Solar Systems
26046 Eden Landing Road
Suite 4
Hayward, CA 94545
Terry Elledge, Mark Perrin (415) 785–0711
Storage tanks & liners

Solar Systems by Sun-Dance, Inc.
13939 N.W. 60th Ave.
Miami Lakes, FL 33014
Thomas L. Abell (305) 947–4456
Flat plate collectors

Solar Tanks and Components
Grand Prairie, TX 75050
Bruce Pratt (214) 255–8453, (214) 225–2765
Storage tanks & liners

Solar Technologies of Florida
Reynolds Industrial Park
P.O. Box 40485
Jacksonville, FL 32203
Jack W. Hoover (904) 269–3264
Absorber plates

Solar Technology Corp.
2160 Clay St.
Denver, CO 80211
Richard Speed (303) 455–3309
Flat plate collectors, greenhouses

Solar Unlimited, Inc.
4310 Governors Drive, W.
Huntsville, AL 35805
Larry Frederick (205) 837–7340
Flat plate collectors, hot water systems

Solar Usage Now, Inc.
Box 306
Bascom, OH 44809
Joe Deahl (419) 937–2226
Absorber coatings, hot water systems, reflective surfaces

SolaRay, Inc.
324 S. Kidd St.
Whitewater, WI 53190
Robert K. Skrivseth (414) 473–2525
Flat plate collectors

Solarex Corp.
1335 Piccard Dr.
Rockville, MD 20850
Bob Edgerton (301) 948–0202
Photovoltaics

Solarics Energy Control Systems, Inc.
P.O. Box 15183
Plantation, FL 33318
Ron Stein (305) 971–0391
Controllers

Solarkit of Florida, Inc.
1102 139th Ave.
Tampa, FL 33612
Wm. Denver Jones (813) 971–3934
Absorber plates, flat plate collectors, hot water systems

Solaron Corp.
300 Galleria Tower
720 South Colorado Blvd.
Denver, CO 80222
Heating systems

Solarsystems Inc.
507 West Elm St.
Tyler, TX 75702
Robert F. Faulkner (214) 592–5343
Flat plate collectors

Solarsystems Industries, Ltd.
5511 128th St.
Surrey, BC V3W 4B5
Erich W. Hoffmann (604) 596–2665
Controllers, flat plate collectors, storage tanks & liners

Solatherm Corp.
1255 Timber Lake Drive
Lynchburg, VA 24502
W. W. Hays III (804) 237–3249
Heat exchangers, storage tanks & liners

Solergy Co.
7216 Boone Ave. N.
Minneapolis, MN 55428
Vince Grimaldi (612) 535–0305
Flat plate collectors, hot water systems

Southeastern Solar Systems, Inc.
2812 New Spring Rd.
Suite 150
Atlanta, GA 30339
Joe Cooper (404) 434–4447
Heating systems, hot water systems

Southwest Ener-Tech, Inc.
3030 S. Valley View Blvd.
Las Vegas, NV 89102
Gary Halderson (702) 876–5444
Flat plate collectors

Specialty Manufacturing Inc.
DBA Insolarator
7926 Convoy Court
San Diego, CA 92111
Frank B. Ames (714) 292–1857
Flat plate collectors

Spire Corp.
Patriots Park
Bedford, MA 01730
Thomas E. Wilber (617) 275–6000
Photovoltaics

S.P.L. Industries Ltd.
400 West Main St.
Babylon, NY 11702
Absorber plates

S.R.D. Corp.
6625 4th St. South
St. Petersburg, FL 33705
Daniel R. White (813) 866–1346
Concentrating collectors

Sun of Man Solar Systems
Drawer W
Bethel Island, CA 94511
Richard Brady (415) 684–3362
Solarsyphon Diode™ (Misc.)

Sun Spot Research
1070 South Leyden
Denver, CO 80224
Harold L. Moses (303) 321–7323
Measurement & data devices

Sun Tech Solar Industries Corp.
P.O. Box 203
Chester, NY 10918
Laurence T. Wansor (914) 469–4212
Heating & hot water systems

Sun Unlimited Research Corp.
P.O. Box 941
Sheboygan, WI 53081
Glenn F. Groth (414) 452–8194
Heating systems

Sunburst Solar Energy, Inc.
123 Independence Drive
Menlo Park, CA 94025
Brian E. Lanhston (415) 327–8022
Absorber plates, flat plate collectors

Sunearth Solar Products Corp.
Box 515SA
Montgomeryville, PA 18936
H. Katz (215) 699–7892
Flat plate collectors, hot water systems

Sunpower Systems Corporation
510 South 52nd Street
Tempe, AZ 85281
Dan Ikeler (602) 894–2331
Concentrating collectors

SunSaver Corp.
Box 276
North Liberty, IA 52317
D. Dunlavy (319) 626–2343
Heating systems

Sunshine Manufacturing Co.
4870 S.W. Main, No. 4
Beaverton, OR 97005
David Carson (503) 642–1123, (503) 643–6172
Concentrating collectors, storage tanks & liners

Sunworks Div., Enthone, Inc.
P.O. Box 1004
New Haven, CT 06508
Floyd C. Perry (203) 934–6301
Flat plate collectors, heat transfer fluids, hot water systems

Swedlow, Inc.
12122 Western Ave.
Garden Grove, CA 92645
W. R. Lee
Fresnel lenses (Misc.)

TA Controls
652 Glenbrook Rd.
Stamford, CT 06906
Leif Eiderberg (203) 324–0106
Controllers

Taco, Inc.
1160 Cranston St.
Cranston, RI 02920
John E. Jessup (401) 942–8000
Pumps

TechniTrek Corp.
1999 Pike Ave.
San Leandro, CA 94577
Vernon Butorovich (415) 352–0535
Flat plate collectors

Telluride Solarworks
Box 700
Telluride, CO 81435
Dean Randle (303) 728–3303
Flat plate collectors

Texas Controls, Inc.
P.O. Box 59469
Dallas, TX 75229
F.W. Schempf (214) 386–5000
Measurement & data devices

Thanor Enterprises, Inc.
817 West St.
Wilmington, DE 19801
Frank W. Arnoth (302) 571–9515
Controllers

Thermacor Process, Inc.
P.O. Box 4529
Fort Worth, TX 76106
Richard B. Bender II (817) 624–1181
Insulation

Thermal Dynamics, Inc.
2285 Emerald Hts. Court
Reston, VA 22091
M.D. Shell (703) 620–3014
Concentrating collectors

Thermal Technology Corp.
P.O. Box 130
Snowmass, CO 81654
Ron Shore (303) 963–3185
Passive solar devices

Thomason Solar Home, Inc.
609 Cedar Ave.
Fort Washington, MD 20022
Jack Thomason, Jr. (301) 839–1738
Flat plate collectors, solar systems

Tranter, Inc.
735 East Hazel St.
Lansing, MI 48909
Robert Rowland (517) 372–8410, ext. 243
Absorber plates

Turbonics, Inc.
11200 Madison Ave.
Cleveland, OH 44102
David R. Essen, Mgr. Sales (216) 228–9663
Heat exchangers

Urethane Molding Inc.
RFD No. 3, Route 11
Laconia, NH 03246
Randy H. Annis (603) 524–7577
Insulation

U.S. Solar Pillow, Inc.
P.O. Box 88
Grand Junction, CO 81501
Dick Carmack (303) 242–2743
Passive solar devices

Vegetable Factory, Inc.
100 Court St.
Copiague, NY 11726
Fred Schwartz (516) 842–9300
Greenhouses

Virginia Solar Components, Inc.
Highway 29 South
Rustburg, VA 24588
Robert Savage (804) 239–9523
Controllers

Vulcan Solar Industries, Inc.
200 Conant St.
Pawtucket, RI 02893
J. Michael Levesque (401) 725–6061
Hot water systems, flat plate collectors

West Wind
Box 542
Durango, CO 81301
Geoffrey Gerhard
Controllers

Western Energy, Inc.
454 Forest Ave.
Palo Alto, CA 94302
Norman Rees (415) 327–3371
Flat plate collectors

Whiteline, Inc.
P.O. Box 3071
Asheville, NC 28802
George T. White (704) 258–8405
Concentrating collectors

Wojcik Industries, Inc.
301 N. Brandon Rd.
Suite 7
Fallbrook, CA 92028
Warren Wojcik (714) 728–0553
Absorber plates

Ying Manufacturing Corp.
1957 W. 144th St.
Gardena, CA 90249
George E. Home, III (213) 770–1756
Flat plate collectors

Zomeworks Corp.
P.O. Box 712
Albuquerque, NM 87103
Steve Baer (505) 242–5354
Passive solar devices

Solar Service Listings

Solar is a relatively new field, and architects, engineers, designers, and consultants experienced in solar are not always easy to find. This listing was composed to help you find services available in your area. It is arranged geographically in four regions—east, midwest, south, west and Canada. The map at the beginning of each region indicates the states included. Many companies provide services throughout the country, and this is indicated in the listing.

Eastern U.S.A.

Adirondack Alternate Energy, Div. of Brownell Lumber
Edinburg, NY 12134

Passive solar house design and alternate energy systems engineering. On-site building supervision and consultation. Energy analysis. Low energy post and beam package houses. **Geographical range:** Adirondacks. **Contact:** Bruce Brownell (518) 863–4338.

AIDCO Maine Corp.
Orr's Island, ME 04066

Designers and installers of complete solar systems of any kind for all applications. Each system custom designed to meet energy demands and economic objectives of client. Special service is "trouble-shooting" of systems to optimize performance and recommend necessary changes. Also federal grant assistance. Current projects include two 100 percent solar residences—one under HUD grant. **Geographical range: unlimited. Contact:** Robert K. Multer (207) 833–6700.

Applied Solar Technology, Inc.
10738 Tucker St.
Beltsville, MD 20705

Consulting engineers offering wide range of professional services. Among these are solar heating and cooling design for residential, commercial and industrial applications, R&D in controllers and solar air conditioning, and instrumentation design. Also present short courses in solar heating fundamentals and currently working on design of solar high school curriculum. Several large scale HUD and ERDA grant projects completed— Garland Lane solar homes, Madison-McCulloh moderate income townhouse project and an office building for Maryland National Capital Park and Planning Commission. **Geographical range:** New Jersey to Virginia. **Contact:** Frank Wilkins (301) 937–7400.

Architects Group
Fern Hill, Harlemville
Ghent, NY 12075

Architectural designers and planners. Two passive houses now under construction and a large integrated industrial installation in planning stages. **Geographical range:** N.Y. and New Eng. **Contact:** G. Raymond de Ris (518) 672–7346.

Barrett Heating and Air Conditioning Co., Inc.
2260 Union Blvd.
Bay Shore, NY 11706

Full engineering, design and installation capabilities. Three years solar experience. Over seventy systems in operation. **Geographical range:** Long Island, N.Y. **Contact:** Gary Shoemaker (516) 665–0940.

Basic Energy Construction Inc.
P.O. Box 2127
Morristown, NJ 07960

Solar and alternative energy consultation, architectural and engineering design, construction and construction management. Experience in domestic hot water, solar space heating and air-conditioning systems and solar process heat systems. **Geographical range:** Northeast. **Contact:** Gregory Egan (201) 540–1148.

Michael Beattie
13 Center St.
Rutland, VT 05701

Integral systems building design with emphasis on low cost, simple methods. Six years experience in design/construction field. Current projects include an active air heating system for steel building, a passive residence and another passive home with attached greenhouse. **Geographical range:** Vt. **Contact:** Michael Beattie (802) 775–4864.

Mark Beck Associates, Inc.
762 Fairmount Ave.
Towson, MD 21204

Provide full architectural design services for planning and installation of solar systems. Active and passive, new and retrofit system and building design. Solar feasibility studies and energy conservation analyses and programs. Total energy system and greenhouse design. Recent projects include the Carroll County (Md.) Environmental Appreciation Area, Patterson residence, Baltimore County, Md.; and a large scale energy conservation analysis for GSA. Have also done monitoring and review of HUD demonstration projects. **Geographical range:** Mid-Atlantic states, D.C. and Va. **Contact:** Peter C. Powell, AIA (301) 828–6000.

Beckman, Blydenburg & Associates
P.O. Box 100
Providence, RI 02901

Architects/engineers specializing in energy conscious design. Experience in passive and active solar and small scale (up to 50 kw) wind energy systems. Have recently completed seven new passive solar houses and an historic retrofit in Rhode Island. Installed a 10kw wind system in Mich. **Geographical range:** unlimited. **Contact:** Ralph Beckman (401) 274–3690.

Richard D. Blazej
14 Elliot St.
P.O. Box 299
Brattleboro, VT 05301

Building construction consultant and troubleshooter. Designer of residential structures with emphasis on alternative energy systems and energy conservation. Twenty-two years experience in general design and construction. Worked on Grassy Brook Village solar condominium project. **Geographical range:** primarily Vt., N.H. and Western Mass., will consider other places in Northeast. **Contact:** Richard Blazej (802) 254–2739.

Blue/Sun, Ltd.
P.O. Box 118
Farmington, CT 06032

Designers of energy balanced houses incorporating active and passive solar collection as well as energy conservation techniques. Specialize in post and beam construction. Over 33 solar houses designed. **Geographical range:** Northeast and Calif. **Contact:** Blue Minges (203) 677–0004.

Parsons Brinckerhoff
One Penn Plaza, 250 West 34th St.
New York, NY 10001

Design/analysis of solar heating and cooling systems—industrial and agricultural applications. Solar insolation field data measurement and economic analysis. TRNSYS computer capabilities. Have done work for the American Can Co.—a DWH system, State of N.H. Dept. of Agriculture—solar heated animal shelter and others. **Geographical range:** Cont. U.S. **Contact:** K. Ramakrishna (212) 239–7925.

B. P. Solar Tech
330 W. 45th Street
New York, NY 10036

Solar consultant providing services at all stages of project—planning, design and construction/installation. Site analysis, product recommendations and evaluations. Computerized surveys available soon. Work includes a solar desalienation plant in the Grand Bahamas and a 780 sq. ft. passive solar retrofit in New York City. **Geographical range:** Primarily N.Y.C., but will consider anything. **Contact:** B. Polak.

Brill Kawakami Wilbourne, Architects
77 Main Street
Cold Spring-on-Hudson, NY 10516

Designers of new and retrofit solar systems for industrial, commercial and residential buildings. Polyglycoat Maintenance Center and private residences among recent efforts. **Geographical range:** unlimited. **Contact:** Ralph Brill (914) 265–2326.

Burt Hill Kosar Rittelmann Associates
400 Morgan Center
Butler, PA 16001

Primarily an architectural firm, but with much wider range of services. Do work in solar engineering/energy management programming, feasibility and energy studies, solar system design, R&D, construction management and proposal preparation, writing and consultation. Have provided architectural and system design for Friendship Federal Bank in Ingomar, Penn. and a "minimum energy building" in Mission Viejo, Calif. Served as system and conservation consultants on St. Charles (Ill.) High School project. **Geographical range:** North America. **Contact:** P. Richard Rittelmann (412) 285–4761.

Carter Engineering, Inc.
1107 Spring St.
Silver Spring, MD 20910

Consulting engineers. Experience includes work on several residential active systems—both air and water. **Geographical range:** unlimited. **Contact:** D. G. Carter.

Cashin Associates, P.C.
499 Jericho Turnpike
Mineola, NY 11501

Consulting engineers, architects, planners. Nassau County Office Building and Hempstead Harbor Park Administrative Building (both on Long Island) are among recent solar heating-energy management projects. **Geographical range:** Northeast. **Contact:** Francis Cashin (516) 248–0690.

Center for Ecological Technology
P.O. Box 427
Pittsfield, MA 01201

Energy audits and consultation. Passive solar and solar greenhouse design. Workshops and lectures. Have designed and built two solar greenhouses, a Trombe wall house and have conducted a county-wide energy audit. **Geographical range:** Western Mass. **Contact:** Ned Nisson (413) 445–4556.

Frank Chapman Architects and Planners
31 Batter Terrace
New Haven, CT 06511

Architectural design incorporating solar. Solar system design. Build low cost hot air systems. Current priority is on design and construction of integral air collector systems for domestic hot water and space heating. Have done several apartment projects under HUD grants—both new and retrofit including the Chesire Elderly Village, the New Heaven (Conn.) Solar Court and the Fitch Warner

Apartments. A dozen new projects now underway. **Geographical range:** N.Y., Conn., Mass., R.I., Vt., N.H. **Contact:** Howard Phillips (203) 776–8600.

CHI Housing, Inc.
Box 566, 68 Main St.
Hanover, NH 03755

Complete energy conserving and solar design and building services for residential and light commercial applications. All projects are custom designed and built with fixed price contract. Most residential projects fall into the $30,000 to $80,000 range for one to five bedrooms. Numerous residential projects completed to date. **Geographical range:** Design—unlimited; Construction—25-mile radius of Hanover, N.H. **Contact:** Douglas Coonley (603) 643–5940.

Clean Energy Systems
63 Maple Ave.
Keene, NH 03431

Registered mechanical engineer, representing Sunworks, will provide advice on applications and installation of solar systems. **Geographical range:** New Hampshire. **Contact:** H. Hamilton Chase (603) 352–0083.

Edward C. Collins II Assoc. Architects
Box 284
Lincoln, MA 01773

Architectural design of solar and energy conscious buildings. Experience in both single and multi-family dwellings. Received American Wood Council "Design for Better Living Award" for solar heated residence. **Geographical range:** N.Y. and New Eng. **Contact:** Keith B. Gross (617) 259–0420.

Community Builders
Canterbury, NH 03224

Design and construction of passive solar and wood heat structures. Both single and multi-family dwellings completed. **Geographical range:** Concord, N.H. area. **Contact:** Don or Steve Booth (603) 783–4743.

Consolar™
P.O. Box 751
Litchfield, CT 06759

Building design consultant for solar applications to new or retrofit construction. Has designed and is in process of building his own passive house as well as doing consultation work for a number of residential projects. **Geographical range:** Western Conn. and Hudson Valley (N.Y.). **Contact:** Walter Gress, Jr. (203) 354–0344.

DAS/Solar Systems
201 Sixth Ave.
Brooklyn, NY 11217

Solar engineering, design and HVAC contracting for domestic hot water and space heating systems. More than fifteen of their systems in operation in metro. New York area. **Geographical range:** East coast from Mass. to Penn. **Contact:** Steven Mueller (212) 636–1471.

DAWN Associates
Box 432
Phoenicia, NY 12464

Energy efficient shelter design incorporating passive and "semi-passive" solar technology. Attention paid to integration of site features into design. Stevenson School solar house in Gardner, N.Y. and the Seven Oaks Community Building in Madison, Va. are examples of firms recent work. **Geographical range:** 150 mile radius of N.Y.C.; 150 mile radius of Washington, D.C. **Contact:** for N.Y.C. area—Jerome Kerner (914) 688–7700; for Washington area—Eugene Eccli; 1448 N. Lancaster St., Arlington, Va. (202) 223–0830.

Design Alternatives, Inc.
1312 18th St., N.W.
Washington, DC 20036

Design work for residential clients in the area of energy conservation and solar energy applications. Wide range of research, program development and educational services in solar energy are offered to clients in both the private and public sectors. Economic and technical residential energy conservation feasibility study conducted for government agency. Several residences designed. **Geographical range:** Design work—Mid-Atlantic region; Consulting—national. **Contact:** Eugene Eccli (202) 223–0830, and (202) 223–6336.

Design Associates
Nettleton Hollow
Washington, CT 06793

Design and/or construction of passive solar heated, energy conserving buildings. Two completed private residences. Solar Merrydale, a three lot subdivision in progress. **Geographical range:** Conn., West. Mass., Eastern N.Y. **Contact:** Forbes Morse (203) 868–2900 or Stephen Lasar (203) 354–0855.

Designers Construction Cooperative
Star Route
New Salem, MA 01355

Design and/or construction of solar and natural climate responsive energy conserving buildings. Consulting services. Self help construction technical assistance and client/firm cooperative contracting. Five years experience in solar and self help. Many designs, structures and research studies completed. **Geographical range:** Design—worldwide; construction—New England and N.Y. **Contact:** Sheldon Klapper (617) 864–0639.

Dublin-Bloome Associates
42 West 39th St.
New York, NY 10018

Consulting engineers and planners. Energy Management consultants. Wide range of services from grant preparation to post-construction/installation analysis. Have been involved or responsible for some of the major active projects of the past few years including the Cary Arboretum Offices and Labs in Millbrook, N.Y. and the George A. Towns Elementary School in Atlanta, Ga. One new project now in planning stage is a solar "new town" in Iran that is planned for 200,000 people. **Geographical range:** world-wide. **Contact:** Fred Dubin (212) 868–9700.

Eco-Energetics
P.O. Box 7
Easthampton, MA 01027

Engineering analysis and design of solar heating systems—active and passive—and domestic hot water systems. Service offered to architects, building contractors and individuals. Prefer small scale projects. Current project is a free-standing, self-sustaining greenhouse at Harmony Farm (Mass.) which will employ passive and active solar, methane generators and a wind turbine. **Geographical range:** Western Mass. **Contact:** Robert Pinkos (413) 527–9456.

Ecos Corporation
P.O. Box 331
Farmington, CT 06032

Design and construction of solar-assisted dwellings using both active and passive systems. Energy management and site analysis. Have completed one solar house with more slated for construction in spring. Energy management surveys conducted for both single and multi-family dwellings. **Geographical range:** Southern New Eng. **Contact:** Thomas Andersen (203) 673–2110.

ECOS, Inc.
21 Imire Rd.
Boston, MA 02134

Shelter Division provides design and consulting services for residential, public and commercial buildings. Special services include feasibility studies, master planning, interior space planning, landscape design and architectural rendering. Emphasis of all services is to integrate all viable energy saving alternatives into architectural designs that are in harmony with local environment. Have done work on a pre-engineered solar house, a SOLTRAN-solar augmented composting toilet facility and a large carriage house solar greenhouse retrofit. **Geographical range:** New England. **Contact:** R. Terry Cline (617) 782–0002.

Energy Design Inc.
1025 Curry Rd.
Rotterdam, NY 12303

Design services for solar systems and component parts. Experience in hot water and hydronic heating systems. **Geographical range:** On-site work—Upper N.Y. and Western New England. Other consulting—will discuss. **Contact:** Mark Urbaetis (518) 355–3322 or 371–9596.

Energy Resource Corp.
88 Main St.
New Canaan, CT 06840

Architectural and engineering design of residential and commercial solar space and hot water heating systems.

Marketing and installation of energy conservation devices. Energy conservation surveys. **Geographical range:** Metro. N.Y.C. area. **Contact:** Alfred Alk (203) 838–7259.

En-Save
117 Partree Rd.
Cherry Hill, NJ 08003

Design of energy efficient space heating, cooling and DWH systems using solar energy and off peak storage. Experience in commercial, institutional and residential applications. **Geographical range:** not given. **Contact:** M. Creedon.

Evog Associates, Inc.
P.O. Box 36
Hebron, NH 03241

Designer and builder of energy efficient homes, offering optimal use of wood heat and passive solar design. New and retrofit construction. Last year completed four passive houses using thermal storage walls, direct gain and greenhouses and two passive retrofit projects. Active solar house in planning stages as speculation project. **Geographical range:** Central N.H. **Contact:** Richard Holt (603) 744–8918.

Walter L. Gottschalk
The Rosemary
198 West Main St.
Orange, VA 22960

Custom design of alternate energy systems—solar, wood, wind, water—for residential and commercial space and domestic hot water heating. Solar air-conditioning/nocturnal cooling systems. Conventional solar manufacturer's components used, but consideration given to use of locally available materials and labor to effect greatest cost reductions possible. Recent projects include a sun tempered/energy conservative greenhouse for the Orange Co. (Va.) School Board and a passive air type system for Orange Co. Historical Society's new headquarters. **Geographical range:** East coast. **Contact:** Walter Gottschalk (703) 672–2731.

B. F. Greene, Consulting Engineers
97–45 Queens Blvd.
Rego Park, NY 11374

Mechanical-electrical engineers doing work in solar system design. Plans and specifications provided. New Paltz (N.Y.) City Hall is among recent projects. **Geographical range:** unlimited. **Contact:** B.F. Greene.

Michael Greene, Designer/Builder
Mount Delight
Deerfield, NH 03037

Complete design and consulting services in direct-gain passive solar residences. Emphasis in design is on site affinity, use of native materials and hand craftsmanship throughout. Several private residences completed, mostly in New Hampshire. **Geographical range:** Northern New England. **Contact:** Michael Green (603) 463–7930.

Keith B. Gross & Associates
153A North Ave.
Weston, MA 02193

Architecture, energy conservation and solar design—passive, active and hybrid—new or retrofit construction. Feasibility studies for solar retrofit. Experience in residential space heating and hot water systems, greenhouse retrofits and multi-family residential retrofit feasibility studies. **Geographical range:** Northeast and Calif. **Contact:** Keith Gross (617) 894–0573.

Hittman Associates, Inc.
9190 Red Branch Road
Columbia, MD 21045

Professional services in the energy and environmental fields: consulting, technical and economic studies, concept analysis and system modelling. Solar energy applications studies done for both private and governmental entities. **Geographical range:** North America. **Contact:** H. M. Curran (301) 730–7800.

Larry Honeywell
R.D.1
Carthage, NY 13619

Design and construction of energy efficient, timber framed houses with integrated hybrid passive systems. Also interested in and capable of working with active systems of all types, wind energy, greenhouses and wood heat. Recent projects have included a Beadwall/trombe wall residential system and a hybrid passive air system. **Geographical range:** Northern N.Y. **Contact:** Larry Honeywell (315) 493–3106.

David F. Jaquith
11 Ober St.
Beverly, MA 01915

Architectural, planning and solar design services. Passive design. Site planning. HUD grant for solar houses completed, New Ipswich, N.H. **Geographical range:** U.S. **Contact:** David Jaquith (617) 927–3745.

Doug Kelbaugh, A.I.A.
70 Pine St.
Princeton, NJ 08540

Architectural design with emphasis in passive solar heating and cooling; Trombe Wall house plan and kit ($35); teaching, speaking, consulting, graphics and conferences. **Geographical range:** nation-wide consulting practice. **Contact:** Doug Kelbaugh (609) 924–2703.

Shannon P. Kennedy, Architect
4101 Kinsway
Baltimore, MD 21206

Architect specializing in design services for incorporating active and passive natural energy systems into residential and commercial projects—new buildings and retrofitting. Experience includes retrofits of historic houses with consideration given to preserving original design while providing for effective solar and energy conservation conversion. **Geographical range:** Md. and Virg. **Contact:** Shannon Kennedy (301) 254–6128.

Sheldon Lazan, P.E.
2255 Center Ave.
Fort Lee, NH 07024

Consulting engineer working in solar. Has designed parabolic solar collector for Solar Sun, Inc. (Cincinnati, Ohio) and various new and retrofit residential projects. **Geographical range:** East coast. **Contact:** Sheldon Lazan (212) 661–4170.

Lombardi and Waldo
89 State St.
Guilford, CT 06437

Architects, engineers and land-use planners specializing in design, construction supervision, program requirements and surveys of existing conditions. Single and multi-family residences and commercial projects. **Geographical range:** Conn. **Contact:** Bernard J. Lombardi.

Mapleleaf Design & Construction, Inc.
17 W. Central Ave.
Paoli, PA 19301

Design and construction of solar and energy efficient dwellings. Background in organic architecture based on experience at Frank Lloyd Wright school. Designed Pleasant Bay Animal Hospital—a burmed building in East Harwick, Mass. **Geographical range:** Design, unlimited; construction, S.E. Penn. **Contact:** T. Victor Stimac (215) 647–3887.

Massdesign Architects and Planners, Inc.
138 Mt. Auburn St.
Cambridge, MA 02138

Architectural, planning and solar design services; energy conservation research and design; solar and energy conservation feasibility studies. Designed Massachusetts Audubon Society headquarters. **Geographical range:** north temperate and cold climates—U.S. and Canada; east coast from Virginia north. **Contact:** Gordon Tully (617) 491–0961.

Mease Engineering Associates
P.O. Box 51
Port Matilda, PA 16870

Engineering company working in alternative energy: solar, wind, methane and water power. Energy efficient and passive solar residential design and construction. Air pollution testing and consulting. Experience includes work in passive construction and design and eutectic salt storage. **Geographical range:** Northeast, Mid-Atlantic. **Contact:** Michael Mease (814) 692–4225.

Millers Run Construction Co., Inc.
RFD
Sutton, VT 05867

Design and construction of solar energy systems and solar and energy efficient homes. Installation of solar domestic hot water systems. Integrative design. Reasonable cost. Several residences completed and a prototypical flat plate collector built. **Geographical range:** Construction—Vt. Design—New England. **Contact:** Robert Starr (802) 626–8045.

Robert Mitchell/Solar System Design
RD 3, Box 147
Selkirk, NY 12158

Design and construction of passively heated structures. Design and installation of active systems. Residential and commercial energy consumption reduction analysis. Highest priority placed on cost effectiveness and material quality. Numerous passive and active projects completed to date. **Geographical range:** Job supervision—Upstate N.Y. Design—unlimited. **Contact:** Robert Mitchell (518) 767–3100.

Moore Grover Harper, PC
Architects and Planners
Essex, CT 06426

Full architectural and planning services for active, passive and hybrid solar heated and energy conserving buildings. Special interest in passive and hybrid systems. Conducting research in areas of building construction for thermal efficiency, heat storage systems and complex climates and building sites. In-house computer analysis of system economics. Winners of thirty-six design awards. Institutional, commercial and residential experience. Recent projects—Armed Forces Reserve Center (Norwich, Conn.) and Guilford (Conn.) Medical Building. **Geographical range:** Residences—U.S. Larger buildings—unlimited. **Contact:** William Grover (203) 767–0101.

National Solar Corp.
Novelty Lane
Essex, CT 06426

Solar system engineers and manufacturers. Provide complete engineering design studies for industrial/commercial and residential projects. In operation since 1975. **Geographical range:** New Eng. and south to N.J. **Contact:** Anthony Easton (203) 767–1644.

The Natural House
271 Washington St.
Canton MA 02021

Design and consulting services in residential solar architecture and integrated energy systems engineering. One hundred percent passive/active house designed for central Maine, other integrated solar residences including single and multi-family. Thirty-three unit condominium village, five-unit townhouse project. **Geographical range:** Northeast. **Contact:** James Serdy (617) 828–7115.

North Country Engineers
P.O. Box 38
Sandy Creek, NY 13145

Consulting engineers. Design and evaluation services for alternative energy and heat recovery systems. Experience includes solar houses, two solar town halls in Wayne County, N.Y. and numerous industrial heat recovery systems around the country. **Geographical range:** Northeast. **Contact:** Daniel French (315) 387–3411.

Northern Owner Builder
RFD 1
Plainfield, VT 05667

Residential architectural design and drafting services with emphasis on energy conservation, solar applications and wood heating. Especially interested in involving the owner in design and construction stages. Have completed several projects—two residences in New England and a greenhouse-workshop at Community College of Vermont. Also distribute books related to all aspects of house building. Catalog of these available for 25¢. **Geographical range:** New Eng. and S.E. Canada. **Contact:** Paul Hanke (802) 454–7808.

One Design Inc.
Mt. Falls Rt.
Winchester, VA 22601

Architectural staff maintained by this manufacturer of passive heating and cooling systems. **Geographical range:** unlimited. **Contact:** Tim Maloney (703) 662–4898.

Russell G. Page
1 Ridgewood Drive
Concord, NH 03301

HVAC and solar designer. **Geographical range:** New Hampshire. **Contact:** Russel Page (603) 228–0111.

Parallax, Inc.
P.O. Box 180
Hinesburg, VT 05461

Architectural and engineering services. Passive, hybrid and active collector system design. Complete design services—residential, commercial and institutional. Solar greenhouses a speciality. Recent projects include the

Enosberg Falls (Vt.) School, Collidge Hall and the Aiken Center for Natural Resources at the University of Vermont, and several houses. **Geographical range:** Worldwide. **Contact:** Robert Holdridge (802)482–2946.

Edwart N. Pedersen
109 Haffenden Rd.
Syracuse, NY 13120

Architectural design services with emphasis on energy conservation and alternative energy technologies. Energy audits and alternate energy system implementation feasibility studies. Current projects include Mirror Lake Estates, a planned solar community in upstate N.Y. and a feasibility study for the village of Pulaski, N.Y. **Geographical range:** Central N.Y., Eastern Mass., Northern N.J. **Contact:** Edward Pedersen (315) 472–5016.

People/Space Company
49 Garden St.
Boston, MA 02114

Architectural services encompassing solar system design. Projects have included new and retrofit, rural and urban residential installations. Progressive Architectural Award for solar residence, 1973. **Geographical range:** U.S. **Contact:** Robert Shannon (617) 742–8652.

William D. Potts, AIA
Providence Church
Glenelg, MD 21737

Architect working in energy efficient building design with special interest in hybrid—passive/active systems. Assessment of energy efficient alterations to existing buildings. Projects include residential, commercial and institutional applications. One of the larger of these is the Upton Multi-Purpose Center in Baltimore, Md. **Geographical range:** Mid-Atlantic states. **Contact:** William Potts (301) 465–8285.

PRC Energy Analysis Company
7600 Old Springhouse Road
McLean, VA 22101

Engineering services in the field of solar-related research and development. Services include economic feasibility analyses, system design and program management. Applications in solar heating and cooling, wind systems, biomass and photovoltaics. Many studies and projects completed including several for Department of Energy and Department of Interior. **Geographical range:** International; 200 offices in 50 countries. **Contact:** E.E. Bean (703) 893–1820.

Princeton Energy Group/Harrison Fraker, Architect
245 Nassau St.
Princeton, NJ 08540

Design, research, and independent consulting services for all types of alternative energy applications, energy conservation and programs of energy management. Architectural and engineering assistance including computer-aided design analysis and laboratory or on-site testing. Emphasis on "low" or "appropriate" technology. Projects have included analysis and design of small scale solar and wind systems for remote villages in Oman, the Princeton Educational Center, Blairstown, N.J. and others. **Geographical range:** travel at customers expense. **Contact:** Lawrence Lindsey (609) 924–7639.

Project Sun—Solar Energy Systems
P.O. Box 93
Mamaroneck, NY 10543

Mechanical design and consulting for solar systems. Grant application preparation, feasibility studies and computer simulations. Installation of systems. Three years experience in field. **Geographical range:** Metro. N.Y.C. area. **Contact:** Bert Siegel (914) 698–9209.

Maurice T. Raiford, Ph.D.
Solar Energy Consultant
4814-B Tower Road
Greensboro, NC 27410

Consultant for all types of solar applications. System design and sizing. Economic and equipment advice. Special interest in passive solar design particularly in agricultural and industrial applications. Experience in new and retrofit projects. Recently completed five nation tour as solar specialist for the Department of State. Available as speaker. **Geographical range:** worldwide. **Contact:** M.T. Raiford (919) 855–8303.

Rockland Solar Energy Co.
6 Beaver Hollow Lane
Monsey, NY 10952

Solar consultant to heating/cooling contractors, architects, builders and government agencies. Several installations completed. **Geographical range:** Rockland and Orange Co. in N.Y. **Contact:** Ronald Cataldo (914) 352–5408.

Roy Larry Schlein and Associates, Inc.
Room 217, 29 Bala Ave.
Bala Cynwyd, PA 19004

Consulting engineers and planners dealing in all phases of mechanical and electrical engineering and plumbing. Solar involvement is in design of solar assisted water-to-water and air-to-water heat pump systems as well as DWH and standard hot water systems. Experience in residential, commercial and institutional applications, including a McDonald's restaurant, work at a community hospital and others. **Geographical range:** Northeast. **Contact:** Roy Schlein in Pa. (215) 667–9516; in N.J. (609) 667–1112.

Shelter Design/Builders
Box 161A, R.D. 1
New Tripoli, PA 18066

Passive solar design and construction management services for residential and light commercial structures. Several residences completed in Pennsylvania and a veterinary clinic. Solar experience since 1975. Sell basic house plans. **Geographical range:** Consulting—unlimited; Design—100 mile radius of Allentown, Pa.; Construction management—40 mile radius of New Tripoli, Pa. **Contact:** Mike Orda (215) 756–6112.

I. Shiffman, P.E.
529 Central Ave.
Scarsdale, NY 10583

Consulting mechanical/electrical engineers working with alternate energy systems and energy conservation for industrial, commercial, institutional and residential facilities. **Geographical range:** not given. **Contact:** I. Shiffman (914) 725–1750.

Robert O. Smith and Associates
55 Chester St.
Newton Highlands, MA 02161

Professional engineering in design of heating systems for buildings and hot water, conventional and solar. Consultation and teaching. Assistance with selection of contractors and suppliers. Heating cost reduction studies. Experience in grant preparation, design of commercial and residential systems—active and passive. Government and private contracts. Many projects completed. **Geographical range:** Northeast. **Contact:** Robert Smith (617) 965–5428.

Solafern, Ltd.
536 Mac Arthur Blvd.
Bourne, MA 02532

Engineering design firm providing both residential and commercial solar systems, as well as testing facilities for air collectors. Several houses completed to date, includes retrofits and new installations. **Geographical range:** as required. **Contact:** Philip Levine (617) 563–7181.

Solar Design Associates
271 Washington St.
Canton, MA 02021

Consulting and design services in solar architecture and integrated energy systems engineering. Several solar residences, some with cooling; three high-rise projects, thirty-three unit condominum village, other multi-family residences, passive greenhouse complex to firms credit. **Geographical range:** Northeast. **Contact:** Steven Strong (617) 828–7115.

Solar Energy Systems
One Olney Ave.
Cherry Hill, NJ 08003

Solar heating, cooling, DWH and process heat system design. Economic analyses. Residential and commercial experience. **Geographical range:** unlimited. **Contact:** N.E. Brussels (609) 424–4446.

Solar Engineering Group
580 Alexander Rd.
Princeton, NJ 08540

Single source responsibility for feasibility studies, design and construction of projects employing energy conservation, solar and wind energy. Commercial, industrial, residential and agricultural. Experience includes several HUD residential systems, a motel/hotel hot water initiative design and a number of commercial and apartment energy conservation projects. **Geographical range:** Mid-Atlantic states. **Contact:** Jim Kopley (609) 921–3330.

Solar Heat Corporation
108 Summer St.
Arlington, MA 02174

Solar system design and engineering. A number of installations completed to date. **Geographical range:** not given. **Contact:** Mark Hyman (617) 646–5763.

SolarTherm
Box 426
Meadville, PA 16335

Consulting services covering all phases of solar-assisted heating, cooling and domestic hot water systems design and installation. All systems are designed for maximum cost-effectiveness by using optimum combination of standard readily available components. Three year solar experience in both new and retrofit installations. **Geographical range:** Continental U.S. **Contact:** Wallace Bixby (814) 724–4258.

Solar USA, Inc.
700 Springfield Ave.
Berkeley Heights, NJ 07922

Solar panel manufacturer offering co-ordinated engineering services. Staff engineer has been involved in solar since 1958. **Geographical range:** Northeast. **Contact:** J. S. Ballantine (201) 464–8870.

Solsearch Architects
1430 Massachusetts Ave.
Cambridge, MA 02138

Offer design, research and development, and construction management of solar and energy conserving projects. Specialize in passive solar and greenhouses, but have experience in active systems as well. Lectures available. Designed the Prince Edward Island and Cape Cod Arks for the New Alchemy Institute and a low energy house in Connecticut. **Geographical range:** Northeast U.S., Maritime Canada. **Contact:** in U.S.: Ole Hammarlund (617) 492–2188; in Canada: David Bergmark, Miller-Solsearch, 126 Richmond St., P.O. Box 2320, Charlottetown, Prince Edward Island, Canada; (902) 892–9898.

Sunnyside Up
14 Elliot St., P.O. Box 299
Brattleboro, VT 05301

Design and energy study services. Design of solar and conservation conscious structures with emphasis on passive systems particularly direct gain, attached greenhouses and waste heat recovery. Energy studies begin with pre-design site analysis to best optimize existing conditions and move on to heat loss, heat gain or annual heat load calculations as well as analyses of most cost-effective equipment for individual projects. Also involved in educational activities. **Geographical range:** Vt., N.H., Mass., R.I. and Conn. **Contact:** Richard Gottlieb (802) 254–2739.

Syska & Hennessy
110 West 50th St.
New York, NY 10020

Consulting mechanical and electrical engineers with capacity to provide solar energy studies for commercial, industrial and institutional applications. In-house computer facility maintains design and energy analysis programs, including one that predicts solar collector performance based on hourly measured solar radiation and weather data. Solar energy system designs include Eastern Liberty Savings & Loan Assn., Washington, D.C. and retrofitting of two GSA buildings in Calif. Feasibility study recently completed for Xerox Corp. headquarters in Stamford, Conn. **Geographical range:** No. Am., Europe, Mid-East, S.A., No. Af. **Contact:** William B. Hankinson (New York) (212) 489–9200. Also offices in Washington, D.C.; Los Angeles, San Francisco, Paris and Tehran.

Torrence, Dreelin, Farthing & Buford, Inc.
P.O. Box 11084
Richmond, VA 23230

Architectural and engineering design of commercial and industrial building incorporating design of solar energy and conservation systems and techniques. Designed Easco Photo building in Richmond, Virg. **Geographical range:** Eastern U.S. **Contact:** Richard Hankins, Jr. (804) 358–9111.

Total Environmental Action, Inc. (TEA)
Church Hill
Harrisville, NH 03450

Passive and active energy conserving building design. Consulting, engineering and research. Also involved in education and information dissemination. **Geographical range:** unlimited. **Contact:** Hilda Wetherbee (603) 827–3361.

Edmund G. Trunk
2424 Hudson St.
East Meadow, NY 11554

Consulting engineer in areas of energy conservation, solar energy and wind power. Has done work in both design and testing of solar equipment. **Geographical range:** L.I. and N.Y.C. **Contact:** Edmund Trunk (516) 785–5622.

Vermont Solar Group, Inc.
P.O. Box 292
Warren, VT 05674

Complete solar design services—active, passive and integrated space heating systems. Engineering and architectural services. Energy conservation consultants. Recent projects have included many residential systems including a HUD Cycle 2 Demonstration house in Waitsfield, Vt. and such commercial installations as the DWH system for the Environmental Resource Group store in Williston, Vt. **Geographical range:** New England and N.Y. **Contact:** Glenn Gazley (802) 496–2041.

Alex Wade
Box 43
Barrytown, NY 12507

Architect will design energy saving and passive solar heated houses. Author of **Thirty Energy Efficient Houses You Can Build.** Designed many small, efficient passive solar houses. **Geographical range:** northeastern U.S. **Contact:** Alex Wade (914) 758–5554.

Donald Watson, A.I.A.
Box 401
Guilford, CT 06437

Architectural design with emphasis in passive solar buildings and owner-built construction. Six completed and occupied solar projects to date. **Geographical range:** current projects in fifteen states from Maine to Wisconsin. **Contact:** Donald Watson.

Wright-Pierce-Barnes-Wyman
99 Main St.
Topsham, ME 04086

Solar energized environmental design. Wastewater treatment. Fuels from bio-mass. Several wastewater treatment plants completed to date with variety of solar applications—active, passive, biomass and combinations. **Geographical range:** unlimited. **Contact:** Stephen Bowers (207) 725–8721.

Midwest U.S.A.

Alternate Energy Research and Development
Box 77
Atlanta, MI 49709

Solar and wood energy consultant. **Geographical range:** N.E. Mich. **Contact:** William Huey.

Arthur Hall Pedersen Design and Consulting Engineers
34 North Gore
Webster Groves, MO 63119

Design, proposal preparation, analysis, inspection and consulting services in areas of solar architecture and mechanical and structural engineering. Residential, commercial and agricultural applications. Among recent projects are a HUD Cycle 3 house incorporating an active collector--water storage system, Aqua-Farm—a catfish production farm using direct solar heating of water recycled and purified through vascular aquatic plants and the Inner City Townhouses employing Trombe-wall and other passive techniques. **Geographical range:** Eastern MO. **Contact:** Arthur Pedersen (314) 962–4176.

Bascom Industries, Inc.
320 E. Tiffin St.
Bascom, OH 44809

Engineering consulting and solar feasibility studies. Sev-

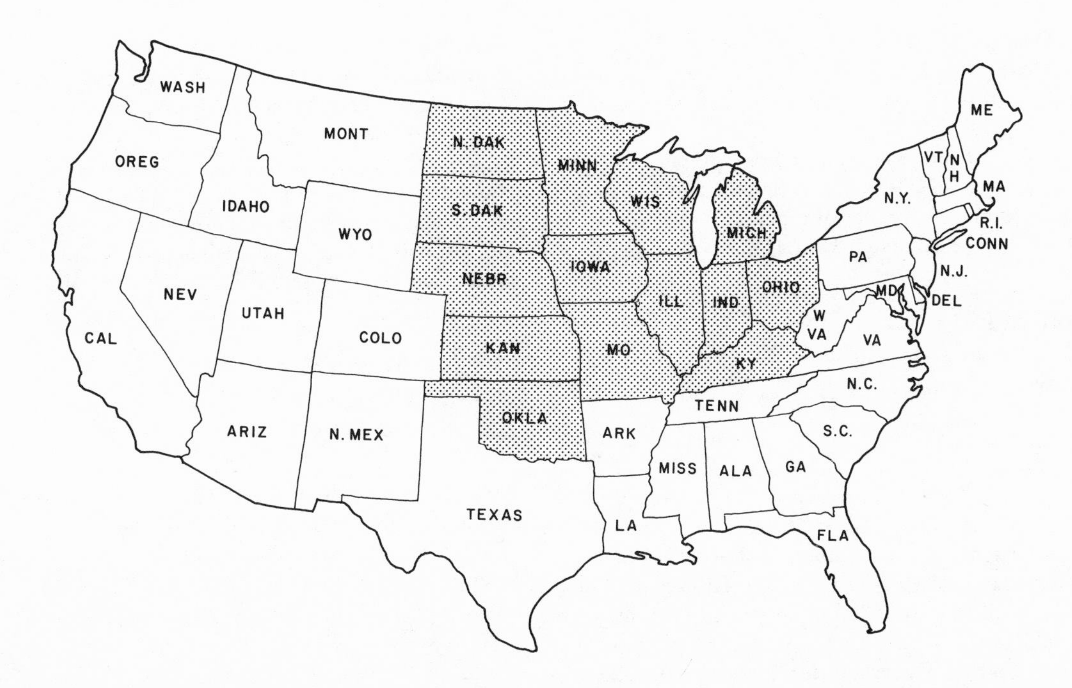

eral pool and hot water heating systems, including a HUD proposal for a nursing home DWH system. **Geographical range:** Northern Ohio. **Contact:** Joseph Deahl (419) 937–2225.

ments (Des Moines) where heat and domestic hot water are provided by an array of Lennox collectors. **Geographical range:** Midwest. **Contact:** Thomas J. Van Hon (515) 244–7167.

Bressler, Armitage & Lunde
1002 Wesley Temple Bldg.
Minneapolis, MN 55403

Architectural and engineering services in the areas of energy conservation, bio-conversion, earth sheltered buildings and various mechanical designs, including solar, heat pump and wood systems. Staff can provide studies, research, development or application design in both commercial and residential fields. Many projects and studies completed to date involving a wide variety of techniques and applications. **Geographical range:** U.S. **Contact:** Martin Lunde (612) 870–7011.

Brooks Borg and Skiles
815 Hubbell Building
Des Moines, IA 50309

Architectural and engineering firm with experience in solar design. Examples of their work are the Raccoon Valley State Bank (Adel, Iowa) which employs a PPG flat plate heating and cooling system, Iowa State Capitol Complex (Des Moines), using Suntec concentrating collectors for heating and cooling and the Carpenter Apart-

The Clark Enersen Partners
1515 Sharp Building
Lincoln, NE 68508

Designers of solar heating, cooling and hot water systems—commercial, institutional and residential. One of their systems has been in successful operation for over two years. Housing Authority Office Building and Hyde Observatory, both in Lincoln, NE are examples of their work. **Geographical range:** Central Plains States. **Contact:** Charles Thomsen.

Christensen Building Services
417 10th
Ames, IA 50010

Architectural design, mechanical engineering and installation services for active and passive systems—commercial, agricultural and residential. Currently building demonstration hybrid house in Ames, Iowa. **Geographical range:** Iowa. **Contact:** Rhys Christensen (515) 233–2058.

Leo A. Daly Company
8600 Indian Hills Drive
Omaha, NE 68114

Architect and engineering firm offers computer-aided design and analysis of solar heating, cooling, and domestic hot water systems. **Geographical range:** U.S. and Alaska. **Contact:** William Brady (402) 391–8111.

Enercon, Ltd.
1603 Orrington Ave., Suite 1390
Evanston, IL 60201

Consulting engineers with emphasis on energy conservation. Feasibility studies, economic analayses and conceptual designs for solar systems to supplement conventional energy sources on various types of educational, institutional, governmental and residential facilities. Recent projects include a solar overview for the Latin School of Chicago, a feasibility study for an apartment complex in Ouray, Colo. and federal grant preparation for an Illinois state office building. **Geographical range:** U.S. **Contact:** David L. Grumman (312) 328–3555.

Ener-Tech, Inc.
1924 Burlewood Dr.
St. Louis, MO 63141

Engineering firm involved in energy management and solar energy consultation. Total energy design service—active and passive residential and commercial applications. Retrofit and new construction experience—hot water, pool heating and space heating. **Geographical range:** Metro. St. Louis. **Contact:** Jim Ray (314) 878–1586.

Environmental Design Alternatives
2011 Rosewood Drive
Kent, OH 44240

Full architectural services for energy efficient building design. Employ both active and passive solar technologies, but emphasis in either is to use simplest means possible. Encourage owner participation in design. Consultation services as well. Have participated in two HUD demonstration cycles. **Geographical range:** Ohio and W. Penn. **Contact:** Douglas Fuller.

Fairbrother & Gunther, Inc.
325 Fuller Ave. N.E.
Grand Rapids, MI 49503

Consulting mechanical and electrical engineers doing feasibility and design of commercial, institutional and industrial solar space heating and process water systems. Four years solar experience. Recent projects include a large flat plate dormitory space and water heating retrofit, a parabolic cylinder concentrator retrofit for heating, cooling and hot water of a college administration building and several flat plate service hot water systems for hospitals. **Geographical range:** Mich., Ill., Ind., and Ohio. **Contact:** Kenneth E. Gunther (616) 451–8476.

The Hawkweed Group, Ltd.
4643 N. Clark
Chicago, IL 60640

Architects and planners involved solely in solar energy. All building design integrates solar concepts with climatic-site considerations. About fifty projects in various stages of completion. Residential, commercial, institutional and industrial. **Geographical range:** virtually anywhere. **Contact:** Rodney Wright (312) 784–5025.

I. E. Associates
3704 11th Ave. South
Minneapolis, MN. 55407

Multi-directional firm offering consulting, engineering design, R&D, and socio/economic studies in areas of energy management, conservation, solar, bioconversion and greenhouses. Also plan and teach short courses in these and related areas. Experience includes design of energy conserving and passive solar greenhouse retrofit, R&D, impact analysis and field evaluation of anaerobic digesters under federal contract and design of digester system and educational package for solar equine system. **Geographical range:** North America. **Contact:** Tom P. Abeles (612) 825–9451.

Interface Design Group
6621 Clayton Rd.
St. Louis, MO 63117

Architects, engineers and economists involved in research through the development stages of solar energy applications. Priorities are towards energy conservation first, then passive and finally active solar if economically justifiable. Have conducted several residential conservation analyses, and planned a solar-assisted hot-water system for a 112-unit apartment building. **Geographical range:** Midwest. **Contact:** Warren Cargal (314) 721–8756.

**Ionic Solar Inc.
8934 "J" St.
Omaha, NE 69127**

Solar system construction management. Active and passive system design. Equipment sales, installation and maintenance. Numerous DHW and space heating systems conpleted—single and multiple family residential applications. **Geographical range:** Western Mo., Western Io., Eastern Neb. and Eastern Kan. **Contact:** Garry D. Harley (402) 339–2420.

**Joseph J. Kawecki
296 Cliffside Dr.
Columbus, OH 43202**

Designer and builder of integrated energy structures. Designs include complete heat loss/gain calculations, estimated costs and paypacks. Experience in a variety of solar and conservation techniques. Current emphasis is on direct gain and underground construction. Designed solar underground house that is open to the public, a low energy house—2000 sq. ft. that uses only 2500 Kw total electric in a typical January and the first passive house in what will be a total solar community in Ohio. **Geographical range:** Central Ohio. **Contact:** Joseph Kawecki (614) 267–8598.

**McKim Solar Energy Systems Inc.
1142 E. 64th St.
Tulsa, OK 74136**

Solar architectural and mechanical design services. Experience in institutional and single and multi-family residential applications—five private houses completed under federal grants. Illustrated applications manual available for $2. **Geographical range:** 250 mile radius of Tulsa, Okla. **Contact:** Richard McKim (918) 749–8896.

**Natural Energy Workshop
Box 130
North Freedom, WI 53951**

Planning, design, drafting, and architectural services; research on low cost energy efficient housing; display and education center. Several projects completed. **Geographical range:** Wisc., northern Ill., and northeast Iowa. **Contact:** Ed Doerr, technical dept.; Russ Kowalski, design dept. (608) 344–3013.

**Northern Solar Power Co.
311 South Elm St.
Moorehead, MN 56560**

Consulting engineering in solar domestic hot water and space heating systems—liquid and air. Installation of commercially available systems. Experience in both new and retrofit installations. Also publish do-it-yourself materials. **Geographical range:** W. Minn., N.D., S.D. and Mont. **Contact:** Bruce Hilde.

**The Schemmer Associates, Inc.
10830 Old Mill Rd.
Omaha, NE 68154**

Solar energy analysis, economic feasibility analysis and design of solar assisted facilities, including architecture, engineering, and planning. Engineering design of sixty solar assisted energy efficient housing units for Walt Disney World, Florida. **Geographical range:** Continental United States and select foreign countries. **Contact:** Roger Wozny (402) 333–4800.

**Schmidt, Garden & Erikson
104 S. Michigan Ave.
Chicago, IL 60603**

Architects and engineers providing building and solar system design. Eighty years experience in planning and design of medical, research, corporate and educational facilities. **Geographical range:** U.S. **Contact:** Louis Michelsen (312) 332–5070.

**Shaffer & Roland, Inc.
20 North Wacker Dr.
Chicago, IL 60606**

Consulting, architectural and engineering services for energy conservation and solar energy harvesting. Specialization in space and domestic water heating systems and in designing and building systems for waste management, including anaerobic digesters. **Geographical range:** not given. **Contact:** John Martin (312) 236–9106.

**Smith, Hinchman & Grylls Associates, Inc.
455 West Fort St.
Detroit, MI 48226**

Architectural, engineering and planning services. Recent

solar projects include the Terraset Elementary School, Reston, Virg.; Sea Loft Resturant, Long Branch, N.J. and the U.S. Army Hospital, Fort Polk, La. **Geographical range:** unlimited. **Contact:** William C. Louie (313) 964–3000.

SolaRay, Inc.
324 S. Kidd St.
Whitewater, WI 53190

Complete professional engineering design of solar systems for residential, commercial and industrial applications. Three years experience in solar, thirty-one in engineering. Recent solar systems design include residences, indoor swimming pool, private camp bath house and commercial store building. **Geographical range:** unlimited. **Contact:** Robert Skrivseth (414) 473–2525.

Solar Energy Engineering
1838 Alverne Dr.
Poland, OH 44514

Consulting engineers in residential, commercial and industrial solar applications. Complete contract installation for turn-key projects available. Services range from energy conservation and heat recovery to solar system design and product development. Several residential projects completed under HUD Cycles 2 and 3. Designed industrial process hot water system incorporating solar assisted heat pump for General Extrusions Inc. in Youngstown, Ohio. **Geographical range:** 500-mile radius of Youngstown, Ohio. **Contact:** Gene Ameduri (216) 757–8687.

Solar Home Systems, Inc.
12931 West Geauga Trail
Chesterfield, OH 44026

Architectural engineering in residential and commercial solar applications. Full services for energy conservation and active or passive solar building design. Six years in solar. Experience in underground construction. **Geographical range:** Ohio, will travel at customers expense. **Contact:** Joseph Barbish (216) 729–9350 or 289–7020.

The Solarway, Inc.
6412 Washington Ave.
Des Moines, IO 50322

Designers and builders of hybrid (passive and active) solar homes. **Geographical range:** Iowa and adjacent states. **Contact:** G.G. Corrigan (515) 277–7760.

Sun Unlimited Research Corp.
P.O. Box 941
Sheboygan, MI 53081

Provide solar engineering and architectural design services oriented primarily toward use of company's Sunstone collectors. Several large institutional projects completed. One—the Howards Grove (Wisc.) Elementary School employs an array of 132 collector panels. **Geographical range:** U.S. and Canada. **Contact:** Glenn Groth.

Systems Technology Inc.
245 N. Valley Rd.
Xenia, OH 45385

Consulting in areas of design and component specification for variety of solar applications. Energy conservation studies, HVAC evaluations, equipment de-bugging. Experience in residential, commercial and institutional facilities. **Geographical range:** Ohio, Ind. and Kentucky. **Contact:** Lloyd Anderson (513) 372–8077.

Darryl Thayer and Associates
2406 Doswell Ave.
St. Paul, MN 55108

Solar system and concentrating collector design. Active and passive consultation. System sizing and cost-effectiveness analysis. Specialize in local area solar data. Construction supervision. Experience in single and multi-family residential systems including a HUD Cycle 3 house. **Geographical range:** Minn., Wisc., N.D. and S.D. **Contact:** Darryl Thayer (612) 644–5060.

Roger G. Whitiner, AIA
2156 N. Cleveland Ave.
Chicago, IL 60614

Architect specializing in energy conscious design. Consulting and design services. Several private residences, all with passive emphasis, completed. Serves on numerous state and professional solar related committees. **Geographical range:** Design—Chicago Metropolitan area. Consulting—nationwide. **Contact:** Roger Whitiner.

Joseph Yohanan, Architect
1041 Cherry
Winnetka, IL 60093

Architectural services with passive solar emphasis. Active experience as well. Evanston (Ill.) Environmental Center and a 15,000 ft. office building in Chicago—passive with active "back-up" are among recent projects. **Geographical range:** Mid-West. **Contact:** Joseph Yohanan.

Southern U.S.A.

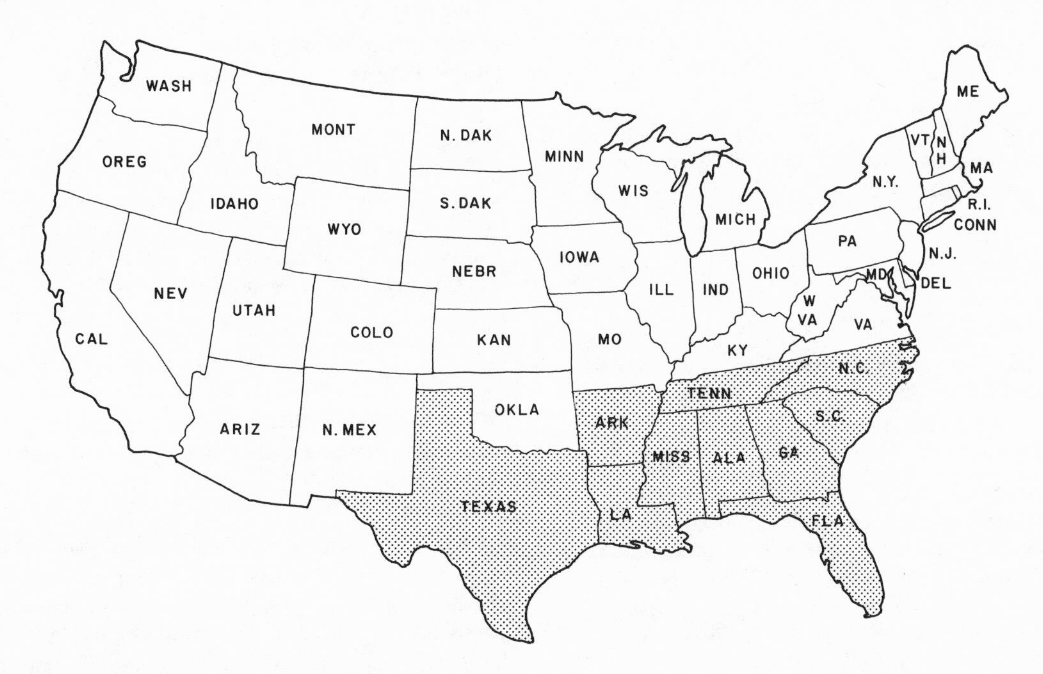

Aries Consulting Engineers, Inc.
115 Northeast 7th Avenue
Gainesville, FL 36201

Consulting engineers with emphasis on solar feasibility studies, and systems and instrumentation design. Coca-Cola U.S.A., Boeing Aerospace and the State of Florida are among recent clients. Services limited to governmental, commercial institutional and industrial applications. **Geographical range:** unlimited. **Contact:** William J. Fielder (904) 372–6687.

Belefant Associates, Inc.
1325 N. Atlantic Ave.
Cocoa Beach, FL 32931

Consulting engineers with solar experience in commercial and residential applications. **Geographical range:** unlimited. **Contact:** Arthur Belefant.

Energy Applications
Route 5, Box 383A
Rutherfodton, NC 28139

Engineering and consulting service for concentrating and tracking solar devices. **Geographical range:** unlimited. **Contact:** Napoleon Salvail (704) 287–2195.

Energy Designs/Architects
201 Woodrow St.
Columbia, SC 29205

An architectural firm specializing in design of energy conserving and solar buildings—both passive and active. Have completed a number of projects including, a solar office building in Chester, S.C. and a residence in Columbia, S.C. In process are a passive home and two HUD grant residences. **Geographical range:** S.C., N.C. and Georgia. **Contact:** Dick Lamar (803) 799–7495.

Patrick N. Espy
10,008 Hickory Hill Lane
Huntsville, AL 35803

Solar energy consultant and instructor. Has designed solar hot water system for a 120 bed hospital in Dallas, Tex. and acted as consultant on use of solar for hot water heating for NASA/MSFC. **Geographical range:** unlimited. **Contact:** Patrick Espy.

Todd Hamilton/Architect
6220 Gaston Ave.
Dallas, Tx 75214

Consultation on passive energy conservation and space and water heating systems. Residential and commercial applications, including Longview (Tex.) Shopping Center. **Geographical range:** Tex. and Okla. **Contact:** Todd Hamilton.

Integrated Energy Systems, Inc.
211 N. Columbia St.
Chapel Hill, NC 27514

HVAC engineering, design and plan energy alternatives, formulate energy conservation policy. Have done work in both active and passive systems for both residential and commercial applications. One interesting project is Windy Harbor—a passively heated shopping center with wind power generators. **Geographical range:** Piedmont N.C. and Virg. **Contact:** Daniel Koenigshofer (919) 942–2007.

Johnson-Dempsey & Associates, Inc.
1800 N.E. Loop 410
San Antonio, TX 78217

Designers of active and passive commercial and residential buildings and systems. Experience in solar goes back to the 1940's and 50's when several projects were built in San Angelo, Tex. More recently have worked on Texas Savings and Loan Branch office in San Antonio and 26 units of HUD sponsored housing under the auspices of the San Antonio Housing Authority. **Geographical range:** 300 mile radius of San Antonio. **Contact:** Bruce B. Johnson (512) 828–6251.

Hugh J. Metz Construction Co.
P.O. Box 843, 800 Oak Ridge Tpk.
Oak Ridge, TN 37830

Designer of and construction consultant for solar heating, hot water and energy conservation systems. Also sells and installs equipment. Numerous installations completed. **Geographical range:** Tenn. **Contact:** Hugh Metz (615) 482–5568.

Solar Development, Inc.
3630 Reese Ave.
Garden Industrial Park
Riviera Beach, FL 33404

Energy products and systems engineering. Computer program for sizing large solar projects. Design and system layout for solar assisted heat pumps and heat recovery systems. Have done projects under both HUD and ERDA grants as well as privately. **Geographical range:** unlimited. **Contact: R.J. Jefferson (305) 842–8935.**

Solar Development Co., Inc.
4000 Old Wake Forest Rd.
Raleigh, NC 27609

Engineering consulting services for solar thermal systems and energy recovery/conservation. Experience is in both residential and commercial applications. Extensive use of parabolic concentrators. One project—a 144,000 sq. ft. building is equipped with a fifty ton chiller and heating and hot water from concentrators. **Geographical range:** Eastern U.S. **Contact:** G.R. Winders (919) 872–6900.

Southeast Design and Development Inc.
P.O. Box 43402
Birmingham, AL 35243

Solar system design with R&D emphasis. DHW, space and pool heating systems for residential, commercial and industrial applications. Passive and liquid systems. Thirteen collector array for residence, 200 collector array for court house (space and water heating) and thirty collector system for bank (space heating) completed in past eighteen months. **Geographical range:** Southeast. **Contact:** Richard Sheley (205) 967–2753.

Southern Energy Conservation Systems
520 N. Spring St.
Pensacola, FL 32501

Consultation and evaluation of energy systems for builders, industries and private individuals. **Geographical range:** Northwest FLa. **Contact:** David Whitfield (904) 432–5158.

Sunshelter Design
610 Glenwood Ave.
Raleigh, NC 27603

Custom residential designers working with passive solar, underground and site-oriented architecture; and alternative back-up systems. Special emphasis placed on well integrated exterior design. Several residences, both single and multi-family, completed or in progress. **Geographical range:** N.C., S.C., Virg. and E. Tenn. **Contact:** John Meachem (919) 832–9919.

Sunspace
Box 71A, Rt. 5
Burnsville, NC 28714

Designers and builders of passive solar buildings. Also provide slide presentations and lectures. Currently conducting a workshop that is building a passive solar house, as well as working on a buried house in North Carolina and a passive one in Georgia. **Geographical range:** Building: Western N.C.; Design: Southeast; Lectures/Workshops: total U.S. **Contact:** Richard Kennedy (704) 675–5286.

Western U.S.A. and Canada

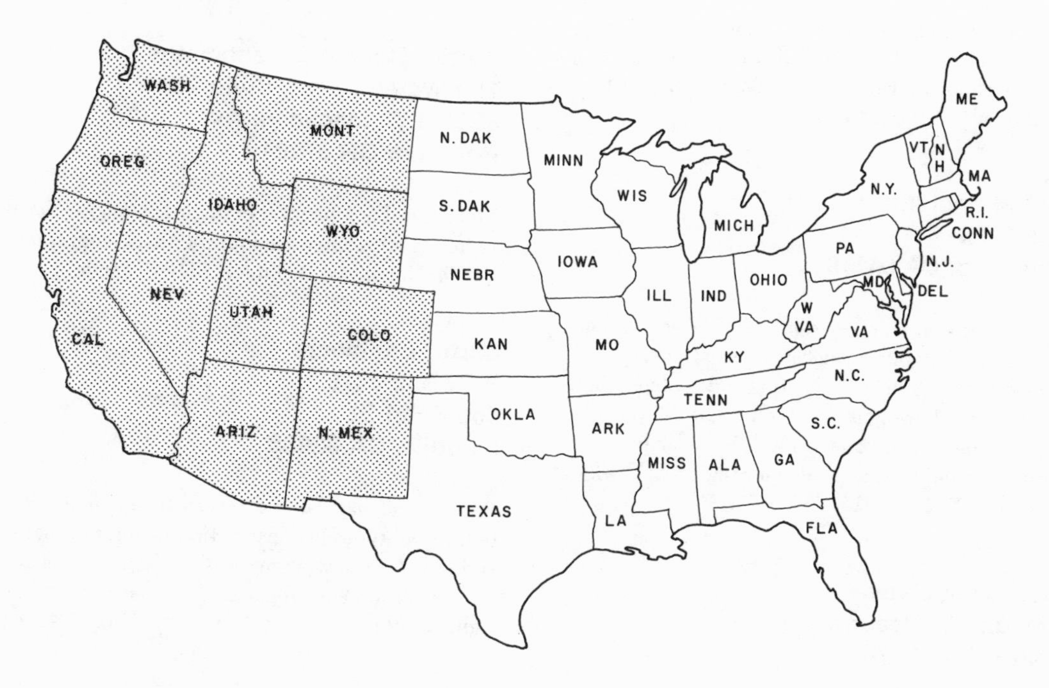

John D. Anderson and Associates, Architects
1522 Blake Street
Denver, CO 80202

Offer full architectural and planning services. Have completed several academic solar buildings in Colorado—Community College of Denver/North Campus in Westminster and the Yale and Jewell Elementary Schools, both in Aurora. Currently working on single family residence and a 240 student dormitory. **Geographical range:** total U.S. **Contact:** John D. Anderson (303) 534–5566.

Ayres Associates
1180 South Beverly Dr., Suite 600
Los Angeles, CA 90035

Consulting engineering firm specializing in design of mechanical and electrical building service systems. Offer computerized techniques for simulation and analysis of regular and solar augmented energy systems. Engaged in R&D of thermodynamically correct system simulation algorithms and the implementing software. Experience in use of such solar programs as TRNSYS, F-CHART, SOLCOST, SHIMSHAC and LASL. Have participated in the design of the National Security and Resource Study Center at Los Alamos, helped in preparation of ERDA Solar Design Manual and worked on the Skytherm Atascader House for ERDA. **Geographical range:** U.S. **Contact:** J. Marx Ayres (213) 553–5285.

Beam Engineering, Inc.
732 N. Pastoria Ave.
Sunnyvale, CA 94086

Consulting electrical and mechanical engineers working on solar collector systems and energy conservation. Solar space heating and hot water systems designed for both residential and commercial applications including a hot water processing system for photographic laboratory at the NASA-Ames Research Center. Site-wide energy survey and audit performed for Stanford Linear Accelerator Center. **Geographical range:** S.F. Bay area. **Contact:** Benjamin Beam (408) 738–4573.

Bechtel National Inc.
P.O. Box 3965
San Francisco, CA 94119

Engineering and construction services in areas of solar thermal power, photvoltaics, ocean thermal energy conversion, desalination and space heating and cooling. Have recently completed designs for both a 10 MWe solar thermal pilot power plant and a 150 kWe solar powered water pumping facility. **Geographical range:** world wide. **Contact:** Ernest Y. Lam (415) 768–6895.

Bran Dem Associates
27048 Mountain Meadow Rd.
Escondido, CA 92026

Qualitative reasearch and writing. Passive solar design. Association management. Feasibility studies conducted for cities of San Diego and Palm Springs, Calif. Several passive water wall residences completed. **Geographical range:** not given. **Contact:** John Brand (714) 749–3083.

Buffalow's Inc., Mechanical Contractors
1245 Space Park Way
Mountain View, CA 94043

Consulting, design, estimate, installation and maintenance services offered for all types of solar systems (active and passive) and applications. Residential, commercial and industrial installations. Four years solar experience. Numerous projects underway. **Geographical range:** No. Calif. and Upper Nev. **Contact:** Henry Buffalow Jr., (415) 961–7550.

Robert H. Bushnell, P.E.
502 Ord Drive
Boulder, CO 80303

Consulting engineer and meteorologist. Consultation, design, specification, construction inspection, testing and evaluation of solar heating systems—both passive and active. **Geographical range:** 100 mile radius of Boulder. **Contact:** Robert Bushnell (303) 494–7421.

Carter, Bringle & Associates
The Water Tower, Suite 255
5331 S.W. Macadam Ave.
Portland, OR 97201

Mechanical engineers specializing in HVAC design. Commercial and residential applications. **Geographical range:** Pacific Northwest. **Contact:** Robert Zanders.

Carter Engineers
P.O. Box 1419
450 Pearl St.
La Jolla, CA 92038

Provide mechanical and electrical engineering consultation for all types of construction including solar—existing and new facilities. Numerous academic projects completed. **Geographical range:** mainly, but not exclusively So. Calif. **Contact:** Richard K. Fergin (714) 459–2341.

Colorado Sunworks
P.O. Box 455
Boulder, CO 80306

Consulting and design services for passive solar heating and drain-down systems for DHW and space heating. Extensive experience with Beadwall® systems and massive construction techniques. Recent projects include both new and retrofit installations such as the Cold Springs solar greenhouse, Maxwell retrofit space heating project and the Hilton Harvest House Hotel—a new DWH project. **Geographical range:** Colo. **Contact:** Paul Shippee (303) 443–9199.

Dr. C. Phillip Colver, P.E.
P.O. Box 4256
Aspen, CO 81611

Professional engineer available for solar and energy management consultation. Teaches short courses and seminars in solar and conservation. Some design work. Has worked in association with ECS, Inc. and Frank Kreith. Designed solar heating system in Duncan, Okla. Currently involved in teaching a number of courses around Western U.S. and writing on energy management. **Geographical range:** Western U.S. **Contact:** C. Phillip Colver.

Crowther Solar Group
310 Steele St.
Denver, CO 80206

Architectural design firm emphasizing development of total natural energy and resource systems. Commercial, residential, industrial and institutional building and system design—active and passive. Perform energy analysis and establish energy budgets. In 30 years of experience, firm has completed 27 passive and active solar buildings. Nine projects currently under construction, 8 others in various stages of design. Crowther and Monigle office buildings (Denver, Colo.) are recent examples of work. Wrote book, **Sun/ Earth-How to Use Solar and Climatic Energies. Geographical range:** total U.S. **Contact:** Richard L. Crowther, AIA (303) 355–2301.

CTA Architects Engineers
P.O. Box 1456
Billings, MT 59103

Architectural and engineering design. Solar feasibility studies. Participated in DOE Commercial Demonstration Program—provided solar heating for Billings (Mont.) office building. **Geographical range:** Mont., Wyo., N.D., S.D., Idaho, Colo. **Contact:** Steven Ottenbreit (406) 248–7455.

L. M. Dearing Associates, Inc.
12324 Ventura Blvd.
Studio City, CA 91604

Design, engineering and cost analysis of solar heating for swimming pools including floating pool blankets, reel handling systems, and locker/benches. Systems have been installed in such places as the University of Western Fla., the Marina Club in Marina del Rey, Calif.; and the Palms Springs (Calif.) Municipal Swim Center. **Geographical range:** unlimited. **Contact:** LeRoy Dearing (213) 769–2521.

Daniel Dixon, AIA
P.O. Box 797
Granby, CO 80446

Architectural and engineering services for both passive and active solar systems. Also specializes in energy conservation design. The Granby solar house, a HUD grant project and an office complex, Solar Plaza East, are two examples of his work. **Geographical range:** Northwest Colo. and Southern Wyo. **Contact:** Daniel Dixon (303) 887–2200.

Econologic Energy Enterprises
3255 Kerner Blvd.
San Rafael, CA 94901

Research, design and manufacturing consultants for alternative energy resources. Services available to architects, engineers, manufacturers and individuals. Recently have done design proposal for Daly City (Calif.) Community Center, R&D of system design for hybrid solar electric-thermal device, engineering feasibility analysis or portable vertical-axis wind generator. **Geographical range:** Pacific states. **Contact:** Ed Quiroz (415) 459–0383.

Ecotope Group
2332 East Madison
Seattle, WA 98112

Solar design and demonstration services for solar greenhouses, residences (passive and active), commercial applications, wood heating and DHW. Also research and development activities in several areas. Several solar house and hot water installations completed. **Geographical range:** Pacific Northwest. **Contact:** Liz Steward (206) 322–3753.

Energy Engineering Group, Inc.
P.O. Box G
Idaho Springs, CO 80452

Engineering consulting firm engaged in design, analysis, R&D, and technical consultation. Services performed in the structural, mechanical, solar energy and energy conservation fields. Feasibility studies conducted for application of solar to a number of residential projects. Designed and built passive test facility under ERDA grant award. **Geographical range:** Colo. and surrounding states. **Contact:** Carl Hocevar (303) 573–5380.

Environmental Concepts
610 South Tejon St.
Colorado Springs, CO 80903

Architectural design, solar consultation, heat flow analysis, duct layout, construction management. Specialize in passive, but experience in active as well. Four years in field. Commercial, residential, agricultural applications. Greenhouses, domestic hot water, space and swimming pool heating systems. **Geographical range:** Rocky Mt. area. **Contact:** Peter Wood (303) 475–0360.

Environmental Energy Consultants
110 Forest Ave.
Fairfax, CA 94930

Design group offering energy and architectural consulting services. Design of solar systems—new or retrofit, energy efficient building design, grant preparation assistance, help with tax and mortgage considerations, environmental impact reports, architectural renderings, life cycle cost analyses and construction management. **Geographical range:** Northern Calif. **Contact:** David Stanley.

Federico Grabiel
1031 Wellesley
Los Angeles, CA 90049

Solar energy on-site use consultant. Efficiency calculations for passive and active systems. Initial solar structural design. Solar tempering service. New and retrofit residential, commercial and ecclesiastical experience. See also listing under Educational section. **Geographical range:** Calif. **Contact:** Federico Grabiel (213) 820–3233.

Hardison and Komatsu Associates
522 Washington St.
San Francisco, CA 94111

Architects of solar heated and energy conservative buildings. Provide full architectural services, including building and system design, construction administration, feasibility studies and research. One interesting current project is a solar heated office building for Pacific Gas and Electric in Vallejo, Calif. **Geographical range:** Calif. and Nev. **Contact:** William S. Taber, Jr. (415) 981–2025.

Hayakawa Associates
1180 South Beverly Drive
Los Angeles, CA 90035

Consulting engineers. **Contact:** Tseng-Yao (213) 879–4477.

Housewarming Development Corp.
Box 8, 1928 Sixth St.
Boulder, CO 80306

Engineering and development, including manufacture, of energy systems. Emphasis placed on use of solar with wood back-up. Several such residences completed. **Geographical range:** Colo. **Contact:** Day Chapin (303) 443–7970.

Hyperion Inc.
7209 Valtec Ct.
Boulder, CO 80301

Manufacturer of collectors offers solar engineering assistance in design, development and building of custom systems and components. Some installation work. **Geographical range:** Central and Greater Southwest U.S. **Contact:** John Eatwell (303) 449–9544.

Interactive Resources, Inc.
117 Park Place
Pt. Richmond, CA 94081

Planning, architecture, engineering, energy and construction management. Richmond (Calif.) Government Service Center, ten Northern California Highway Patrol stations and Stanford University Food Service Center, all with solar space and hot water heating systems are examples of recent work. **Geographical range:** unlimited (travel at clients expense). **Contact:** Dale Sartor (415) 236–7435.

Joint Venture Inc.
1406 Pearl Street
Boulder, CO 80302

Architects with emphasis on solar and/or energy responsive design. Will entertain any size or type of project. Previous and ongoing work includes multi-family housing projects in Boulder, Colo. and Des Moines, Iowa and a new manufacturing facility for the Celestial Seasonings Tea Co. in Boulder. **Geographical range:** Rocky Mts. and Mid-West. **Contact:** Alan Brown (303) 444–1752.

Kastek Corporation
P.O. Box 8881
Portland, OR 97208

Engineering, consulting, designing and installation of solar heating systems. Three domestic hot water systems completed last year. **Geographical range:** Northwest. **Contact:** William Brown.

Dr. Jan Kreider
1929 Walnut St.
Boulder, CO 80302

Solar engineering consultation in areas of economic analysis, computer modeling, system mechanical design, critical design review. Has worked on numerous projects including the RTD Maintenance facility—called the largest solar heating system in the world. Author of two books on solar energy and editor of another, all published by McGraw-Hill. **Geographical range:** No. Am., Europe, East Asia. **Contact:** Dr. J. Kreider (303) 447–2218.

James P. Leshuk
P.O. Box 12833
Salem, OR 97309

Consulting mechanical engineer providing design and development services for solar energy conversion, energy conservation and resource management projects. Has participated in various energy audits and surveys including one for the Portland (Ore.) Energy Conservation Demonstration Project under HUD contract, an ERDA funded Solar Pond Life Cycle analysis for Industrial and Agricultural Process Heat and an audit of energy use and needs for fourteen nursing homes. **Geographical range:** Pacific N.W., No. Calif. and Nev. **Contact:** Jim Leshuk (503) 362–7117 or 981–6377.

MacDonald Budget Energy Systems
Bootlegger Trail
Great Falls, MT 59401

Dealer in solar and wind energy hardware. Designers and engineers of low-cost solar residential heating systems. Currently working on system incorporating hot air and solar water-to-air heat pump system in own home as state-funded demonstration project. **Geographical range:** North-Central Mont. **Contact:** L. Clark MacDonald (406) (452–5967).

Manock Comprehensive Design
P.O. Box 4192
Stanford, CA 94305

Architectural planning and design engineering of energy efficient dwellings with emphasis on passive solar design. Product design of solar-related equipment. Research in passive solar techniques. Markets "solar design aid"—extra large computer generated plots of solar altitude and bearing vs. site standard time for exact building locations. Charts are corrected for longitude and equation of time, i.e., the earth's varying rate of rotation. **Geographical range:** Continental U.S. and Hawaii. **Contact:** Jerrold C. Manock (415) 328–1086.

Matrix
P.O. Box 4883
Albuquerque, NM 87106

Architectural design, passive solar consulting, computer simulations and research. Also passive workshops. Numerous new and retrofit installations to date including residential, commercial and educational facilities. **Geographical range:** unlimited. **Contact:** Edward Mazria.

James E. Molle
P.O. Box 5
Marylhurst, OR 97036

Design residential, commercial, industrial or utility photovoltaic power systems. Three years previous experience in field. Seven systems currently in progress. **Geographical range:** unlimited. **Contact:** James Molle.

More Combs Burch, P.C.
3911 East Exposition Ave.
Denver, CO 80209

Architectural/engineering design and advisement for active and passive systems—heating, cooling, how water—commercial, institutional and residential. Service includes life cycle cost analysis. The Boulder (Colo.) Post Office and the Parker (Colo.) Junior High School are examples of the firm's work. **Geographical range:** Advisory service, U.S.; Design service, Rocky Mt. region. **Contact:** Donald More (303) 744–3157.

Ralph A. Morrill & Associates, Inc.
P.O. Box 1382
Corvallis, OR 97330

Consultants in resource and environmental sciences with emphasis on development of solar and wind energy systems for households. Provide complete analytical design and system optimization and, through associates, installation. Evaluation, sizing, cost estimates, feasibility studies for solar, wind and water power. **Geographical range:** Pacific N.W. and Calif. **Contact:** Ralph Morrill (503) 757–7223.

Mountain Mechanical Sales, Inc.
5270 Broadway
Denver, CO 80216

Design and engineering service for solar domestic hot water and space heating. Manufacture systems. Forty-five solar jobs in past eighteen months. **Geographical range:** Colo., Mont., Wyo., N.M., Utah, Idaho. **Contact:** Paul Williams (303) 534–3000.

Natural Heating Systems
2417 Front Street
West Sacramento, CA 95691

Offer a wide range of services including design of solar heating systems, feasibility studies, computer simulation of system performance, architectural consulting, management of grant proposals and construction management. Designed over 60 solar heating systems for new subdivision in Davis, Calif. Have reviewed grant proposals for the State of California and done feasibility study for North Hollywood, Calif.-based, United Rent-All Corp. **Geographic range:** Western U.S. **Contact:** Mikos Doka-Suna (916) 372–2993.

Northwest Solar Systems Inc.
7700 12th Ave., N.E.
Seattle, WA 98115

Design, consulting and marketing firm utilizing solar energy as the primary design factor. Engineer, architect and solar consultant work as team on each project. Market Lennox and Rheem/Rudd solar heating equipment. Developing own air-collector. Both commercial and residential experience—new and retrofit. **Geographical range:** Washington. **Contact:** Stanley Gustafson (Consulting) or Jerome Gustafson (Marketing) (206) 523–3951 or Fredrick Ritchie (Production) (206) 524–6585.

Pacific Sun, Inc.
540 Santa Cruz Ave.
Menlo Park, CA 94025

Engineering consultants for the designers of solar/mechanical systems with specialization in commercial and industrial energy systems. Also provide instrumentation, data acquisition and data interpretation services. Current projects include solar heating and cooling system for City of Mt. View (Calif.) Fire and Police Administration Building, an active/passive heating and cooling system for Oakmead Solar Industrial Buildings in Santa Clara, Calif. and a solar heated municipal swimming pool system in Palo Alto, Calif. **Geographical range:** Calif., Nev., Arz., Colo. **Contact:** Harry T. Whitehouse (425) 328–4551.

Professional Design Builders, Inc.
P.O. Box 275
Loveland, CO 80537

Consulting engineers for solar design—passive and active. Analyze heat loss, thermal gain, thermal storage characteristics, structure strength, site orientation. **Graphical range:** Front range Colorado and Wyoming. **Contact:** Ivar Larson (303) 669–2650.

Rho Sigma
11922 Valerio St.
North Hollywood, CA 91605

Manufacturer of solar controls offers engineering services. **Geographical range:** National. **Contact:** Preston Welch (213) 982–6800.

Sennergentics
18621 Parthenia St.
Northridge, CA 91324

Design service for solar heating and cooling systems. Several residential installations completed. **Geographical range:** Continental U.S. **Contact:** James Senn (213) 885–0323.

Sharpe Solar Systems
2114 Woolard Drive
Bakersfield, CA 93305

Solar-related architectural service offered by retail sales company. **Geographical range:** Bakersfield and environs. **Contact:** Charles Sharpe (805) 831–6611.

Solar Energy Research Corp.
701B South Main Street
Longmont, CO 80501

Consultant for residential, industrial, commercial solar heating system design. Computer analysis of solar heating system performance. Preparation of reports, plans and specifications for solar equipment. Product evaluation, development, testing and business consulting. Five years in solar. Holds Patent on Thermo-Spray™. **Geographical range:** will travel. **Contact:** James Wiegand (303) 772–8406.

Solar Power Supply
12520 West Cedar Dr.
Lakewood, CO 80228

Design and consulting firm specializing in passive solar buildings and greenhouses. Past experience in installation, but current focus on design and consultation. Recent projects include, Rick's Cafe in Denver where solar hot water system and greenhouse were installed, the Ossgood house, Indian Hills, Colo., design and installation of active system and attached greenhouse and the Larned house in the same town where an integrated active/passive system with greenhouse was planned and built. **Geographical range:** 200 mile radius of Denver. **Contact:** Malcolm Lillywhite (303) 988–3055.

Solar Systems, Inc.
Box 931
Livermore, CA 94550

Solar system and instrumentation design. Passive heat load calculations. Computer modelling. General solar consulting. Experience in design of space and hot water heating systems, pool heating and a large system for a solar office building. **Geographical range:** San Francisco Bay area. **Contact:** George Bush (415) 447–9286.

Solar Systems West
1570 Linda Way
Sparks, NV 89431

Solar system design and consulting. Available to mechanical engineers and other professional system designers only. Manufacturer's representative for Sunworks and Solar Control Corp. Have recently worked as subcontractor on a number of military base installations including a feasibility study and final design for a barracks domestic hot water system. **Geographical range:** Nev. and Calif. **Contact:** Sheldon S. Gordon (702) 331–2595. In California contact: Solar Industrial Inc., 4500 Campus Dr., Ste. 207; Newport Beach, CA 92660; (714) 979–3311.

Solar Technology Corp.
2160 Clay St.
Denver, CO 80211

Engineering design and backup, and installation of solar heating and domestic hot water systems; manufacture solar greenhouse. **Geographical range:** Colorado and adjacent states. **Contact:** Richard Speed (303) 455–3309.

Southwest Energy Management, Inc.
8290 Vickers St., Suite B
San Diego, CA 92111

Solar and energy efficient design, analysis, engineering, construction and project management. Three years in the solar field. Over 300 systems installed. More than 50 design/feasibility studies. **Geographical range:** Design, engineering and consultation—unlimited: Construction—Calif., Ariz., Nev. **Contact:** Brian Langston (714) 292–5185.

Suncraft Company
Rt. 4, Box 90
Golden, CO 80401

Consulting on and design and construction of energy efficient dwellings. Stress passive solar and underground housing. Principal of company is products editor for **Alternative Sources of Energy.** Experience in both new and retrofit construction. **Geographical range:** Construction: Gilpin County, Colo. until 9/78, then Bitterroot Valley, Mont.; Consulting: anywhere. **Contact:** Steve Coffel (303) 582–5719.

Telluride Designworks
Box 700
Telluride, CO 81435

Architectural, engineering and planning firm specializing in solar and alternative energy structures and systems. Design work is done in conjunction with Telluride Sunworks, an associated solar components manufacturer and installer. Experience mainly in residential applications—active, passive and wind electric. **Geographical range:** Rocky Mt. States. **Contact:** Eric Doud (303) 728–3303.

Thacher & Thompson
215 Oregon St.
Santa Cruz, CA 96060

Design and construction of energy efficient homes blending active and passive heating and hot water applications into traditional homes. All houses designed to fit site both aesthetically and in terms of local energy requirements. Several residences completed to date. **Geographical range:** Santa Cruz, Calif. **Contact:** Richard Rahders (408) 426–4683.

Thermal Technology Corporation
Box 130
Snowmass, CO 81654

Solar design and engineering, a solar radiation field station, solar R&D and product development are among the services provided by this firm whose specialty is passive systems. Over 45 solar heated structures completed including the Aspen (Colo.) air terminal. **Geographical range:** Rocky Mt. region. **Contact:** Ronald Shore (303) 963–3185.

Walton-Abeyta & Associates, Inc.
1221 South Clarson St.
Denver, CO 80210

Engineering firm with extensive experience in solar. Design both active and passive systems with wide range of applications: heating, cooling, domestic hot water and swimming pools. Conduct economic and feasibility studies and life-cycle cost analyses. Over twenty commercial and residential projects completed. Among them are the Burdines Mall (Miami, Fla.) solar energy study, Robert Redford's home in Sundance, Utah and the Corbin Retail Building in Evergreen, Colo. **Geographical range:** total U.S. **Contact:** Monty Abeyta, Denver office or Neal McAbee, Glenwood Springs office (address: 2404 Glen Ave., Glenwood, CO).

West Wind
Box 542
Durango, CO 81301

Electrical engineering consulting for wind and solar applications. Design and construction of solar residences. Design and testing of active solar collectors. **Geographical range:** West and Southwest. **Contact:** Geoffrey Gerhard.

Orme and Levinson, Architects and Planners
1043 Johnson St.
Victoria, B.C., Canada V8V 3N6

Architects involved in integral solar design. Over past two years have planned several apartment building and duplex projects and done feasibility studies for others. **Geographical range:** British Columbia. **Contact:** Ben Levinson (604) 382–5125.

Solar Applications and Research Ltd.
3356 W. 13th Ave.
Vancouver, B.C., Canada V6K 2R9

Solar and energy conservation consultants. Design and analysis of passive and active solar space and water heating systems, photovoltaic system design, solar home and greenhouse design. Solar seminars. Have done work in both single and multi-family residences using variety of techniques including trombe walls and thermo syphoning. **Geographical range:** B. C. And Alberta, Can. **Contact:** Chris Mattock (604) 733–5631.